AF883062

DEFENSE, SECURITY AND STRATEGIES

# U.S. COUNTERINSURGENCY GUIDE AND TACTICS

# DEFENSE, SECURITY AND STRATEGIES

Additional books in this series can be found on Nova's website
under the Series tab.

Additional E-books in this series can be found on Nova's website
under the E-books tab.

DEFENSE, SECURITY AND STRATEGIES

# U.S. COUNTERINSURGENCY GUIDE AND TACTICS

TIMOTHEUS ANZELMO
EDITOR

Nova Science Publishers, Inc.
*New York*

Copyright © 2012 by Nova Science Publishers, Inc.

**All rights reserved.** No part of this book may be reproduced, stored in a retrieval system or transmitted in any form or by any means: electronic, electrostatic, magnetic, tape, mechanical photocopying, recording or otherwise without the written permission of the Publisher.

For permission to use material from this book please contact us:
Telephone 631-231-7269; Fax 631-231-8175
Web Site: http://www.novapublishers.com

**NOTICE TO THE READER**

The Publisher has taken reasonable care in the preparation of this book, but makes no expressed or implied warranty of any kind and assumes no responsibility for any errors or omissions. No liability is assumed for incidental or consequential damages in connection with or arising out of information contained in this book. The Publisher shall not be liable for any special, consequential, or exemplary damages resulting, in whole or in part, from the readers' use of, or reliance upon, this material. Any parts of this book based on government reports are so indicated and copyright is claimed for those parts to the extent applicable to compilations of such works.

Independent verification should be sought for any data, advice or recommendations contained in this book. In addition, no responsibility is assumed by the publisher for any injury and/or damage to persons or property arising from any methods, products, instructions, ideas or otherwise contained in this publication.

This publication is designed to provide accurate and authoritative information with regard to the subject matter covered herein. It is sold with the clear understanding that the Publisher is not engaged in rendering legal or any other professional services. If legal or any other expert assistance is required, the services of a competent person should be sought. FROM A DECLARATION OF PARTICIPANTS JOINTLY ADOPTED BY A COMMITTEE OF THE AMERICAN BAR ASSOCIATION AND A COMMITTEE OF PUBLISHERS.

Additional color graphics may be available in the e-book version of this book.

**Library of Congress Cataloging-in-Publication Data**

U.S. counterinsurgency guide and tactics / editor, Timotheus Anzelmo.
p. cm.
Includes index.
ISBN 978-1-60876-121-0 (hbk.)
1. Counterinsurgency--Handbooks, manuals, etc. 2. Integrated operations (Military science)--Handbooks, manuals, etc. 3. Counterinsurgency--United States. I. Anzelmo, Timotheus. II. United States. Dept. of the Army.
U241.U88 2011
355.4'25--dc23
2011028027

*Published by Nova Science Publishers, Inc. † New York*

# CONTENTS

# PREFACE

This book examines American counterinsurgency practices and tactics. At its heart, a counterinsurgency is an armed struggle for the support of the population. This support can be achieved or lost through information engagement, strong representative government, access to goods and services, fear, or violence. This armed struggle also involves eliminating insurgents who threaten the safety and security of the population. However, military units alone cannot defeat an insurgency. Most of the work involves discovering and solving the population's underlying issues, that is, the root causes of their dissatisfaction with the current arrangement of political power.

Chapter 1 - Insurgency is the organized use of subversion and violence to seize, nullify or challenge political control of a region. As such, it is primarily a political struggle, in which both sides use armed force to create space for their political, economic and influence activities to be effective. Insurgency is not always conducted by a single group with a centralized, military-style command structure, but may involve a complex matrix of different actors with various aims, loosely connected in dynamic and non-hierarchical networks. To be successful, insurgencies require charismatic leadership, supporters, recruits, supplies, safe havens and funding (often from illicit activities). They only need the active support of a few enabling individuals, but the passive acquiescence of a large proportion of the contested population will give a higher probability of success. This is best achieved when the political cause of the insurgency has strong appeal, manipulating religious, tribal or local identity to exploit common societal grievances or needs. Insurgents seek to gain control of populations through a combination of persuasion, subversion and coercion while using guerrilla tactics to offset the strengths of government security forces. Their intent is usually to protract the struggle, exhaust the government and win sufficient popular support to force capitulation or political accommodation. Consequently, insurgencies evolve through a series of stages, though the progression and outcome will be different in almost every case.

Chapter 2 - This manual gives the US Army a common language, concept, and purpose to fight and achieve success in a counterinsurgency. COIN is a complex subset of warfare that encompasses all military, paramilitary, political, economic, psychological, and civic actions taken by a government to defeat an insurgency at the company, battalion, and brigade levels. To do this, the manual merges traditional approaches to COIN with the realities of the current operational environment.

In: U.S. Counterinsurgency Guide and Tactics
Editor: Timotheus Anzelmo
ISBN: 978-1-60876-121-0
© 2012 Nova Science Publishers, Inc.

*Chapter 1*

# U.S. GOVERNMENT COUNTERINSURGENCY GUIDE

*Timotheus Anzelmo*

## EXECUTIVE SUMMARY

Insurgency is the organized use of subversion and violence to seize, nullify or challenge political control of a region. As such, it is primarily a political struggle, in which both sides use armed force to create space for their political, economic and influence activities to be effective. Insurgency is not always conducted by a single group with a centralized, military-style command structure, but may involve a complex matrix of different actors with various aims, loosely connected in dynamic and non-hierarchical networks. To be successful, insurgencies require charismatic leadership, supporters, recruits, supplies, safe havens and funding (often from illicit activities). They only need the active support of a few enabling individuals, but the passive acquiescence of a large proportion of the contested population will give a higher probability of success. This is best achieved when the political cause of the insurgency has strong appeal, manipulating religious, tribal or local identity to exploit common societal grievances or needs. Insurgents seek to gain control of populations through a combination of persuasion, subversion and coercion while using guerrilla tactics to offset the strengths of government security forces. Their intent is usually to protract the struggle, exhaust the government and win sufficient popular support to force capitulation or political accommodation. Consequently, insurgencies evolve through a series of stages, though the progression and outcome will be different in almost every case.

Counterinsurgency (COIN) is the blend of comprehensive civilian and military efforts designed to simultaneously contain insurgency and address its root causes. Unlike conventional warfare, non-military means are often the most effective elements, with military forces playing an enabling role. COIN is an extremely complex undertaking, which demands of policy makers a detailed understanding of their own specialist field, but also a broad knowledge of a wide variety of related disciplines. COIN approaches must be adaptable and agile. Strategies will usually be focused primarily on the population rather than the enemy and will seek to reinforce the legitimacy of the affected government while reducing insurgent influence. This can often only be achieved in concert with political reform to improve the quality of governance and address underlying grievances, many of which may be legitimate. Since U.S. COIN campaigns will normally involve engagement in support of a foreign government (either independently or as part of a coalition), success will often depend on the willingness of that government to undertake the necessary political changes. However great its know-how and enthusiasm, an outside actor can never fully compensate for lack of will, incapacity or counter-productive behavior on the part of the supported government.

This guide employs a COIN model that comprises five main functional components:

The political function is the key function, providing a framework of political reconciliation, and reform of governance around which all other COIN activities are organized. In general, a COIN strategy is only as good as the political plan at its heart.

The economic function seeks to provide essential services and stimulate long term economic growth, thereby generating confidence in the government while at the same time reducing the pool of frustrated, unemployed young men and women from which insurgents can readily recruit.

The security function is an enabler for the other functions and involves development not just of the affected nation's military force, but its whole security sector, including the related legal framework, civilian oversight mechanisms and judicial system. Establishing security is not a precursor to economic and governance activity: rather security, economic and governance activity must be developed in parallel.

The information function comprises intelligence (required to gain understanding), and influence (to promote the affected government's cause). It is essential that the influence campaign is in tune with the strategic narrative, resonates with the relevant audiences, is based on genuine resolve by the affected government and that physical actions match. What makes COIN different from other stabilization and humanitarian tasks is that both elements of the information function will be conducted in stark competition with the insurgents' own information functions.

These four functions contribute to the overall objective of enabling the affected government to establish control, consolidating and then transitioning it from intervening forces to national forces and from military to civil institutions.

The imperative to achieve synergy among political, security, economic and information activities demands unity of effort between all participants (the affected government, USG agencies and coalition partners). This is best achieved through an integrated approach to assessment and planning. A common interagency assessment of the insurgency establishes a deep and shared understanding of the cultural, ideological, religious, demographic and geographical factors that affect the insurgency. Such understanding provides the foundation for policy formulation when the risks and costs of intervention are weighed against U.S. interests in determining whether to become involved and what form that involvement should take. This decision should not be taken lightly; historically COIN campaigns have almost always been more costly, more protracted and more difficult than first anticipated. Much will hinge on the degree to which policy makers consider the affected government to be receptive to assistance, advice and reform; it is folly to intervene unless there is a reasonable likelihood of cooperation. If the USG does decide to become involved, then policy makers should seek a careful balance which employs the most appropriate, most indirect and least intrusive form of intervention yet still gives a high probability of achieving the necessary effect. The sovereignty of the affected government must be maintained and too high a U.S. profile may be counter-productive (historically, some of the most successful U.S. engagements have been indirect and low key). Once U.S. assistance is committed, a COIN strategy must be devised, ideally in collaboration with the affected government and other coalition partners, since their early inclusion can help mitigate the effects of operational level differences in goals, capabilities and culture. Detailed, integrated planning then follows and a process of continuous monitoring, evaluation and assessment is used to measure progress and identify where changes in approach are necessary to achieve success.

Success in COIN can be difficult to define, but improved governance will usually bring about marginalization of the insurgents to the point at which they are destroyed, co-opted or reduced to irrelevance in numbers and capability. U.S. intervention may cease when success is assured but before it is actually achieved. Ultimately, the desired end state is a government that is seen as legitimate, controlling social, political, economic and security institutions that meet the population's needs, including adequate mechanisms to address the grievances that may have fueled support of the insurgency.

# 1. Theory and Principles

## Part A: Insurgency

### Definition

Insurgency can be defined as 'the organized use of subversion and violence to seize, nullify, or challenge political control of a region.'

U.S. Army General Samuel Sumner meets with the Sultans of Bayang and Oato, Philippines, 1902.

Insurgents seek to subvert or displace the government and completely or partially control the resources and population of a given territory. They do so through the use of force (including guerrilla warfare, terrorism and coercion/intimidation), propaganda, subversion and political mobilization. Insurgents fight government forces only to the extent needed to achieve their political aims: their main effort is not to kill counterinsurgents, but rather to establish a competitive system of control over the population, making it impossible for the government to administer its territory and people. Insurgent activity is therefore designed to weaken government control and legitimacy while increasing insurgent control and influence.

## Characteristics

Many of the more renowned insurgencies of the 20th Century followed the Maoist 'Protracted Warfare' model; being monolithic organizations with a centralized, hierarchical command structure, clearly defined aims and a sequenced approach to achieve them. However, modern insurgencies are increasingly being recognized as complex matrices of irregular actors with widely differing goals. They often lack a centralized command structure but typically are linked by dynamic, flat networks (often significantly enabled by modern communications systems). Motivations within this eclectic mix may vary from religious extremism to pure criminality and many groups may not themselves intend to become the governing authority. Often, the only common factor will be a desire to achieve local freedom from control by the government and its international supporters.

## Ideology

During the Twentieth Century, insurgents were often motivated by Marxism, religion or nationalism (or a combination of these). Insurgencies were often led by university educated 'intellectual elites' whose personal circumstances were sometimes far removed from those of the rank-and-file insurgents that they inspired.

Modern insurgencies are often more complex matrices of irregular actors with widely differing goals. At least some of the principal actors will be motivated by a form of ideology (or at least will claim to be), but that ideology will not necessarily extend across the whole insurgent network. Modern insurgencies are typified by the points below.

- The charisma of insurgent leaders can sometimes be more important than ideology in convincing others to join their movement;
- Some insurgent actors will be more interested in financial reward than ideology. This applies from the unemployed youth getting paid to fight to the criminal gang leader exploiting a state of lawlessness;
- The basic wants, needs and grievances of the population may have little to do with the intellectual ideology of insurgent leaders, but may still be exploited to generate support;

- Even those fighters, sympathizers, and supporters who justify their actions with the rhetoric and symbols provided by insurgent propagandists may not be fully conversant with the ideology;
- Hatred that emerges during armed conflict, through atrocities and dispossession, often overshadows the initial motivators that drove individuals and community groups to join the insurgency or support the government;
- Players in pre-existing local conflict may draw on the insurgents (or the government) as an external ally to help them;
- In tribal societies (as found in parts of South and Central Asia, the Middle East and Africa) the support of one tribe or faction for the government may often predispose tribal rivals to support the insurgents, and vice versa.

## Building Networks

Insurgents require supporters, recruits, safe havens, money, supplies, weapons and intelligence on government actions. A robust insurgency can be waged with the support of just a small percentage of a given population. From the remaining majority, insurgents require only compliance (acquiescence or inaction). The position of an active individual within an insurgent network will be determined by the combination of a number of factors including:

- The level of respect and trust they hold within a community;
- Their reputation established through previous insurgent actions;
- Their degree of motivation, ideological or otherwise;
- Their perceived loyalty to other network members;
- Their level of expertise in a particular field;
- Their access to resources, human or otherwise;
- The degree of risk they are prepared to accept.

Insurgent networks provide life support for the movements they support, but they also entail vulnerability. Command and support networks establish lines between isolated cells whose operational security may otherwise be impeccable. Some key functions may be deliverable only by individuals with dubious loyalty, for example criminal smugglers enabling logistics supply or personnel movement. Modern information infrastructure including mobile phones and the internet provide means of rapid communications and networking between insurgents, but are also open to exploitation.

The most secure insurgent networks involve small numbers of active personnel who are trustworthy and employ tight operational security. However, the insurgent dilemma is that in order to promote the insurgency and exploit success, these small networks have to expand, exposing themselves to action by government security forces. Effective interdiction can lead to a cycle of expansion and contraction of insurgent networks as security and trust is repeatedly built up and then lost.

## Funding

To fund their activities, insurgents may foster an illicit economy, sometimes of international scope, eluding government monitoring, taxation and interdiction. Such illicit financial activities diminish government revenues, increase corruption among local officials, and weaken the control and legitimacy of the government. Criminal activities may include theft, extortion, trafficking (of narcotics, arms and people), money laundering, piracy, document fraud, bribery, kidnapping and black market activity. These funding streams will often drive insurgents into alliances of convenience with organized crime. In some cases, long-standing insurgencies morph into gangs or organized criminal networks that are motivated by profit and economic self-interest, rather than ideology.

Funding may also be obtained through donations from sympathetic foreign governments, diaspora groups and individuals. Such funding streams may be simple and direct or complex and masked dependent on the efforts being taken internationally to interdict them. In extreme cases, funding may be channeled through a third party organization purportedly conducting charitable work.

## Trans-National Dynamics

Most insurgencies need a physical safe haven, and may find it in neighboring countries. Moreover, contemporary insurgencies are often supported or driven by transnational networks with access to satellite communications, the Internet, global media and transnational banking systems. International support may be leveraged from diaspora or émigré communities, international institutions, friendly foreign governments and populations, or the international media. If other countries give support to the affected government, the insurgents may directly target public opinion there, pressuring them to cease their assistance. Such pressure may be exerted from the affected territory through the kidnap, torture and murder of intervening civilian nationals, often broadcast internationally to reach the population of origin. Alternatively, more direct effect may be achieved through terrorist attacks launched within the intervening country itself (perhaps facilitated by immigrant or other sympathetic community groups). Policy makers must therefore take into account regional and global dynamics as well as the internal situation of the nation most directly affected.

## Insurgent Political Strategy

To gain the support (or at least the acquiescence) of the people, insurgents may apply a combination of persuasion, subversion and coercion.

- Persuasion will often involve the promotion of insurgent ideology, but it can also include the provision of money, basic social services, control of land, or positions of authority. Insurgents may appeal to the self-interest of constituencies through alliances with political parties, tribal leaders, ethnic or religious groups, warlords, organized crime networks, and local bandits. To do so, they will exploit societal trends and popular grievances or needs, manipulating elements of religious, tribal, ethnic or local identity that resonate with some subset of the target population. These trends do not necessarily have to be linked directly to the ideology of insurgent leaders to be exploited by them in the mobilization of support. Such partnerships may exacerbate localized conflict, perpetuate instability and help mobilize support for the insurgency.

Propaganda is a key element of persuasion and is used at the local, national and often international levels to influence perceptions of potential supporters, opinion leaders, and opponents in the favor of the insurgents; promoting the insurgent cause and diminishing the government's resolve. More specifically, propaganda may be designed to control community action, discredit government action, provoke overreaction by security forces, or exacerbate sectarian tension.

- Subversion is the action used by insurgents to penetrate, manipulate, undermine or disrupt government institutions and organizations. At the same time, insurgents may exploit competing power structures, such as tribal hierarchies, clerical authorities or criminal networks that challenge the authority and reach of control of the central government. In doing so, the insurgents seek to 'out administer' the local authorities. Clearly, these techniques will be particularly effective in areas where government services are weak or absent.
- Coercion can either augment or replace persuasion as a means to mobilize support, depending on the ability of government forces to protect the population. Insurgents seek to intimidate government supporters or collaborators, and force community leaders to take sides in the conflict. This often entails the use of violence as follows:
    - Insurgents can use violence to intimidate or eliminate those who oppose their aims. In particular, insurgent attacks against government infrastructure and personnel will undermine the government's morale, confidence and capability, weakening its authority and control over affected areas;
    - This effect is magnified by the depiction of such violence in propaganda, portraying the government as weak and the insurgents as strong, and exacerbating local grievances. Propaganda is sometimes the primary aim of insurgent violence;
    - Targeting members of different ethnic or sectarian groups may engender a sense of social identity, solidarity and alienation from the government;
    - By creating violent instability, insurgents may be able to encourage people to turn to them in preference to the government to 'restore' public order;
    - If insurgents can provoke excessive government action against a population, then death, injury, mistreatment, or dishonor can become a powerful motivator for retributive action against the government.

## Challenging Government Security

Insurgents usually have less conventional military capacity than the government (at least in the early stages of insurgency) and so tend to use guerrilla tactics to inflict damage without allowing their fighters to be engaged by equal or larger government forces. Tactics such as raids, ambushes, assassinations, sabotage, booby traps, and improvised explosive devices take advantage of mobility, stealth, deception and surprise to weaken, discredit, or paralyze the less agile government security forces. Insurgents try to manage the tempo and intensity of their activities to permit a level of effort they can sustain indefinitely. By prolonging the conflict, they hope to exhaust the opposition, seeking to impose unsustainable costs on the government to force capitulation. Although the permutations of insurgent activity are context-driven, historical analysis shows that insurgents typically apply four basic tactics, or variations of them, to defeat stronger security forces:

- Provocation: Insurgents often commit acts (such as atrocities) that are intended to prompt opponents to react irrationally, in ways that harm their interests. For example, government forces, frustrated by their inability to distinguish fighters from non-combatants, may be provoked into indiscriminate reprisals or harsh security measures that alienate parts of the population. Alternatively, one tribal, religious, ethnic or community group may be provoked into attacking another in order to create and exploit instability.
- Intimidation: Insurgents intimidate individual members of the government (especially police and local government officials) to dissuade them from taking active measures against the insurgents. They will also publicly kill civilians who collaborate with government or coalition forces, thereby deterring others who might seek to work with the government.
- Protraction: Insurgents seek to prolong the conflict in order to exhaust opponents, erode their political will, and avoid losses. Typically insurgents react to government countermeasures by going quiet (reducing activity and hiding in inaccessible terrain or within sympathetic or intimidated population groups) when pressure becomes too severe. They then emerge later to fight on.
- Exhaustion: Insurgents conduct activities such as ambushes, bombings, attacks on government facilities, economic assets and transport infrastructure that are designed to compel security forces to undertake numerous onerous, high-cost defensive activities that expend scarce resources without significantly advancing the counterinsurgents' strategy.

## Stages of Insurgency

Every insurgency develops differently, but some general patterns can be observed. Insurgencies may evolve through some or all the stages of subversion and radicalization, popular unrest, civil disobedience, localized guerrilla activity, and widespread guerrilla warfare to open, semi-conventional armed conflict. Alternatively, they may wither away to dormancy if they are effectively countered or if they fail to capture sufficient popular support. One or more different stages may appear simultaneously in a country or region affected by insurgency. An insurgency may actually succeed in overthrowing the government (historically a rare event), may force the government into political accommodation (a more common outcome), may be co-opted by the government and cease fighting (also common), or may be crushed. Insurgencies may be co-opted by domestic or trans-national terrorist groups, morph into criminal networks, or wither into irrelevance. Measures that succeed against incipient insurgencies often differ greatly from those that are effective against mature or declining insurgencies. Thus, planners and decision-makers must clearly understand the stage the insurgency has reached, to develop appropriate responses.

# PART B. COUNTERINSURGENCY

## Definition

Counterinsurgency may be defined as 'comprehensive civilian and military efforts taken to simultaneously defeat and contain insurgency and address its root causes'.

Best practice COIN integrates and synchronizes political, security, economic, and informational components that reinforce governmental legitimacy and effectiveness while reducing insurgent influence over the population. COIN strategies should be designed to simultaneously protect the population from insurgent violence; strengthen the legitimacy and capacity of government institutions to govern responsibly and marginalize insurgents politically, socially, and economically.

## Characteristics

COIN is a complex effort that integrates the full range of civilian and military agencies. It is often more population-centric (focused on securing and controlling a given population or populations) than enemy-centric (focused on defeating a particular enemy group). Note that this does not mean that COIN is less violent than any other conflict: on the contrary, like any other form of warfare it always involves loss of life. It is an extremely difficult undertaking, is often highly controversial politically, involves a series of ambiguous events that are extremely difficult to interpret, and often requires vastly more resources and time than initially anticipated. In particular, governments that embark upon COIN campaigns often severely underestimate the requirement for a very long-duration, relatively high-cost commitment (in terms of financial cost, political capital, military resources and human life). The capabilities required for COIN may be very similar to those required for peacekeeping operations, humanitarian assistance, stabilization operations, and development assistance missions. However, the intent of a COIN campaign is to build popular support for a government while marginalizing the insurgents: it is therefore fundamentally an *armed political competition* with the insurgents. Consequently, control (over the environment, the population, the level of security, the pace of events, and the enemy) is the fundamental goal of COIN, a goal that distinguishes it from peace operations or humanitarian intervention. Within these broad characteristics, the specific nature of any particular COIN campaign arises from the complex interaction of three key factors: the characteristics of the environment (physical, economic, political and human) in which it takes place; the nature of the insurgent group (or groups); and the nature of the counterinsurgent government and its security forces.

## Different Forms of COIN

- Domestic COIN versus Overseas COIN: A nation faces very significant conceptual and practical differences between conducting COIN within its own national boundaries and intervening in a foreign country in support of another government. Intervention to conduct COIN in a foreign country is often a discretionary activity whereas internal/domestic COIN is usually not a matter of choice. Forces operating in another government's territory are vulnerable to the insurgents' "home ground" advantage: the insurgents live in the country and never plan to leave, whereas the intervening force must eventually plan on transition and departure. The population knows this and is therefore less likely to support it. A government conducting COIN in its own territory will generally have greater strategic patience to stay the course of a protracted struggle. Another "home ground" advantage is the detailed understanding of the geography, culture, history, sociology and politics of the affected country which insurgents will already possess but the intervening country will usually have to learn.
- Since the United States presently faces no credible internal insurgency, all U.S. COIN campaigns are likely to be external interventions in support of a foreign government (or in failed/collapsed states). Intervention to support COIN merits careful consideration of a range of factors that are addressed in detail in Chapter 4 (Assessment and Planning).
- Bilateral Versus Multilateral COIN: The United States may not be the only foreign country prepared to assist the affected nation in countering insurgency. There are significant differences between campaigns supported by a single intervening power and those involving an intervening coalition or United Nations force. Coalition COIN will often be seen as a more legitimate endeavor than a U.S. only intervention, but it requires significant alliance management and coordination and is inherently less efficient than unilateral COIN. This is covered in more detail in Chapter 4.
- Different Levels of Consent: Not all COIN interventions will have the full consent of the affected government. There will be major differences between campaigns conducted with full consent, partial consent, or where there is no effective government. A final variation (recognized as especially difficult) is where an insurgency follows a conventional war in which an invading power or coalition overthrows the existing government, then builds a new government from scratch (or radically reforms an existing structure) while increasingly being opposed by insurgents. In this scenario, the challenges of conducting COIN may not have been fully anticipated or considered during the original decision to invade. However, by the time the insurgent threat is manifested, intervening governments may have little choice but to remain committed to a protracted and costly COIN campaign. It is important to recognize that the decisions to intervene in Afghanistan and Iraq were not taken as decisions to undertake COIN campaigns. The strategic drivers and the factors considered were very different. Use of these examples in the study of COIN intervention could therefore be misleading and policy makers should exercise caution in drawing parallels.

## Purpose

COIN differs from other civil-military operations both in the methods employed and in the purpose of the undertaking. The purpose of COIN is to build popular support for a government while suppressing or co-opting insurgent movements.

## Approaches

COIN is a response to a specific set of insurgent threats and requires in-depth assessment of that threat matrix, based on solid understanding of the relevant social, cultural, economic, political and security conditions, along with detailed knowledge of insurgent motivations, goals, organization, and methods. Central to gaining the confidence and support of the population is to improve the quality of governance through political reform, strengthening the rule of law and conducting economic development as appropriate. Simultaneously, a mix of diplomacy and negotiation, police methods, intelligence work, military combat and non-combat activities should be used to destroy, contain, marginalize or co-opt the insurgents. Effective COIN therefore involves a careful balance between constructive dimensions (building effective and legitimate government) and destructive dimensions (destroying the insurgent movements). As noted, there are two basic approaches to COIN strategy:

- The enemy-centric approach conceptualizes COIN as a contest with an organized enemy, and focuses COIN activity on the insurgent organizations. This approach emphasizes defeat of the enemy as its primary task and other activities as supporting efforts. There are many variants within this approach, including "soft" vs. "hard," direct vs. indirect, violent vs. non-violent, and decapitation vs. marginalization strategies. This approach can be summarized as "first defeat the enemy, and all else will follow."
- The population-centric approach shifts the focus of COIN from defeating the insurgent organization to maintaining or recovering the support of the population. While direct military action against the insurgent organization will definitely be required, it is not the main effort; this approach assumes that the center of gravity is the government's relationship with and support among the population. It can be summarized as "first protect and support the population, and all else will follow."

A purely enemy-centric approach might work against incipient insurgencies that are led or centrally controlled by a particularly charismatic or powerful individual. However, historical experience has shown that against mature insurgencies and complex, non-hierarchical insurgencies, population-centric approaches have a higher likelihood of success.

In reality, COIN campaigns will rarely be purely enemy-centric or population- centric, but will generally include elements of both, with the relative balance changing over time. Winning the support of the population for the legitimate government will often entail a breaking of the ideological, financial or intimidation linkages between insurgent leaders and their supporters, perhaps one family grouping or village at a time. Counterinsurgents should always be looking for potential fracture lines where the COINcidence of interests between the ideological leadership and a particular part of the insurgent network is weakest. A wedge may then be created through the use of carrot (political, economic & development benefits) and stick (detention & disruption) operations. Key 'bridging' individuals (insurgents, who by personal connections link whole tribes or other groupings to the insurgent leadership) should be a priority for reconciliation or detention, but to achieve this, a deep understanding of regional sociology and relative motivations is critical. This could be viewed as an enemy centric approach, since it focuses on the insurgent network, but its ultimate objective is to win over whole sectors of population to the government cause.

## Primacy of Non-Military Means

While the provision of security is a necessary activity in COIN, it will not defeat an insurgency on its own. When possible, civilian and military measures should be applied simultaneously to achieve success in an integrated strategy that delegitimizes and undermines the insurgency, builds government control and strengthens popular support. In counterinsurgency, military forces are, in a sense, an enabling system for civil administration; their role is to afford sufficient protection and stability to allow the government to work safely with its population, for economic revival, political reconciliation and external non-government assistance to be effective.

## Unity of Effort

COIN usually involves the efforts of multiple stakeholders. Unity of effort is highly desirable at the national level (among the various agencies of the affected government) and at the international level (between the affected nation and all supporting nations). This is not easily achieved, especially in the context of a coalition intervention operation. Hence, robust command-and-control (C2) mechanisms, as well as diplomatic efforts to maintain coalition cohesion and support, are fundamentally important. They must be considered in detail at the outset of a campaign, and given adequate resources and attention during it. In more complex coalition scenarios, the appointment of a single overall strategic advisor to an affected government (ideally with the mandate and responsibility to coordinate all international civil and military assistance) may be the only means to ensure unity of effort.

## COIN as a Special Environment

For military forces, COIN often involves a wider range of tasks and capabilities than are required in conventional conflict. Armed forces that are optimized for major combat operations will usually require specific training (and perhaps even structural reorganization) to meet the unique requirements of COIN. For civilian agencies engaged in diplomatic, development and information activities, COIN is less about performing a different set of tasks than about operating in a different kind of environment; often conducting their activities in chaotic, unstable, or actively hostile environments. Tasks will often need to be carried out in dramatically different ways, generally requiring specialist training and sometimes requiring development of new capabilities. However it is achieved, the targeted application of diplomatic, development and information effects in a conflict situation is fraught with the risk of unintended consequences and requires a sophisticated understanding of the local context.

### Success in COIN

A COIN effort may be deemed successful when the following conditions are met:

- The affected government is seen as legitimate, controlling social, political, economic and security institutions that meet the population's needs, including adequate mechanisms to address the grievances that may have fueled support of the insurgency;
- The insurgent movements and their leaders are co-opted, marginalized, or separated from the population;
- Armed insurgent forces have dissolved or been demobilized, and/or reintegrated into the political, economic, and social structures of the country.

It should be noted that it is usually far harder to totally eliminate insurgent activity than it is to reduce it as a significant threat to the affected government. Insurgents can become extremely difficult to identify, track and interdict once their manpower and activity drop below a critical mass. Consequently, low level insurgencies may persist for many years after the main threat has been broken. International support in COIN will usually be consensually withdrawn once insurgency can be comfortably contained by the affected government. The combination of these factors means that a COIN intervention in an affected country may end several years before the insurgency there ceases.

## 2. Components of COIN Strategy

To be effective, officials involved in COIN campaigns must address two imperatives — political action and security operations—with equal urgency, recognizing that insurgency is fundamentally an armed political competition and that effective security operations, though unlikely to deliver success by themselves, will almost always be a prerequisite to political resolution. Security operations, conducted in support of a political strategy, coordinated with economic development activity and integrated with an information campaign, will provide human security to the population and improve the political and economic situation at the local level. This should increase society's acceptance of the government and, in turn, popular support for the COIN campaign. COIN functions therefore include informational, security, political and economic components, all of which are designed to support the overall objective of establishing and consolidating control over the environment, then transferring it to effective and legitimate local authorities.

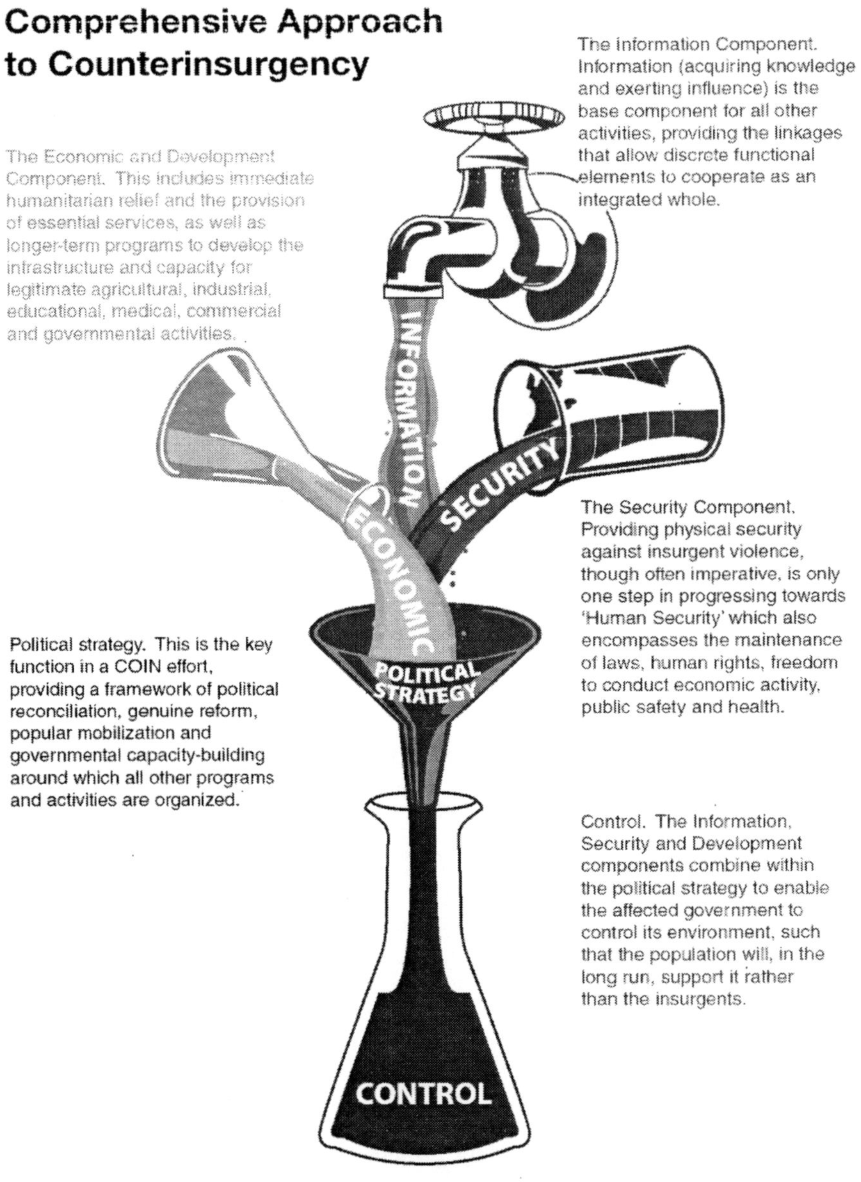


The diagram on the preceding page is designed to help policy makers visualize the interaction of COIN components by illustrating the key functions of a comprehensive COIN framework. This approach builds on classic COIN theory but also incorporates best practices that have emerged through experience in numerous complex operations over recent decades.

The diagram is a visualization tool, not a template for action. It is intended to demonstrate to policy makers and program implementers where their efforts fit into a COIN strategy or campaign, rather than telling them what to do in a given situation (it is an aid to collaboration, not an operational plan). The functions are linked to one another primarily through the information function, which underpins and integrates the whole effort, and through the common campaign intent embodied in the control function.

## Information

Information is the foundation for all other activities, and provides the linkages that allow discrete functional elements to cooperate as an integrated whole. The collection, formulation, storage and dissemination of information are crucial in shaping perceptions of the conflict by all stakeholders.

In COIN, the information flow can be roughly divided into that information which we wish to assimilate in order to inform our approach; i.e. knowledge and that information which we wish to disseminate in order to influence populations. At the same time, as counterinsurgents we are also interested in impeding the information flow of insurgent groups; both their intelligence collection and their ability to influence.

- Knowledge: In COIN, decisions at all levels must be based on a detailed understanding and awareness of the environment. No COIN strategy can be better than the degree of understanding on which it is based. The information required to engender this understanding encompasses a far broader range of subjects than would normally fall under the auspices of military intelligence. In conventional warfare, decision makers mostly require intelligence about the enemy, but in COIN they primarily need intelligence about the population. COIN intelligence must therefore incorporate the spectrum of characteristics of a nation's system of systems, including political, military, economic, sociocultural, infrastructural, informational and environmental knowledge.
  At the strategic level, understanding is required of the population factors behind the insurgency, its stage of progression, the reforms required to address its causes, and the willingness and ability of the affected government to make those reforms and the implications of foreign intervention.
  At the operational level, understanding is required of the strengths and vulnerabilities of the insurgent strategy, the strengths and weaknesses of the affected government and the requirements of the population. Continuous feedback on the degree of success of ongoing COIN efforts is also critical.
  At the tactical level, understanding is required of the identity of active insurgents, their networks, logistics, capabilities and intent. It is also very useful to understand the views, concerns and sympathies of non-combatant civilians in order to influence them, gain additional intelligence and further isolate the insurgents. Almost all forms of intelligence collection have a role in COIN, but historically, intelligence gathered from human sources (including civilians, agents and captured/reconciled insurgents) has made the greatest contribution to success.
  One of the most critical yet pervasive shortcomings that interagency operations face is the failure to manage and share knowledge. This is especially true during COIN, when a common base of understanding is key to effective collaboration.
- Influence: Effective COIN requires the shaping of opinions among several different population groups through messages and actions:
  - Affected Nation: The fundamental influence aim in COIN is to build confidence in the government while diminishing the credibility and influence of the insurgents. All actions and messages should support this aim.

- U.S. Population: Where the United States is conducting a direct intervention in support of an affected nation, the costs involved (financial and human) will often be high and prolonged. Understanding and support in the U.S. will be key if the nation is to remain engaged long enough to achieve decisive effect.
- Neighboring Countries: Many insurgencies depend on safe havens in countries adjacent to the affected nation. Sanctuary may be giving willingly or may be beyond control of the government there. Even in non-democratic nations and ungoverned spaces, there may be merit in efforts to influence populations in these areas.
- Coalition Nations: Political resolve will rarely be consistently robust across a coalition. U.S. policy makers should be cognizant of the difficulties some coalition members will face in maintaining popular support for their participation.
- Diaspora Communities: As previously discussed, diaspora communities can play a significant role in supporting or opposing insurgencies. Positive influence here can pay dividends.

Clean separation of messages to these various populations is rarely possible and a high degree of coordination is required to allow messages to be tailored to different audiences without contradiction.

The influence strategy must cascade down from a set of strategic narratives from which all messages and actions should be derived. The narratives of the affected government and supporting nations will be different but complimentary. Messages and actions must address ideological, social, cultural, political, and religious motivations that influence or engender a sense of common interest and identity among the affected population and international stakeholders. They should also counter insurgents' ideology in order to undermine their motivation and deny them popular support and sanctuary (both physical and virtual). In doing so, counterinsurgents should seek to expose the tensions in motivation (between different ideologies or between ideology and self-interest) that exist across insurgent networks.

To enhance the legitimacy of the affected government, messages aimed at their population should be closely coordinated with and ideally delivered by their own officials. Themes and messages should be simple and memorable, and must resonate with the population. This requires detailed understanding of the COIN environment which must be continually updated. Detailed target audience analysis is required for each separate population group and reliable measures of effectiveness must be sought to assess the success of messages and if necessary recalibrate them.

Messages are delivered partly through media operations, but more prolifically and often more credibly through the thousands of daily interactions between the population, the government and the security forces. Every action in COIN sends a message, which means that words and deeds *must* be synchronized. Messages cannot simply be spin, they must be grounded in truth and reflect a genuine willingness on the part of the affected government to undertake real reforms that address its people's needs. Failure to honor promises is usually extremely counter-productive, so officials should be cautious in making promises, and should track any that are made, with the aim of meeting or exceeding the expectations of the population.

- The Information Contest: The flow of information (intelligence and influence) is as important to insurgents as it is to counterinsurgents. A COIN campaign should seek to limit the intelligence available to the insurgents through use of counterintelligence, deception and where possible their physical separation from the populace. Similarly, the ability of the insurgents to exert influence should be restricted by physical separation and by the pre-emption and timely countering of their messages.

Influence activities (actions and messages) can be proactive or reactive. Being proactive gives a significant influence advantage, since the first impression or report of an event that reaches a population will often receive the widest exposure and will subsequently be most resistant to alternative accounts. The insurgents that are most effective in this field design whole operations to support their influence themes. Successful COIN requires an approach to influence which is similarly proactive, entrepreneurial and prolific in 'selling' messages.

That said, the imperative to counter insurgent messages demands a reactive element to our influence activities. Speed is of the essence. The longer it takes for a rebuttal, denial or counter-message to be released, the less relevant and effective it will be. Cumulatively, whichever protagonist (insurgent or counterinsurgent) is fastest at processing the cycle of messaging will have a significant advantage in gaining influence. Some of the counterinsurgents' delay in response will be derived from the need to investigate events and establish facts (a constraint from which insurgents are often exempt), though a holding response is generally preferable to silence. Less justifiable is the delay inherent in lengthy approvals processes.

The time sensitivity of reactive influence requires counterinsurgents to employ delegation of authority, universal understanding of the narrative and a degree of risk-tolerance throughout the command chain. This has strong parallels to the military concept of the OODA loop (Observe – Orient – Decide – Act) and the theory of the mission oriented approach.

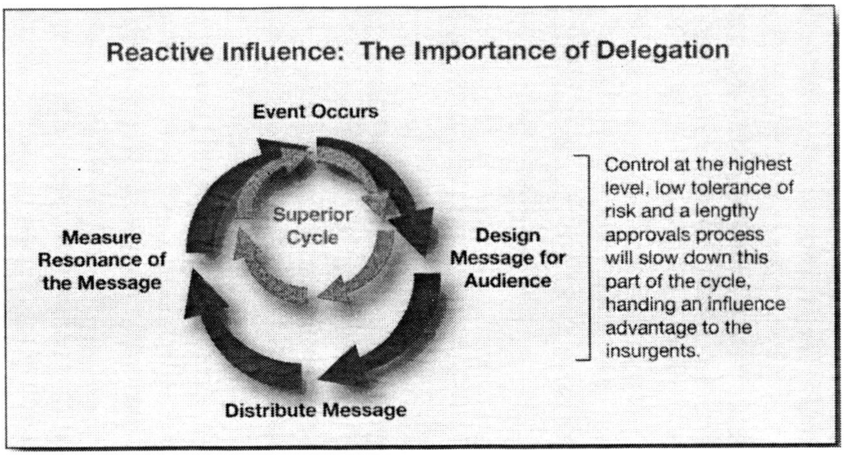


## Security

In COIN scenarios, the term security is frequently used to refer to the degree to which the government can suppress insurgent activity in an area. However, the concept of 'Human

Security' is a more complex metric which can only be measured through the collation of individual perceptions across a community. The paramount concern is the absence of physical violence, but other relevant factors include maintenance of laws, the protection of human rights, freedom to conduct economic activity, public safety (fire, ambulance, etc.) and public health (safe drinking water, sanitation, etc.). COIN emphasis on physical security is not necessarily an indicator that the wider concept of human security is not important, but more a case of imperative and sequencing. The end state of providing human security should be implicit in the wider efforts to improve the standard of governance down to the local level.

Physical security efforts must not focus too greatly on strengthening the military and police forces of the affected nation. Such capacity building should only be part of a broader process of Security Sector Reform (SSR) in which the whole system is developed, including the civil institutions that oversee the security forces and intelligence services, the legal framework and the justice institutions (prosecution services, judiciary and prisons) that implement it. It is particularly important that a sense of civil ownership and accountability should extend to the local level and that all elements of the security apparatus should be trusted by the population. Taking this broader view of security is very useful to countries engaged in counterinsurgency, since it links the reduction of violence to the improvement of many of the issues that are most important to a population. Indeed, effective SSR (especially reform of the justice sector) may address many of the grievances that initially fueled the insurgency.

SSR should be conducted in a balanced and carefully sequenced way, or it may have negative unintended consequences. For example dramatic improvement in police capability will only cause demoralization if the judicial system remains inherently corrupt and fails to prosecute those arrested. There may then be a tendency for the police to take justice into their own hands in the form of 'extra-judicial killings'. Similarly, increasing military capacity and competence may risk a coup if the civil control mechanisms are not in place and the government is seen by the military to be corrupt or ineffective. Once again, a sound understanding of context is critical.

In many stable democracies, the military forces are primarily responsible for defense against external threats while the police are responsible for maintaining internal law and order. However, COIN situations often arise because the police are incapable of maintaining order (whether through lack of capacity, lack of capability, corruption or active bias) and so military intervention is often necessary. Maintenance or prompt restoration of police primacy is highly desirable as it reinforces the perception of insurgents as 'criminals' rather than 'freedom fighters' and highlights the government's commitment to uphold its peacetime legal framework. It is also undesirable for the military to take on too great a role in delivery of local civil services as this may unbalance the distribution of power within government and increase the risk of a military coup.

The U.S. Government has significant capacity, through the Department of Defense, to develop the military forces of an affected nation. However, U.S. ability to assist in other areas of SSR is currently limited by a shortfall of deployable capacity to assist in law-making, judiciary and policing. Other countries within a coalition may be able to make a significant contribution in this area.

Unsecured areas provide particular challenges to many of the actors who are best able to remedy political and developmental deficiencies, but this does not mean that establishing security must be seen as a necessary precursor to economic and governance activity: rather, security, economic and political efforts should ideally be developed simultaneously. The common terminology "clear-hold-build" is very useful, but is probably interpreted as more of a set sequence than is always necessary or advantageous.

## The Political Strategy

Political strategy is the key function in a COIN effort, because it provides a framework (of political reconciliation, reform, popular mobilization and governmental capacity-building) around which all other programs and activities are organized. In general terms, the progress of a COIN campaign is only as good as the political strategy it supports. Where the political strategy is vague, unrealistic, or lacking in support from domestic or international stakeholders, the campaign is unlikely to succeed, whatever the efficiency of individual programs. An effective political strategy focuses on strengthening the government's capability and capacity to respond—*and be seen to be responding*—to the needs of its people. Almost by definition, a government facing insurgency will require a degree of political "behavior modification" (substantive political reform, anti-corruption and governance improvement) in order to successfully address the grievances that gave rise to insurgency in the first place. Supporting nations may be able to assist in these reforms.

Where the security environment prevents U.S. civilian agencies from operating freely, the U.S. military may be required to provide extensive support to political, economic and governance efforts in their stead. This will be the case during the "establishment of control" phase in every COIN campaign and, in many cases, throughout the campaign. Given the difference in risk acceptance and the large and enduring resource imbalance between civilian and military agencies this is simply a fact of life: *officials and policy makers must plan for it accordingly.*

## The Economic and Development Function

The economic and development function in COIN includes immediate humanitarian relief and the provision of essential services such as safe water, sanitation, basic heath care, livelihood assistance, and primary education, as well as longer- term programs for development of infrastructure to support agricultural, industrial, educational, medical and commercial activities. It also includes efforts to build the absorptive capacity of local economies and generate government and societal revenues from economic activity (much of which may previously have been illicit or informal). Assistance in effective resource and infrastructure management, including construction of key infrastructure, may be critically important to COIN efforts. It must be tailored to the affected government's willingness to undertake key reforms, capacity to absorb support, and ability to manage its outcomes.

The first principle for most development agencies is to ensure that their activities 'do no harm' and cultivate sustainable benefits in the regions concerned. They seek to ensure this through an in depth assessment of the background situation followed by the application of program management tools to give continuous evaluation and adjustment.

A COIN scenario presents particular challenges for the conduct of economic and development activity. For example:

- Security: There is often a geographical correlation between areas of highest threat from insurgents and areas most in need of development. Development agencies and non-government organizations (NGOs) may be specifically targeted by insurgents, restricting their ability to operate independently and generating more reliance on military forces to undertake this role.
- Time-scale: Counterinsurgents often seek to generate a rapid and obvious improvement in quality of life and economic opportunity in areas where security operations are suppressing insurgent activity. The aim is for the population to see clear benefit in government control. However, quick impact projects will usually lack the depth of assessment that precedes more deliberate programs and may therefore have a higher risk of unintended consequences. They also give less sense of long term economic optimism and commitment from both the affected government and its international supporters.
- Parity: U.S. assistance can sometimes influence a particular local leader or population element. Giving assistance to any one part of a population however, may be seen as preferential and discriminatory by other groups and may actually exacerbate underlying sectarian, regional or tribal tensions. This is especially so when development has actually been targeted for political effect. Using development to buy allegiance may be immediately effective in reducing violence, but in the long term it may foster corruption and reduce the credibility of the affected government and its international supporters. Development resources should be allocated in a balanced way and must not be seen to be given 'as a reward for bad behavior'.
- Corruption: The requirement in COIN to build the credibility and effectiveness of the affected government can imply that development resources should be channeled through its ministries. Governments facing insurgency will often have endemic corruption and may therefore be deemed unfit to handle U.S. financial assistance, yet bypassing those ministries might only further enfeeble them. Corruption prevention measures should be implemented within a wider program of human capacity development, but will often take a lengthy time period to become effective. Sensitivity may be necessary in defining "corruption" in the context of other cultures or established norms. Similarly, the decision over when or whether to attempt change should be taken carefully, weighing the impact in terms of potential for success and desired outcome.
- Oversight: NGOs will often undertake a significant proportion of development activity in COIN scenarios, yet they will not always heed any form of direction from the affected nation or intervening U.S. officials. This reduces the ability of the counterinsurgents to anticipate and prevent the unintended consequences of development activity.

## Control

The four functions (information, political, security and economic) contribute to the overall objective of enabling the affected government to control its environment. This implies the ability to contain insurgent activity (the tempo of operations, level and intensity of violence, and degree of instability that it engenders) such that the population will, in the long run, support the government against the insurgents — noting that this "balance" can differ from one society to the next. The progress of control in a successful COIN campaign often develops in three overlapping phases: establishment, consolidation and transfer (or transition):

- Establishment of control: During the initial phases, a government conducting COIN seeks to establish control over the environment. This requires controlling terrain, key infrastructure, economic production assets, population movement, resources and information flow. In the initial stages there will almost always be a need to catch up with insurgents who have gained the initiative, made inroads into popular support and undermined the government. Because the population is often fearful of the insurgents, the use of force ("kinetic" or lethal security operations to kill or capture insurgents) is almost always a significant feature of this phase of a campaign. Typically, only when the population sees insurgents losing engagements against the government, and sees the death or capture of insurgent leaders who previously intimidated the population, will its support begin to swing behind the government. However, establishing control requires not only the military defeat of insurgents, but also their political marginalization and the provision of economic and governance benefits to the population in order to reduce the insurgents' appeal.
- Consolidation: Once control has been established in one area, the government seeks to extend and consolidate it (in geographic, demographic and functional terms). This phase is typically the longest in duration, lasting years or even decades. In COIN, establishing control over population groups and population centers is more important than the control of territory. Consolidation involves replacing insurgent networks with pro-government ones, rooting out insurgent underground cells and infrastructure, eliminating illicit governance structures and cleaning up illegal economic activity that supports insurgents.

  The key indicator of success is the degree of government control in each district and not the level of insurgent violence, since the latter tends to be low in areas that are fully controlled by either side but high in contested areas. This phase involves substantially less kinetic force than the initial phase, with the military "holding the ring" while police, intelligence, governance, information and economic programs assume the lead, and political leaders work to resolve key grievances and mobilize popular support for ending the insurgency. During this phase there are often outbreaks of insurgent violence, large-scale provocations or insurgent atrocities that seek to derail the process of consolidation. Consequently the entire theater (or parts of it that have already been cleared) may slip back under insurgent control.

  The process of consolidation must be carefully managed to ensure that it does not progress too slowly (leading to popular frustration and regression to insurgent control in key areas) nor too quickly (leading to premature fielding of inadequate local security forces, or premature transition before control is fully consolidated). In a U.S. intervention, the tendency of officials to seek an early handover to elected local leaders needs to be balanced against the requirement to consolidate control, so that those leaders can assume responsibility for a stable functioning system.

- Transition: 'Transition' is used to describe not just the transfer of authority from an intervening nation to the host nation but also the handover from indigenous military forces to local civilian authority (an essential step in normalizing the environment and ending insurgency). Although described last in this theoretical sequence, transition occurs throughout a campaign, as indigenous civil structures become sufficiently legitimate, effective and credible to take the reins. In an intervention, the external force's "exit strategy" timeline will depend almost entirely on the rate at which indigenous capacity can be built and rendered effective and legitimate. This requires considerable time and resources and must be carefully planned for from the outset. The process, pace and sequence of the process requires considerable judgment and is one of the most difficult decisions that officials directing a counterinsurgency campaign must undertake. Examples abound of COIN campaigns where ill-judged or hasty transition created opportunities for insurgent comebacks. Conversely, too slow a transition risks the loss of domestic political support for the campaign, as tangible signs of progress may be hard to perceive. It also risks creating a dependency culture in which any appetite for autonomy dwindles.

### Relationship between Functions

Maintaining an agile and flexible balance among the key counterinsurgency functions is difficult but extremely important. For example, economic assistance programs with inadequate security presence could simply create an array of soft targets for the insurgents. Security assistance in the absence of capable political leadership and oversight could create more—and more capable—armed groups outside the control of the government. Moreover, while an action may fall within one function, it often has immediate effects in the other functions. Efforts must be integrated because effects are impossible to segregate and are often complementary. Maintaining a balance between functions requires an integrated conflict management system (which may be based on a joint committee system, an integrated command model, a consultative alliance process or a combination of measures) that enacts the overall COIN strategy and coordinates the activities of key agencies (civil, military, affected nation and external/coalition).

## 3. Actors

James Kunder (on right), Deputy Administrator of USAID, and Members of a Provincial Reconstruction Team discuss development issues with an Iraqi Sheikh.

## The Affected Government

The government threatened by a nascent or active insurgency is the most important actor in COIN. However great its know-how and enthusiasm, an outside actor can never fully compensate for lack of will, incapacity or counter-productive behavior on the part of the supported government: any COIN campaign is only as good as the political strategy which the affected nation adopts. The U.S. Government perception of the role of each actor in a COIN struggle (including its own role) may not COINcide with the perception of other actors or the population. Additionally, the insurgents may portray the U.S. role as one of foreign aggressor, which may well resonate with the affected population. That population may or may not support and appreciate the security provided by an external force, but they will certainly see it as foreign and temporary; their long-term allegiance will tend to swing toward whichever local actor (government or insurgent) best aligns with their needs and political identity, or best provides security. Effective COIN therefore requires that the major effort is (and is seen by the local population to be) led by the indigenous government. Under ideal conditions, foreign forces do not operate independently of the affected government, nor are political, economic or other development assistance activities undertaken except at the request of the affected government.

But real world conditions are never ideal. Effective, legitimate governments that meet the needs of their people and are capable of managing internal security threats are, almost by definition, unlikely to require external COIN assistance. Governments that do require such assistance almost always need encouragement and assistance to address governance shortfalls, corruption, and the real or perceived issues which insurgents use to mobilize the population. Similarly, in cases where an insurgency develops after forcible regime change, the affected government will be undergoing significant reform or even institution building from scratch. An intervening nation or coalition will almost always need to co-opt, persuade or occasionally pressure the local government to give up counter-productive behaviors, take genuine steps to reform its actions, win the support of its people and demonstrate effectiveness and legitimacy. Four traits that can be used to gauge the legitimacy of a state are:

- The degree to which it honors and upholds human rights and fundamental freedoms;
- The degree to which it responds to the opinions of its citizens;
- Whether it exercises effective sovereignty;
- The degree to which it provides reasonable limits on the power of government over individual rights.

The affected government may have the desire to do only the minimum necessary to defeat an insurgency before returning to business as usual. This would bring about tension between the affected government's desire and an intervening government's aspiration for wholesale reform and institution building to prevent a recurrence of unrest. Because leaders in an affected nation almost always emerge through the same institutional processes that gave rise to the insurgency, they may be unable or unwilling to undertake reforms; conversely, intervening forces may lack the local knowledge and situational awareness to judge what reforms are possible and understand how to undertake them effectively.

Cooperation between affected nation and intervening force, however difficult, is absolutely essential for effective COIN.

Relations between the U.S. Government and the affected nation will be dynamic, especially when elections or other changes affect the composition of its government. Any sovereign government may exercise its autonomy in ways that are in opposition to U.S. interests. A quandary may arise between the U.S.'s desire to reach the end-state (a fully functional, independent and legitimate nation state) and its protection of the very U.S. interests which prompted engagement in the first place.

It is important to recognize the distinction between national level government and local or provincial government. Provincial governments are usually responsible for managing resources on behalf of the national government, but some countries that have fairly competent central governments are still unable to extend their authority and legitimacy into outlying provinces, especially where those provinces cross a social, ethnic, religious or economic fault line. In many regions of the world, the local delivery of administrative services is traditionally far more important than central government in a distant capital. Local government entities which are perceived as illegitimate, corrupt, oppressive, or non-inclusive will provide fertile ground for an insurgency to develop and operate, but are the lowest level through which the national government can deliver security and other public services. The perceived capacity of local government to provide for the population is critical to national government legitimacy.

## The U.S. Country Team

All United States Government COIN strategies, plans, programs, and activities are undertaken in support of the affected government and managed through the U.S. Mission's Country Team, led by the Chief of Mission (COM) in coordination with the Department of State. As the U.S. Ambassador, the Chief of Mission is the President's personal representative to the host nation and is responsible for recommending and implementing U.S. policy regarding that country, as well as overseeing all executive branch employees there and the activities of such employees with limited exceptions. Appointed by the President and confirmed by the Senate, the Ambassador has extraordinary decision-making authority as the senior U.S. official present during periods of instability and crises. Where a confirmed ambassador is not present, the Charge d'Affaires represents the Secretary of State as the senior diplomat accredited to the foreign government.

The Country Team is the central element of interagency coordination and execution in the foreign country. When an insurgency can be identified in its early stages, the Chief of Mission and his or her senior staff may encourage the affected government to take preventive action through the use of informational, security, political and economic measures. The Chief of Mission will also bring the matter to the early attention of decision makers in Washington D.C., his or her local knowledge and situational awareness often providing the most incisive and realistic source of advice. In coordination with policy makers in Washington, the Chief of Mission may also request foreign assistance for the affected nation, help to mobilize international support, and engage non-governmental organizations. In the future, the Chief of Mission may be able to call upon elements of the Civilian Response Corps, which is currently being developed within the Department of State and which aims to provide a pool of civilian specialists and experts in reconstruction and stabilization able to respond rapidly to countries in crisis. These actions may help to address the causes of unrest before the crisis escalates and limits political alternatives to the use of force.

The applicable U.S. geographic Combatant Commander, a four star general or admiral, will be in contact with the Chief of Mission and will be able to assist in pre-empting nascent insurgencies by providing military advice and supporting security enhancement programs. If the United States decides to deploy U.S. combat forces to assist an affected government, the Combatant Commander will be tasked to plan and execute the military aspects of that support.

U.S. efforts must be designed and executed in such manner to increase both the legitimacy and effectiveness of the threatened government in the eyes of its population. COIN strategy requires a tailored approach that captures and integrates the range of U.S. Government agency and department capabilities. The U.S. Government may also elect to provide operational assistance to indigenous forces; such assistance will be the product of deliberate foreign policy decisions taken in Washington D.C., and will be subject to the oversight of the Chief of Mission. Representative capabilities of U.S. Government agencies relevant to COIN are listed in Appendix A.

## The International Community

Considerable attention should be paid to the role of the international community in the development of a COIN strategy. The formation of a sanctioned multinational mission offers four principal areas of advantage to a COIN campaign:

- Legitimacy: When actions taken in support of a COIN campaign are consistent with applicable international law and are supported by international entities (which simultaneously condemn the insurgents) it will bolster support for intervention in the affected nation, the U.S. and the wider international community.
- Capacity: A multinational coalition will be able to call on a larger number of troops and more financial resources than the U.S. can muster on its own. This may have a positive effect on U.S. popular support for the campaign when U.S. taxpayers see other countries participating in and funding part of the counterinsurgency effort. The troops of other nations may be limited in capability or by political restrictions, but by undertaking suitable tasking they will still free up U.S. troops for missions in which they have a comparative advantage.
- Specialist Capabilities: Many U.S. allies and coalition partners have a comparative advantage in deployable capabilities relevant to COIN, such as developing national police forces, promoting economic growth or developing the administrative capacity of local officials in high threat or remote environments.
- Regional Effects: Regional partners can help prevent the establishment of external sanctuaries, prevent or slow the spread of the conflict to other areas, and provide local expertise, basing, and possibly even security assistance.

As well as these advantages, coalition operations bring many additional challenges such as differences in goals, training, capabilities, equipment, logistics, culture, doctrine, intelligence and language. They require early and close coordination of effort to best integrate their capabilities and expertise in the achievement of common political, economic, security and informational objectives. However, the importance of international solidarity and legitimacy means that coalition operations will remain the preferred path for U.S. COIN engagements in the foreseeable future.

## Inter-Governmental Organizations (IGO)

Inter-Governmental Organizations (IGOs), formed when two or more national governments sign a multilateral treaty to form such a body and finance its operations, possess legal personality in international law and their staffs enjoy diplomatic status. Most IGOs are regionally focused, and as such when IGO member states could be adversely affected by an insurgency in their region, the organization may act collectively to deny legitimacy, sanctuary, and support to insurgents. IGOs can also play an important role in humanitarian assistance and development.

## Non-Governmental Organizations (NGO)

NGOs are private, self-governing, non-profit organizations. Their activities (a direct function of the interests of their donors) are very diverse, but include interests such as education, health care, environmental protection, human rights, conflict resolution and similar issues. Some NGOs are implementing partners for U.S. foreign assistance, but these are a special case and will be discussed separately. Since their aims are often complimentary to the COIN effort in meeting the needs of a local population, they should, as far as possible, be consulted to minimize duplication or gaps in assistance. However, in order to secure freedom of movement, including access to semi-permissive environments, NGOs generally strive to be independent, politically neutral, needs-driven organizations. Consequently, they often try to minimize contact with uniformed military personnel or other governmental actors, seeking 'humanitarian space' in which to operate. In an attempt to address this difficult dynamic, the U.S. Institute of Peace (USIP) has created a set of guidelines (referenced in Appendix C) for relations between U.S. armed forces and nongovernmental humanitarian organizations. This requirement for separation is not absolute: despite their best efforts, NGOs will not be seen as neutral by some insurgents and where security is problematic, NGOs may rely on government resources – military and otherwise—for transportation and protection. COIN planning should take NGO activities into account, although it is unwise to rely on NGOs to provide key elements of any plan due to their independent status.

Some NGOs may actually promote aims which conflict with or are counter to the COIN strategy and COIN planners should take care to establish their motivations and activities before engagement. NGOs that oppose the affected government or actively forge links with insurgents will create security problems and risks, but they may also provide opportunities to open channels for negotiation or mediation with insurgents. Criticism by NGOs can be useful in drawing attention to those issues where reform actions are necessary. Finally, as an independent and often credible source of 'ground truth' about the areas in which they work, they will be an important source of information to many interested parties.

## Diaspora Groups

Exile or diaspora groups from the affected nation, whether relocated in the U.S. or elsewhere, will sometimes attempt to influence insurgency situations. They may sympathize with the insurgents and therefore oppose any intervention on behalf of the affected nation, or they may offer assistance to the intervention; looking to play a role in the planning or implementation of a COIN strategy. If they succeed in having any influence over the situation then it is important for COIN practitioners to have an accurate picture of their motives, capabilities and relationship with the affected government, since these are often misunderstood or misrepresented.

## Media

Media is a key actor in a successful information strategy in any COIN campaign. The U.S. Government is accustomed to interaction with western media groups, but there is often a weakness in the relationship with regional media in overseas insurgency situations. This can result in missed opportunities to influence key stakeholders.

## The Private Sector

### *Multinational Corporations*

Multinational Corporations usually become involved in counterinsurgency when their corporate interests (financial interests, foreign based personnel, or infrastructure, etc.) are threatened, or when a financial advantage is perceived. COIN planners will not always have the ability to influence the activities of multi-national corporations in an affected country, but may find that their interests complement the COIN effort. In particular, firms in extractive sectors (oil, mining, etc.) have large initial investment and long production cycles, which mean that withdrawal from a country could result in significant financial cost. For such companies, investment in local stabilization activities (from micro-loans to security sector reform activities) makes economic sense.

### *Contractors*

Contractors are frequently used by the affected nation and supporting countries to provide a wide range of functions of which security is only one. Policy makers should be aware that the dynamics of contracting arrangements often lead to “contractor mission creep” over time, resulting in expanded costs and responsibilities for contractor personnel. Policy makers should also be cognizant of the fact that U.S. hired contractors, especially security contractors, are often viewed by the local population as U.S. Government representatives and any negative behavior or interaction with the local population can have an adverse impact on COIN efforts. They should therefore ensure that armed contractors (including security personnel, facility guards, trainers and advisers) are subject to robust, well-designed rules for the use of force, and that their relationship with the affected government is ethical and transparent. Despite these concerns, contractors (across all functions) are now so essential to

U.S. Government overseas operations that large scale COIN intervention is probably inconceivable without them.

### *Grantees*

The U.S. Government generally delivers development assistance through grants to non-governmental organizations and private individuals or through contracts. U.S. Government grant managers oversee the activities of grantees to ensure that funds are used for the intended purpose. Generally, they can only change the nature of the grant once it has been authorized if the grant is written to provide for termination or modification in the event of changed circumstances. Grantees are often subject to the same 'humanitarian space' considerations as NGOs and frequently prefer to be separated from military activity. COIN planners must therefore maintain awareness of U.S. Government-funded grantee activities in order to protect U.S. investment, avoid duplication of effort, and prevent military compromise of their status. Grantees may require armed security personnel who would be subject to the same concerns as armed security contractors.

## 4. Assessment and Planning

A USAID officer works with her Afghan counterparts.

Involvement in countering an insurgency overseas is not an endeavor that the U.S. Government should embark on lightly. As this guide has illustrated thus far; insurgencies often arise from a deep rooted social grievance which the affected government may be reluctant or unable to address. Moreover, COIN campaigns are generally protracted affairs entailing significant costs in lives and resources. U.S. decision making on whether and how to assist a country affected by insurgency should follow the sequence below:

- Assessment of the situation
- Formulation of policy
- Development of strategy
- Integrated planning
- Implementation
- Continuous monitoring, evaluation and assessment

This Guide is aimed at U.S. Government policy makers and will therefore give greatest emphasis to their role in the 'formulation of policy' stage, in particular determining *whether* the U.S. should become engaged in a COIN campaign overseas and if so how.

## Assessing the Situation

A whole-of-government approach to a COIN engagement begins with a strategic- level interagency analysis of the conflict. This should be conducted by an interagency team comprised of all relevant agencies with core competencies needed to counter the insurgency. As much as possible within release constraints, the relevant knowledge and understanding of the affected state and strategic environment should be shared across all participating agencies. The effort may be conducted at the direction of the National Security Council (though it may be recommended by the Chief of Mission or the State Department regional Assistant Secretary) and will usually be led by the U.S. Agency for International Development or by the Department of State. It may involve extensive field evaluation activities conducted through the U.S. Embassy in the affected country, or it may be conducted entirely through remote assessment methods.

Insurgencies are frequently described in terms of sources and root causes, parties and actors, and drivers and triggers. The sources and root causes of insurgency can be described in terms of the stakeholders' frustrated needs and grievances. The drivers of insurgency can be expressed in terms of the dynamics among the stakeholders; in particular the way in which dissatisfaction is harnessed, channeled and directed by opinion leaders. To fully understand these factors may require detailed analysis of regional history, ethnicity, culture, politics and religion.

The interagency assessment process should yield a comprehensive picture of the environment and a common understanding of the nature of the problem. It will provide the first step towards developing a whole-of-government COIN strategy and supporting plans.

At the strategic level, analytical tools such as the Interagency Methodology for Analyzing Instability and Conflict, the Interagency Conflict Assessment Framework (ICAF), and the Global Forecasting Model of Political Instability (the last from the Political Instability Task Force, a U.S. Government-sponsored grouping of researchers and scholars from a number of American universities) may be useful aids to the development of situation-specific information for policy-level strategic planning. The ICAF, the Tactical Conflict Assessment Framework, and similar modeling tools can be used to inform programmatic, operational and tactical level plans. Use of these frameworks, must incorporate (or at least consider) any pre-existing analysis and data, especially that which has been used in support of previous country plans. A deliberate analysis undertaken carefully will be of enduring utility in providing deep understanding. See Appendix B for web links to these and other assessment and modeling tools.

## Forming U.S. Government Policy

The assessment phase described above lays the foundation for the formulation of U.S. Government policy; most critically whether the U.S. should engage with the affected government, and if so, what form that engagement should take. Only with a full understanding of the causes, nature and maturity of the insurgency, along with knowledge of the applicable international and domestic legal frameworks, can policy makers balance U.S. interests against likely costs and risks of becoming involved in what could prove to be a prolonged and expensive conflict.

### *Deciding Whether to Engage*

It is folly to become engaged with counterinsurgency in a foreign country unless there is a reasonable likelihood that the affected government will introduce necessary reforms and will demonstrate adequate willpower and capacity to defeat insurgents (or at least be willing to accept advice as well as assistance). Before deciding to provide overseas COIN assistance, U.S. officials must determine how likely it is that the local government will cooperate and how willing it is to undertake necessary reforms. For this reason, the following key characteristics of the affected nation must be examined in depth during the assessment phase:

- Character of the affected government: Supporting an oppressive, authoritarian or abusive government against an insurgency is highly problematic, not only from an ethical standpoint but also in terms of the practical likelihood of success. Such a government is unlikely to develop the necessary legitimacy to succeed in COIN. At the same time, the credibility and moral authority of the United States may be tarnished or compromised by too close a relationship with such a regime. In deciding whether or not to become engaged in a given circumstance, policy makers must consider the degree to which the insurgency derives from feckless administration on the part of the affected government, and therefore the degree to which the insurgents represent legitimate grievances. A government that lacks capacity and capabilities for COIN but is open to international community assistance and has a fundamentally democratic and responsible character is more likely to benefit from assistance than a government whose political or moral character is fundamentally unsound. The latter type of government is rarely a good candidate for engagement, regardless of its perceived geo-strategic importance.
- Government bias: Insurgencies that occur along ethnic or sectarian lines frequently derive from genuine sectarian or ethnic bias in the government's administration of its population (though this is often manipulated by extremist groups from outside the affected society). Such biased governments may require wholesale reform, including

changes in the demographic recruiting base for soldiers, police and civil service, and changes in the political orientation of key leaders. This is a costly, time-consuming and often politically controversial process. Planners must assess the likelihood that the government can be sufficiently reformed to meet the needs and address the legal and human rights of its entire population. They should conduct a detailed assessment of requirements for reform, and seek a firm commitment to specific reforms from the affected government. Continued leverage may be necessary to maintain that commitment, so policy makers may decide to tie continued assistance to measured progress in meeting reform benchmarks. Without effective reform, intervention may stabilize the government in the short term, but may simply enable continued behavior by officials that renders long-term success unlikely.

- Rule of law: Most countries affected by insurgency do not have robust, transparent and effective rule of law systems. Indeed, read or perceived inequalities in the administration of the law and injustices are often triggers for insurgency. Consequently, building the government's legitimacy and effectiveness often requires the wholesale reform of rule of law systems. Planners must make a judgment about how eroded or ineffective those systems are, including judiciary and legislative processes, court and prison systems, police, prosecutors, defense attorneys and legal record-keeping systems. This will indicate the amount of effort required to assist and the likely degree of success.
- Level of Corruption: Many countries affected by insurgency exhibit pervasive problem of government and security force corruption. This creates grievances which insurgents exploit, and places great friction and cost on international assistance. In some cases this may simply be a factor for planners to take into account, but in others policy makers may decide to seek a commitment to specific anti-corruption programs as a prerequisite for assistance. In judging the importance of corruption, planners should note that the forms of corruption that are most relevant in an insurgency scenario are those that alienate the people from the government or that lead to waste and inefficiency in government programs.
- Civil-military relations: Many insurgency affected governments have weak institutions, including military and police forces and civil administration. In deciding to become involved, U.S. planners must consider whether assistance to the military and police is likely to alter the balance between military and civilian power in the country. Large-scale assistance to militaries in the absence of matching assistance to civil administration (or without military leaders making firm commitments to civilian control and democracy) may increase the risk of a coup d'état in the affected country, either during or after the phase of international assistance. Such an outcome would ultimately harm the affected country and would undermine the moral authority of the international community. Policy makers need to be encouraged to plan for civil-military relations as an integral part of security sector reform, establishing safeguards to mitigate the risk of coups.
- Economic viability: Many insurgent fighters at the local level, regardless of rhetoric, are motivated by economic factors: youth unemployment and lack of economic opportunity. Insurgents often pay local fighters (or allow them access to profits from illegal activities) to gain their support. Planners must therefore judge the likelihood that key economic infrastructure and systems can be put in place to generate

alternative livelihoods and make the affected country economically viable over the long term.

- Presence of terrorist or transnational criminal groups: The presence of adversaries to the global interests of the United States, such as international terrorist or transnational organized crime groups, is a significant but complex consideration. Where terrorist groups are present, policy makers may be highly motivated to engage, in order to prevent the emergence of transnational threats from under-governed or insurgent-controlled areas. However, large-scale or clumsy intervention in such areas may actually lead to a backlash from local people who are alienated by increased government presence. International involvement in a conflict that does not currently include a transnational element may give a foothold to extremists from outside the affected country to exploit, internationalizing the conflict from both the government and insurgent sides. Policy makers should be very cautious about such escalation and should seek to assist in the lightest and least intrusive manner possible, working by, with and through the local government wherever possible. If this is not feasible due to the scale of the problem, policy makers should carefully weigh the risks of inaction against the costs and benefits of involvement.
- Border security/ungoverned spaces: An affected government that cannot control its borders, has large areas of ungoverned space near its frontiers, or faces an active insurgent sanctuary in a neighboring country will be particularly challenged in conducting COIN. Policy makers must judge the likelihood that areas of ungoverned space can be brought under government control. They must also take a regional view, considering whether neighboring countries can be persuaded to play a constructive role (or at least be dissuaded from undermining the affected government). Assisting an affected country without an effective strategy for border security, reduction of ungoverned space and denial of cross-border insurgent sanctuaries is highly unlikely to succeed over the long term.

Clearly, a country that scores well on each of the factors listed above, and is therefore a good candidate for assistance, is by definition unlikely to need that assistance in the first place. Countries that are candidates for U.S. engagement in the real world therefore usually score badly on several of these considerations. Hence, for each factor listed here, planners and policy makers should not necessarily expect to find a good current situation, but rather should consider the potential long-term viability. An affected country with sound political, economic and social fundamentals but poor current conditions is much more likely to respond well to assistance than one where fundamentals are poor, even if current conditions are not so bad.

Unfortunately, there will inevitably be occasions when the assessment of the insurgency situation will weigh heavily against U.S. involvement, but specific U.S. national interests will drive policymakers towards engagement. However, this does not negate the value of thorough assessment. On the contrary; it means that the decision will have been made with a good understanding of the inherent risks and the challenges that will need to be overcome. It may also prompt caution over the *form* of engagement to be used, perhaps encouraging a more limited involvement from which a subsequent exit can be made with less political consequence.

***Deciding How to Engage***

Depending upon the strength, legitimacy and effectiveness of tools available to the affected government, the U.S. Government may play a subtle role in countering an incipient insurgency or may intervene more forcefully. For reasons of cost, to minimize any backlash from the population against foreign presence, and to protect the sovereignty of the affected government, policy makers should select the most appropriate, most indirect and least intrusive form of intervention that will still have a high probability of achieving the necessary effect. Counter-intuitively for some planners, it is often the case that the less intrusive and more indirect the approach selected, the more likely it is to succeed, though this may be dependent on the maturity of the insurgency.

Insurgencies evolve in stages, and the nature of the U.S. response will often be dependent on the stage of development of the insurgency at the point when the U.S. decides to engage. There is a significant difference between responding to an incipient insurgency and responding to a full-blown insurrection where a well-developed (though not necessarily effective) counterinsurgency program is being implemented by the host nation government. An incipient insurgency can often be more easily addressed by a small scale U.S. response than a well developed one. However, most affected nations will only seek U.S. assistance when the insurgency has developed sufficient maturity to pose a real threat, by which time the smaller scale response options may no longer be effective.

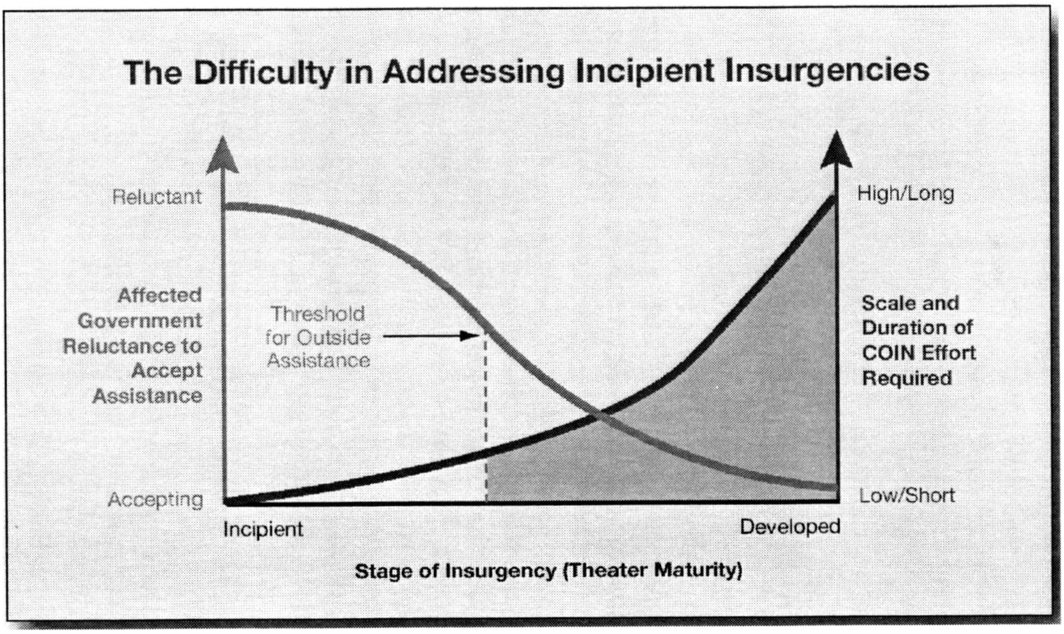


From least to most intrusive, forms of intervention include:

- Mission Augmentation: The mission augmentation approach involves the deployment of a specialist team to augment the U.S. Embassy in the affected country and/or the U.S. Consulate in an affected region of the country. An example of this approach was the Joint Strategic Assessment Team (JSAT) deployed to reinforce U.S. Mission Iraq at the start of 2007. Further examples are the State Department's

Foreign Emergency Response Team (FEST) that can deploy to support embassies experiencing an emergency situation and the intelligence community's Rapid Analytic Support and Expeditionary Response (RASER) teams. The augmentation team may operate on a temporary duty basis, or may be assigned directly to the embassy staff. It includes a team leader well versed in all civil and military aspects of COIN, and team members selected in consultation with the embassy for specialist skills relevant to the needs of the affected government. The team should be as small as possible and would have minimal direct interaction with the affected government or population. Instead, it performs its function by training, advising, supporting and assisting the U.S. Country Team in its role of providing advice and support to the Ambassador. Assistance to the affected government is then carried out by the Country Team in the normal manner. This is a low cost, low profile, small footprint approach which is appropriate early in the development of an insurgency, or as a short term surge at other stages in a campaign. It is sustainable over very long periods due to its low cost. It may also be the chosen approach in situations when U.S. intervention would be extremely politically sensitive.

- Single Expert Advisor: The single expert advisor approach involves the seconding of one advisor, who may be either a civilian or a military officer, directly onto the staff of the affected government. The advisor should be placed, in close consultation with the affected government, in a position to advise, train and assist its elected leaders and officials in dealing with the insurgency. He or she assesses the situation, develops plans and capabilities in support of the affected government, and advises on the placement and tasking of additional U.S. assets if deployed. They will usually maintain a close relationship with the U.S. Ambassador and Country Team and may be supported by an embassy augmentation team or by specialist capabilities that can be called forward as needed. In order to achieve the necessary influence, the advisor must have appropriate rank, status and freedom of action as well as a diplomatic approach to his or her work. The advisor should take a low key, backstage role and support the affected government as its leads the COIN effort, and avoid even the appearance of taking on the lead. This approach is relatively low profile, low cost and sustainable yet it has historically been extremely effective. It is most successful when the selected advisor possesses cultural and language skills appropriate to the affected nation, is paired with an effective indigenous leadership team, and deploys for a long duration.
- Civil-Military Assistance: The civil-military assistance approach involves the deployment of a specialist team, potentially operating as a Joint Interagency Task Force (JIATF), to work with civilian and military agencies of the affected government. Team members or detachments may be embedded in key positions in the affected government to provide support, advice, technical assistance, education and training. Team members do not normally engage in direct activity against the insurgency. Through the civil-military assistance program they may help develop an Internal Defense and Development (IDAD) strategy in consultation with the affected government, and become a conduit for international community assistance. The team operates as a separate entity from the U.S. Mission, but normally works under Chief of Mission authority. If a military Joint Task Force is deployed, or if the security threat is especially high, the team may operate under military authority (as Provincial

Reconstruction Teams do in Afghanistan). However, unlike Foreign Internal Defense, it remains civilian-led and military-supported. This approach is relatively low- cost and sustainable over the long term, but has a higher profile than the two previous methods discussed above. The size of the civil-military assistance team should be kept as small as possible, and the duration of deployment rotations as long as possible, to increase its cost-effectiveness.

- Foreign Internal Defense (FID): Foreign internal defense is defined as the participation by civilian and military agencies of a government in any of the action programs taken by another government or other designated organization, to free and protect its society from subversion, lawlessness, and insurgency. The FID approach involves the deployment of military teams, often originating from the U.S. Special Operations Command, to support the affected government. It differs from civil-military assistance in that it is normally military- led, but still includes very substantial interagency input and support. FID is described in detail in U.S. Army Field Manual 31-20-3 and in Joint Publication 3-07.1. It varies in scope, cost and intrusiveness depending on the nature of the insurgency and the capabilities of the affected government, but is usually more intrusive than the models discussed above (though significantly less intrusive than direct COIN intervention).
- Direct COIN Intervention: Direct intervention in a COIN campaign may follow previous attempts to handle an insurgency using the approaches discussed above, or it may be the initial engagement. The current campaigns in Iraq and Afghanistan are not standard examples of direct COIN intervention, since troops were initially deployed to bring about regime change. The military role in direct COIN intervention is described in detail in Army Field Manual 3-24/Marine Corps Warfighting Publication 3-3 3.5 and in the Defense Department's forthcoming Joint Publication on Counterinsurgency 3-24, as well as being discussed elsewhere in this Guide.

When considering options, it must be remembered that every insurgency is different and will require a carefully tailored response. The approaches outlined above should therefore be seen as broad categories and not specific models.

It should be noted that there is a tendency for assistance to creep incrementally from small scale and less intrusive forms to ever larger and more obvious assistance. This is clearly illustrated by the history of U.S. involvement in Vietnam. The danger of this type of escalation is that the in-depth assessment and policy evaluation that occurred for the initial decision to assist may not necessarily be repeated for every increment and the government may find itself enmeshed in a scale of effort which was not reached by logical deliberation.

The risk of escalatory involvement should be considered during the formulation of policy. If the assessment of the situation is thorough enough and accurate, then the level of engagement chosen should be sufficient to address the problem. However, most countries significantly underestimate the scale of effort required to defeat an insurgency. If escalation of involvement does occur, then a full reappraisal of the situation and policy response should be conducted prior to each and every increment of involvement.

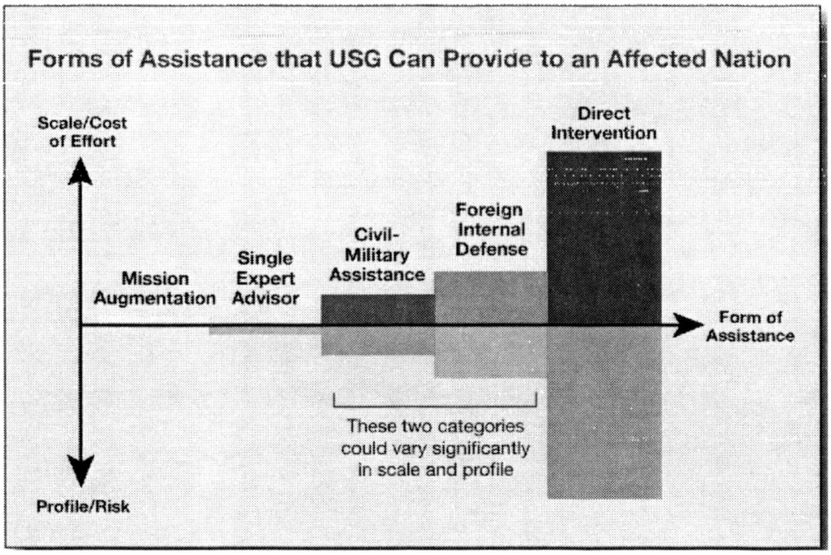


## Developing a Strategy

Regardless of the model of engagement selected, the policy decision to engage requires the development of a detailed framework. The more detailed framework for the U.S. response, in which objectives are determined and resources matched to their achievement, is achieved through strategy development. The components of a COIN strategy have already been outlined in detail during Chapter 2. Ideally, the overall COIN strategy should be devised by the affected nation, as their understanding of it and their commitment to it will be key to success. If possible, the role of the U.S. should therefore be one of advising and assisting the affected nation to improve its strategy (if it already has one) or to help it write one from scratch (if it does not). If the latter is the case, the U.S. should also strive to build up the strategy development capacity of the affected government.

As previously discussed, the affected government may not be particularly eager to address some of the underlying causes of insurgency and so may find U.S. suggestions unpalatable. If so, the U.S. will need to work with the affected government to encourage it to take the necessary steps.

Once the affected nation has a viable COIN strategy, the U.S. should determine where its own resources and actions can best be applied to contribute to the affected nation's strategy.

## Integrated Planning

The planning process to put a COIN strategy into effect will seldom be simple. To be effective, it must be integrated in two dimensions: internationally and between U.S. Government agencies.

- International Integration: By the very fact that the U.S. is assisting another nation to conduct COIN, the planning process must be at least bilateral. If the U.S. is involved as part of a coalition, then planning will require a multinational approach if the capabilities of other nations are to be integrated to best effect and the gains in legitimacy are to be preserved. In such a coalition, the degree of influence should be proportional to the degree of investment that each nation is willing to make in support of the affected nation.

As with the strategy, the plan should ideally be devised and owned by the affected government whose legitimacy and credibility are central to the campaign. If its competency and capacity to conduct COIN is limited, then initial planning will require a greater proportion of outside assistance, but supporting nations should recognize (indeed welcome) the increasing autonomy of the affected government in planning and conducting COIN as it grows in competence, capacity and confidence. While such independence may create conflicts of interest with the supporting nations, it represents progress towards the desired end-state.

- Interagency Integration: COIN planning by the affected nation should integrate civilian and military capabilities across each of the four COIN strategy functions of security, politics, economics and information. This requires the affected nation to conduct 'whole-of-government' planning to synchronize and sequence each department's activities towards achieving the objectives of the COIN strategy. The synergies achieved will be key to exercising control over the environment through the strengthening of legitimate and effective government institutions.

  The U.S. Government also needs to take a 'whole-of-government' approach to its support of the affected nation. It must employ a tailored approach that captures and integrates the range of capabilities that U.S. Government departments and agencies offer, so as to best support the affected government.

  A whole-of-government plan should specify:

  - The over-arching goal to be achieved;
  - Critical facts and assumptions about the environment;
  - Critical facts and assumptions about cause and effect;
  - Major mission elements necessary and sufficient to achieve the goal;
  - Essential task areas within each major mission element (tailored to the unique context and with possible consequences weighed against the desired end-state);
  - Sequencing of essential tasks;
  - Resources available to support the plan (skilled people, relationships, expert knowledge, money, materials, and time);
  - Metrics to assess progress towards the overarching goal and major mission elements;
  - The applicable international and domestic legal constraints;
  - The structure and business rules for contributing, storing and sharing all relevant information.

The outputs of whole-of-government assessment and planning should include detailed descriptions of:

- Dynamics driving the conflict, including those that create and support the insurgency, and those that might mitigate the conflict and defuse the insurgency;

- Primary actors and factors, including opinion leaders and identity groups (legitimate government leaders and their constituencies, insurgent groups and their supporters,), identity issues around which the actors coalesce (ideologies or other organizing principles), the degree to which the insurgency has subverted or penetrated the legitimate government, indigenous and external support to the insurgency, and vulnerabilities of the insurgent movement;
- Purpose of engagement (the "what" or mission statement for the COIN campaign);
- Major mission elements and essential tasks (the "how" for COIN operations);
- Resources required; detailed description of how the programs will be funded and managed by each U.S. Government department and agency; and resource shortfalls;
- Measures of effectiveness and performance indicators for each component of the plan and for the overall strategy;
- Key legal requirements, constraints and redlines;
- Coordination and synchronization mechanisms—"business rules" for integration of activities across departments and agencies, including clear lines of authority, command, and communication;
- Incorporation, where appropriate and possible, of other national, IGO, and NGO capabilities into plans and operations.

In summary, the success of the USG in helping other nations to defeat insurgencies will often be dependent on its proficiency at coordinating all committed agencies and resources (including its own, those of the affected nation, and those of international partners) towards a common objective. The first requirement for the U.S. is that it must synchronize its own agencies in a 'whole-of-government' understanding and approach. The second requirement is that it exercise sufficient diplomatic skill to coax, guide and assist the affected nation through the necessary steps of planning and execution to regain legitimacy and control. In situations where other coalition partners are involved, that diplomatic acumen must extend to maintaining the coalition and ensuring that partner efforts are woven as effectively as possible into the overall COIN strategy.

The 'Principles of the U.S. Government Planning Framework for Reconstruction, Stabilization and Conflict Transformation' document can be a useful tool for strategy development and planning. See Appendix C for web links to this and other planning tools.

## Implementation

The requirement for integration does not end with planning; but should carry over into the execution of the plan. Unity of command may seldom be achievable, but a common sense of purpose and teamwork (between U.S. agencies, with the affected government and with other coalition players) will greatly increase the probability of success.

## Continuous Monitoring, Evaluation and Assessment

Counterinsurgency situations are typically dynamic; insurgencies evolve and mature, affected governments (especially democratic ones) will alter in composition, competency and stance and the opinions held by populations will change. Concurrently both insurgents and counterinsurgents will evolve and adapt their strategies and tactics in a Darwinian struggle to outmaneuver each other. Under such dynamic conditions, it is not sufficient for assessment and planning to occur once, when the decision to become involved is taken. Planning should be adaptive and flexible, though for unity of effort and continuity the main themes should be maintained wherever possible. The situation should be continuously reassessed and the relative success of insurgent and counterinsurgent should be evaluated. Humanitarian and development activities should be monitored and evaluated according to international standards and best practices. Evaluations are often best achieved by an independent team of experts reporting directly to the senior U.S. official. The U.S. military frequently applies this concept, using retired military commanders and diplomats. The views of the local population and non-U.S. Government entities should always be sought. Where the situation has changed significantly, the counterinsurgents (the affected nation, the U.S. and partners) must be prepared to review the strategy to determine whether it is still valid.

# CONCLUSION

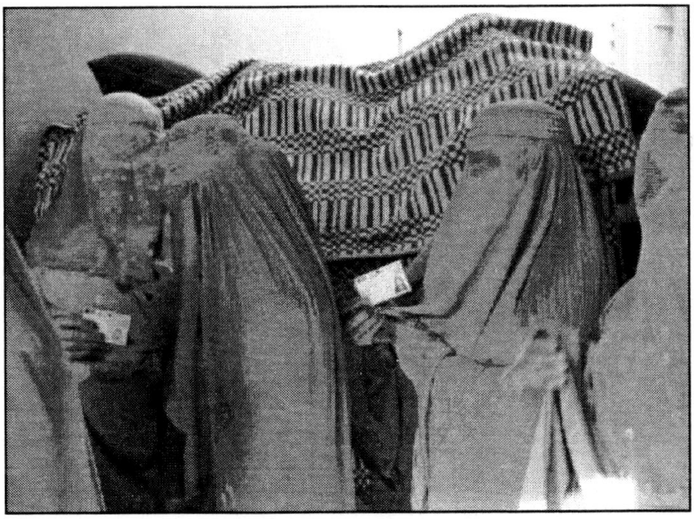

Afghan women waiting in line to vote at their local polling place.

Effective counterinsurgency requires multi-faceted and integrated operations that apply civilian and military capabilities across information, security, political and economic functional areas. The goal of intervention in a COIN campaign is to help an affected government achieve control over its sovereign territory by establishing, developing, and consolidating legitimate, effective government institutions.

The U.S. can assist an affected government with strategies that combine information, security, political and economic elements. Before committing to engagement, careful consideration must be given to the affected government's legitimacy, its willingness to reform and its general political and economic viability. Approaches ranging from augmentation teams, through advisory support, civil-military assistance, Foreign Internal Defense and direct COIN intervention are available and historically proven. Diplomatic efforts in COIN, which shape the international environment as well as helping the affected government to reform, mobilize support, marginalize insurgents and extend its control throughout its territory, are led by the Department of State. Development efforts help the affected government to meet essential needs, develop infrastructure and build economic capacity and are led by the U.S. Agency for International Development. Both diplomacy and development are enabled by and contribute to security activities, which are led by the Department of Defense. The complex nature of insurgency also requires the integration of capabilities extant in a number of other U.S. Government agencies and departments, as well as those of other partner nations, inter-governmental and non-governmental organizations, and the private sector.

Insurgencies, and thus COIN strategies, can vary significantly from one situation to another. COIN efforts succeed if they result in a political resolution acceptable to the parties involved. Diplomacy, development and defense are interdependent at every level of a COIN effort, and civil-military integration is required at the strategic, theater/operational and local/tactical levels. Most successful COIN campaigns have achieved this unity of effort through unified authority.

This Guide serves to synthesize counterinsurgency theory with the recent experience of officials across U.S. Government departments and agencies working in this field. It deliberately focuses at the broad national level so as to develop civilian literature on counterinsurgency to complement existing military doctrine. As the first serious U.S. effort at creating a national counterinsurgency framework in over 40 years, this Guide is intended to provide the basis for continued discussion among and feedback from practitioners. The ultimate intent of this effort is to develop our national capability to support the counterinsurgency efforts of legitimate and responsible governments that respond to the needs of their people.

## APPENDIX A: U.S. GOVERNMENT ROLES IN COIN

### National Security Council

The National Security Council (NSC) is the President's principal forum for consultation with senior advisors and cabinet officials on national security and foreign policy matters. The NSC staff provides advice to the President with respect to the integration of domestic, foreign, and military policies and manages the processes through which the President's polices are coordinated and implemented. Due to COIN's inherent requirement for a whole of government approach, the NSC is uniquely positioned to guide COIN policy development and implementation.

## Intelligence Community

Intelligence is central to any COIN campaign; it is the basis upon which a precise and deep understanding of the nature of insurgency, its context, and its remedies are based. The U.S. and international intelligence communities are indispensable contributors, providing intelligence support to policy makers, including indications and warning; conflict assessment tools; deployable support, including Rapid Analytic Support and Expeditionary Response (RASER) teams; collaborative tools; and dedicated support to planning staffs.

## Department of State

The Department of State, through its bureaus, offices, and missions overseas, leads and oversees U.S. Government support to COIN efforts. The relevant regional bureau will normally direct primary policy regarding U.S. engagement in or operations in support of other governments. Several functional bureaus and offices, including the Bureau of Political-Military Affairs, the Office of the Coordinator for Reconstruction and Stabilization, the Bureau of International Narcotics and Law Enforcement, the Bureau of Democracy, Human Rights, and Labor, the Bureau of Population, Refugees, and Migration, the Bureau of Intelligence and Research, and the Legal Adviser's Office will have substantive roles in the development and execution of COIN strategy. The functional bureaus within State have the capability to design and execute full-spectrum assistance programs in the security sector to include counter-narcotics, anti-corruption, and police and non-military security forces. The Office of the Coordinator for Reconstruction and Stabilization has been tasked to develop a Civilian Response Corps, to provide a pool of civilian expertise in reconstruction and stabilization able to respond rapidly to countries in crisis. Chiefs of U.S. Missions will oversee official U.S. Government operations in the countries to which they are accredited.

## Department of Defense

The Department of Defense and U.S. military forces provide a broad range of capabilities to support an integrated U.S. counterinsurgency effort. These may include advising and training foreign military, internal security, and police units; planning and conducting security operations in support of indigenous security forces; intelligence, communications, and logistical support; public affairs and military information operations; medical assistance; civil affairs support; and infrastructure repair and construction. *Army Field Manual 3-24/Marine Corps Warfighting Publication 3-33.5, Counterinsurgency,* and U.S. military joint doctrine describe U.S. military COIN capabilities and operations in detail.

## U.S. Agency for International Development

USAID can assist U.S. COIN efforts by fostering economic growth, promoting human health, providing emergency humanitarian assistance and enhancing democracy in developing countries. This is achieved through a spectrum of actions from policy reform to community level programs. USAID has extensive experience in developing and implementing programs with national governments and has field offices in 100 developing countries, working in close partnership with private voluntary organizations, indigenous groups, universities, American businesses, international organizations, other governments, trade and professional associations, faith-based organizations, and other U.S. government agencies. USAID programs are designed to enhance institutional capacity and ameliorate the root causes of conflict; community-level programs in particular have a good track record in addressing the grievances that fuel insurgency. The large numbers of foreign service nationals that make up the professional cadre of field staff provide a unique understanding of the local situation, while the range of sectors and levels of activity allow USAID great operational flexibility and agility to both implement and track the effectiveness of COIN operations.

## Department of Justice

The Department of Justice, through its constituent agencies (the Federal Bureau of Investigation, the Drug Enforcement Administration, the U.S. Marshals Service, and the Bureau of Alcohol, Firearms, Tobacco and Explosives) and components, works with other nations to combat transnational crime and international terrorist activities, including financial and operational support that may buttress insurgency operations. Justice also has offices devoted exclusively to providing overseas technical assistance that are highly relevant to COIN: the International Criminal Investigative Training Assistance Program, which develops police and corrections institutions; and the Office of Overseas Prosecutorial Development, Assistance and Training, which develops prosecutorial and judicial institutions.

## Department of the Treasury

The Department of Treasury's Office of Terrorism and Financial Intelligence (TFI) plays an important role within the U.S. Government with the twin aims of safeguarding the financial system against illicit use and combating rogue nations, terrorism facilitators, weapons of mass destruction (WMD) proliferators, money launderers, drug kingpins, and other national security threats by disrupting and dismantling terrorist and insurgent financial networks as well as building partner nation capacity. Both of these aims are highly relevant to COIN. TFI's components include the Offices of Terrorist Financing and Financial Crime, Intelligence and Analysis, Foreign Assets Control, and the Financial Crimes Enforcement Network. The Office of International Affairs, through its Office of Technical Assistance, works directly with foreign governments to support their efforts to improve their financial systems.

## Department of Homeland Security

The Department of Homeland Security (DHS) was formed by merging 22 separate and distinct federal agencies. The Department's overarching purpose is to govern domestic security operations; however, several component agencies and offices operate overseas, interacting and cooperating with host nation government agencies. A significant by-product of that interaction and cooperation is improved U.S. and host government capabilities to provide security and safety for their populations. DHS component agencies and offices with capabilities most relevant to COIN are U.S. Customs and Border Protection (CBP), U.S. Immigration and Customs Enforcement (ICE), the Transportation Security Administration (TSA), U.S. Citizenship and Immigration Services (USCIS), the U.S. Coast Guard (USCG), the U.S. Secret Service, the Office of Intelligence and Analysis, the Office of International Affairs, and the Office of Operations Coordination.

## Department of Agriculture

Many insurgencies occur in countries where the majority of the population is dependent upon agriculture, and where unemployed or underemployed rural youth are considered prime candidates for recruitment. Development of the agricultural sector and its institutions helps facilitate trade and increase incomes, reducing recruitment and support for insurgency. The Foreign Agricultural Service (FAS) executes the U.S. Department of Agriculture's (USDA) international programs, including market development, trade agreements and negotiations, and the collection and analysis of statistics and market information. The FAS delivers training and technical assistance, and collaborates with developing and transitional countries to facilitate trade and promote food security. The goals of USDA's international work are to help ensure that countries critical to U.S. national security strengthen their institutions, policies, and market-based agricultural systems, thereby contributing to long-term economic and political stability; adopt regulations consistent with international standards to increase trade, resulting in economic growth and stability; and employ agricultural practices that will reduce instability, increase regional cooperation, and ensure an adequate resource base for future generations.

## Department of Transportation

Transportation infrastructure (roads, rail, ports, air and pipeline) is a critical component of the economic health and development of countries, factors that mitigate conditions that encourage insurgency. All facets of commerce, trade, travel, and quality of life depend on mobility. Transportation facilitates a government's ability to provide its population with basic services and security, thereby reducing the appeal of insurgency within the most likely populations of potential recruits. Transportation can support COIN efforts by helping countries to strengthen their institutions, policies, and intermodal transportation systems, contributing to long-term economic and political stability; adopt regulations consistent with international standards to increase trade and safety, resulting in economic growth and stability; and adopt transportation practices that promote infrastructure development for local, regional and international movement of people and commerce.

## Appendix B: Website Links to Assessment and Modeling Tools

[1] Global Forecasting Model of Political Instability http://globalpolicy.gmu.edu/pitf/PITF global.pdf
[2] USAID—Conducting a Conflict Assessment: A Framework for Strategy and Program Development http://www.usaid.gov/our_work/cross-cutting_programs/conflict publications/docs/CMM_ConflAssessFrmwrk_May_05.pdf
[3] USAID Conflict Mitigation and Management Policy http://www.usaid.gov/our_work /cross-cutting_programs/conflict publications/docs/USAID_Conflict_MM_Policy.pdf
[4] USAID Community-Based Development in Conflict-Affected Areas http://www.usaid .gov/our_work/cross-cutting_programs/conflict publications/docs/CMM_CBD_Guide _May_2007.pdf
[5] Principles of the USG Planning Framework for Reconstruction, Stabilization and Conflict Transformation http://www.crs.state.gov/index.cfm?fuseaction=public .display &shortcut=49Q9
[6] Graphical Overview of Whole-of-Government Planning Framework and Process for Reconstruction and Stabilization http://www.crs.state.gov/index.cfm?fuseaction =public .display&shortcut=49QF
[7] S/CRS Triggering Mechanisms for "Whole-of-Government" Planning for Reconstruction, Stabilization and Conflict Transformationwww.crs.state
[8] S/CRS Interagency Management System for Reconstruction and Stabilization www.crs .state
[9] OECD DAC Handbook on Security System Reform (SSR) (includes a framework for SSR assessment) http://www.oecd.org/dac/conflict

## Appendix C: References

[1] Baker, Jim Systems. (2006). *Thinking and Counterinsurgencies Parameters Winter, 07*, 26-43.
[2] Baltazar, Thomas & Elisabeth Kvitashvili "Combating Terrorism. (2007). *The Role of USAID and Development Assistance*" Marine Corps Gazette Web Article Apr.
[3] Blaufarb, Douglas S. & George K. (1984). *Tanham Fourteen Points: A Framework for the Analysis of Counterinsurgency* Washington: BDM.
[4] Cassidy, Robert M.( 2004)."*Back to the Street without Joy: Counterinsurgency Lessons from Vietnam and Other Small Wars*" Parameters Summer. 73-83.
[5] Cassidy, Robert M. (2006)."*The Long Small War: Indigenous Forces for Counterinsurgency*" Summer.
[6] Childress, Michael. (1995). *The Effectiveness of U.S. Training Efforts in Internal Defense and Development: The Cases of El Salvador and Honduras Santa Monica*: RAND.
[7] Cohen, Eliot, et al. (2006). "*Principles, Imperatives, and Paradoxes of Counterinsurgency*" Military Review March-April. 49-53.

[8] Bensahel, Nora. (2006). "*Preventing Insurgencies after Major Combat Operations" Defense Studies* Vol. *6*, No. 3 Sep. 278-291.

[9] Byman, Daniel Going to War with the Allies You Have. (2005). Allies, Counterinsurgency, and the War on Terrorism Carlisle: SSI.

[10] Echevarria, Antulio J. (2005). II. Fourth-Generation War and Other Myths. Carlisle: SSI.

[11] Galula. (1964). *David Counterinsurgency Warfare: Theory and Practice*. London: Praeger.

[12] Gompert, David C. (2007). "*Heads We Win: The Cognitive Side of Counterinsurgency*" RAND Occasional Paper.

[13] Grdovic, Mark. (2004). "*Understanding Counterinsurgency*" Special Warfare, Dec, 5-8.

[14] Hammes, T.X. (2004). *The Sling and the Stone: On War in the 21st Century*. Osceola: Zenith Press.

[15] Hoffman, Bruce. (2004). "*Insurgency and Counterinsurgency in Iraq*" RAND Occasional Paper.

[16] Hoffman, Bruce. (2006). "*From the War on Terror to Global Counterinsurgency*" Dec.

[17] Jureidini, Paul A. et al. (1962). *Casebook on Insurgency and Revolutionary Warfare*: 23 Summary Accounts. Washington: SORO.

[18] Kilcullen, David. (2004). "*Countering Global Insurgency*".

[19] Killebrew, Robert, ed. (2006). *The Country Team in American Strategy*.

[20] Komer, Robert W. Bureaucracy Does its Thing: Institutional Constraints on U.S.- GVN Performance in Vietnam. Washington: RAND, 1972.

[21] Kiszely, John. (2006). "*Learning About Counter-Insurgency*". RUSiJournal Dec, 16-21.

[22] Ladwig, Walter C., III. (2007). "*Managing Counterinsurgency: Lessons from Malaya*". Military Review May-June, 56-66.

[23] Lawrence, T. E. (1917). "*The 27 Articles of T.E. Lawrence*". The Arab Bulletin Aug. 20.

[24] Long, Austin On "Other War". (2006). *Lessons from Five Decades of RAND Counterinsurgency Research. Santa Monica*: RAND.

[25] Marks, Thomas A. (2005). "Insurgency in a Time of Terrorism" *JCOA-LL Bulletin,* Vol. *II*, No. 4, Sep, 33-43.

[26] Metz, Steven Learning from Iraq. (2007). *Counterinsurgency in American Strategy*. Carlisle: SSI.

[27] Muckian, Martin. (2006-07). "*Structural Vulnerabilities of Networked Insurgencies: Adapting to the New Adversary*" Parameters Winter. 14-25.

[28] Nagl, John A. (2005). *Learning to Eat Soup with a Knife: Counterinsurgency Lessons from Malaya and Vietnam Chicago*: University of Chicago Press.

[29] Record, Jeffrey. (2006). "*External Assistance: Enabler of Insurgent Success*" Parameters Autumn, 36-49.

[30] Record, Jeffrey. (2006). "*The American Way of War: Cultural Barriers to Successful Counterinsurgency*".

[31] Rosenau, William. (2007). "*Subversion and Insurgency*" RAND Occasional Paper.

[32] Schwarz, Benjamin C. (1991). American Counterinsurgency Doctrine and El Salvador: The Frustrations of Reform and the Illusions of Nation Building. Santa Monica: RAND.

[33] Sepp, Kalev I. (2005). "*Best Practices in Counterinsurgency*". Military Review May-June. 8-12.

[34] Stewart, Richard W. (2006). "*CORDS and the Vietnam Experience: An Interagency Organization for Counterinsurgency and Pacification*" National War College Research Paper, 1 May.

[35] Ucko, David. (2007). "Countering Insurgents through Distributed Operations: Insights from Malaya 1948-1960" *The Journal of Strategic Studies* Vol. *30*, No. 1, Feb, 47-72.

[36] United States Departments of the Army and Navy (2006). *Field Manual 3-24/ Marine Corps Warfighting Publication 3-33.5 Counterinsurgency.*

[37] United States Department of the Army (2008). *Field Manual 3-07 Stability Operations.*

[38] United States Department of Defense DOD (2005). *Directive 3000.5 Military Support for Stability, Security, Transition, and Reconstruction (SSTR) Operations.*

[39] United States Department of State *Lessons Learned: Disarmament, Demobilization and Reintegration (DDR) in Reconstruction and Stabilization Operations— Guide for U.S. Government Planners.*

[40] United States Institute for Peace (2008). *Guidelines for Relations Between U.S. Armed Forces and Non-Governmental Humanitarian Organizations in Hostile or Potentially Hostile Environments.*

[41] United States Institute for Peace *Guide for Participants in Peace, Stability and Relief Operations.*

[42] United States White House (1962). *United States Overseas Internal Defense Policy.*

[43] United States White House (2005). *NSPD-44 Management of Interagency Efforts Concerning Reconstruction and Stabilization.*

## APPENDIX D: ACRONYMS AND ABBREVIATIONS

| | |
|---|---|
| CAF | Conflict Assessment Framework, USAID |
| CCDR | Combatant Commander, DOD |
| COCOM | Combatant Command |
| COIN | Counterinsurgency |
| COM | Chief of Mission, State |
| DOD | Department of Defense |
| Justice | Department of Justice |
| State | Department of State |
| FAS | Foreign Agricultural Service, USDA |
| FM | Field Manual, DOD |
| IGO | Inter-Governmental Organization |
| MCWP | Marine Corps Warfighting Publication, DOD |
| MNC | Multinational Corporation |
| NGO | Non-Governmental Organization |
| NSC | National Security Council |
| U.S. | United States |

| | |
|---|---|
| USAID | U.S. Agency for International Development |
| USDA | U.S. Department of Agriculture |
| USG | United States Government |

In: U.S. Counterinsurgency Guide and Tactics
Editor: Timotheus Anzelmo
ISBN: 978-1-60876-121-0
© 2012 Nova Science Publishers, Inc.

*Chapter 2*

# TACTICS IN COUNTERINSURGENCY

***Headquarters, Dept. of the Army***

## INTRODUCTION

This manual gives the US Army a common language, concept, and purpose to fight and achieve success in a counterinsurgency. COIN is a complex subset of warfare that encompasses all military, paramilitary, political, economic, psychological, and civic actions taken by a government to defeat an insurgency at the company, battalion, and brigade levels. To do this, the manual merges traditional approaches to COIN with the realities of the current operational environment.

The US Army thinking and doctrine on COIN tactics since the end of World War II have focused on the conduct of counterguerrilla operations in the later stages of insurgency. The Army has seen itself as defeating guerrilla forces—usually communist forces—rather than defeating an entire insurgency. It saw success as something it could achieve by using the force of arms directly against guerrilla forces. This doctrine of COIN began to take shape shortly after World War II in manuals such as FM 31-20, *Operations against Guerrilla Forces* (1951) and later in FM 31-15, *Operations against Irregular Forces* (1961). The Army refined its counterinsurgency doctrine during Vietnam in FM 31-22, *US Counterinsurgency Force*, FM 31-16, *Counterguerrilla Operations* (both published in 1963) and in FM 31-23, *Stability and Support Operations* (1972). After Vietnam, the Army split COIN doctrine off from conventional "high intensity" operations in FM 100-20, *Military Operations in Low Intensity Conflict* (1990) in which the "light" forces owned counterinsurgency, and FM 90-8, *Counterguerrilla Operations* (1986), where the focus remained on defeating the guerrilla force. This manual is the historical successor to FM 90-8. In addition, parts of FM 100-20 have been integrated into this FM, as have the Army's concept of full-spectrum operations and all elements of COIN operations.

At its heart, a counterinsurgency is an armed struggle for the support of the population. This support can be achieved or lost through information engagement, strong representative government, access to goods and services, fear, or violence. This armed struggle also involves eliminating insurgents who threaten the safety and security of the population. However, military units alone cannot defeat an insurgency. Most of the work involves discovering and solving the population's underlying issues, that is, the root causes of their dissatisfaction with the current arrangement of political power. Dealing with diverse issues such as land reform, unemployment, oppressive leadership, or ethical tensions places a premium on tactical leaders who can not only close with the enemy, but also negotiate agreements, operate with nonmilitary agencies and other nations, restore basic services, speak the native (a foreign) language, orchestrate political deals, and get "the word" on the street.

Today's counterinsurgent battlefield is increasingly cluttered with US, Host Nation, and other coalition forces, each with its own strengths and limitations. In addition, multiple insurgent groups, nongovernmental organizations, armed contractors, and a local population divided into several ethnic groups add to this clutter. A counterinsurgency long-range plan for a tactical unit combines offensive, defensive and stability operations. To achieve the appropriate ratio between these and accomplish unity of effort among diverse units and actors, units must build long-term plans around the seven counterinsurgency lines of effort: establish civil security, establish civil control, support Host Nation forces, support to governance, restore essential services, support economic and infrastructure development, and conduct information engagement.

All seven lines of effort are critical to establishing *unity* of *effort* for actions conducted by US units, Host Nation security forces, and the Host Nation government. These actions can range from killing or capturing an insurgent cell known to emplace IEDs, to solving unemployment in an area, to publicizing the opening of a water treatment facility. Without unity of effort over time, the tactical unit's long-range plan will face challenges in securing the population, gathering the population's support, and defeating the insurgency.

Counterinsurgency is an iterative process. Tactical units can conduct a wide variety of operations. These can include anything from a combined cordon and search operation with Host Nation security forces, to a medical operation to inoculate a hamlet's children against disease, to a road project to connect a village to the highway, to a loudspeaker broadcast to inform a village about a recent council meeting. Regardless of the mission, successful tactical units learn and adapt as they discover more about their own strengths and limitations—and the strengths and limitations of the Host Nation government, the populace, and the insurgents. This manual furthers FM 3-24's theory that "in COIN, the side that learns faster and adapts more rapidly—the better learning organization—usually wins."

Each counterinsurgency is unique. Leaders must always execute good judgment, tactical patience, and innovation to defeat an insurgency. As the US Army continues its lengthy battles against insurgency around the world, tactical units must continue to focus on securing the support of the population, achieving unity of effort, and learning and adapting faster than the insurgents do.

## 1. Operational Environment of Counterinsurgency

*"Learn all you can about your Ashraf and Bedu. Get to know their families, clans and tribes, friends and enemies, wells, hills and roads."*

T. E. Lawrence, The 27 Articles of T.E. Lawrence.

FM 3-0 defines an operational environment as "...a composite of the conditions, circumstances, and influences that affect the employment of capabilities and bear on the decisions of the commander." The particular operational environment heavily influences both the nature of insurgent warfare and the methods of counterinsurgency operations. This chapter helps units define the operational environment of COIN by using operational variables, employing mission variables; understanding the effects of the operational environment; and comprehending the importance of cultural awareness.

## Section I—Overview

For more than two centuries, the United States military has been called upon to defeat insurgencies like the Whiskey Rebellion in the eastern United States, the Native Americans on the western plains of the United States, the Boxer Rebellion in China, Pancho Villa in Mexico, Augusto Sandino in Nicaragua, and the Viet Cong in Vietnam. Although the Army does have historic examples of COIN operations, our doctrine and COIN skills atrophied between Vietnam and the invasion of Afghanistan. In addition, the world is increasingly shaped by population explosion, urbanization, globalization, technology, religious fundamentalism, resource demand, climate change and natural disasters, and proliferation of weapons of mass destruction. The increasing complexity of the world has made it more challenging for governments to maintain order and satisfy the rapidly growing needs of their populations. As these governments try to maintain their tenuous hold on power, dissatisfied portions of their population have, like dissatisfied groups for thousands of years, turned to violence to achieve political goals. Using violence to achieve political goals is known as *insurgency*. As a result, US forces have conducted counterinsurgency operations around the world in Colombia, Somalia, Kosovo, Afghanistan, the Philippines, and Iraq. Before developing a better understanding of the operational environment (OE), it is important to understand what insurgencies, counterinsurgencies, and the influences shaping the OE are—

### *Insurgency*

This is an organized movement aimed at the overthrow of a constituted government through use of subversion and armed conflict (JP 1-02). The key distinction between an insurgency and other movements is the decision to use violence to achieve political goals. An insurgency is typically an internal struggle within a state, not between states. It is normally a protracted political and military struggle designed to weaken the existing government's power, control, and legitimacy, while increasing the insurgency's power, control, and legitimacy.

The majority of insurgencies have been limited to local regions or specific countries. However, today's instant communications allow insurgent groups and leaders to communicate worldwide to find support for their cause, and to support causes they view as compatible with their own goals. External forces, including nation-states, may support an insurgency for their own benefit. They may also oppose a competing nation-state that supports the existing government. As a result, modern insurgencies can often cross multiple countries.

An insurgency is made up of components (the five insurgent elements, the eight dynamics, and one or more of the six insurgent strategies) and manifestations (tactics, strengths and vulnerabilities). (For a greater understanding of insurgencies, see Chapter 2.)

### *Counterinsurgency*

COIN involves all political, economic, military, paramilitary, psychological, and civic actions that can be taken by a government to defeat an insurgency (JP 1-02). COIN operations include supporting a Host Nation's military, paramilitary, political, economic, psychological, and civic actions taken to defeat an insurgency. Avoiding the creation of new insurgents and forcing existing insurgents to end their participation is vital to defeating an insurgency. COIN operations often include security assistance programs such as foreign military sales programs, the foreign military financing program, and international military training and education programs.

Counterguerrilla operations, on the other hand, focus on detecting and defeating the armed insurgent or guerrilla, without solving the society's underlying problems. Military efforts alone, however, cannot defeat an insurgency. (For a better understanding of counterinsurgency, see Chapter 3.)

### *Influences on Current Operational Environments*

Eight forces are shaping current operational environments:

#### Population Explosion

Population explosion is the rapid growth of the world's population over the last six decades. In 1950, the world's population was 2.5 billion people; in 2008, the population is estimated at 6.5 billion people; and in 2050, the population is expected to reach 9 billion. The population explosion provides more opportunities for the insurgent to hide within the population and places a premium on winning the struggle for the populace's support. As the population continues to grow, governments will struggle to provide their people with food, water, and power, giving potential insurgent groups an opportunity to exploit a vulnerable population.

#### Urbanization

Urbanization is the growth of urban areas due to both a population surge and migration. In 1950, 29 percent of the world's population lived in urban areas; in 2008, almost 50 percent of the population lives in urban areas; and by 2050, it is estimated that 60 percent of the population will live in urban areas. This rapid growth of urban areas indicates that there is a greater potential that future insurgencies will be fought in urban areas.

#### Globalization

Globalization is a combination of the technological, economic, social, cultural, and political forces that are bringing nation-states and the people of the world closer together. These forces are making the world more interconnected and economically linked. Positively, it has reduced poverty in nations like China and India. It has increased the gap between rich and poor nations, caused an increased demand for resources, and may be affecting the climate.

### Technology

Technological developments such as the computer, the internet, the digital camera, and satellite television have transformed the world since 1950. Information can be exchanged around the world in less than a second. This has also increased the reach, impact, and influence of the media to the insurgent and the counterinsurgent. Additionally these same technologies, along with advanced weaponry, have dramatically changed the battlefield. Both insurgents and counterinsurgents will continue to innovate and adapt these technologies to the battlefield of today and the future.

### Religious Fundamentalism

Religious fundamentalism is defined as a belief in the infallibility of holy scriptures, absolute religious authority, and strict adherence to a set of basic religious principles without any compromise with modern life. As nation-states struggle to provide for their people, some of the dissatisfied population, as a backlash against globalization, will turn to religious fundamentalism to provide those needs that the nation-state cannot. (This is a primary insurgent ideology, and is further discussed in Chapter 2.)

### Resource Demand

Demand for energy, water, and food for growing populations will increase competition and, potentially, conflict.

### Climate Change and Natural Disasters

Climate change and natural disasters will compound already difficult conditions in developing countries and have the ability to cause humanitarian crises, driving regionally destabilizing population migrations and raising potential for epidemic diseases.

### Proliferation of Weapons of Mass Destruction and Effects

Proliferation of weapons of mass destruction and effects will increase the potential for catastrophic attacks, especially if used by failed states or terrorist organizations.

## Section II—Operational and Mission Variables

A thorough analysis of the population, the insurgency, and the counterinsurgent using the eight operational variables and the six mission variables is critical to developing a counterinsurgency plan that can defeat the insurgency. Even, a tactical unit will use the operational variables as a way to define their operational environment, which often corresponds to their area of interest (AI).

### *Operational Variables*

Army doctrine uses eight interrelated operational variables to analyze the operational environment. Known as PMESII-PT, the eight operational variables are—

### Political

The political variable describes the distribution of responsibility and power at all levels of government. Since an insurgency is fundamentally a struggle for political power, the political environment in the HN country is critical. Attention should be paid not just to the formal political system (such as political parties and elected officials) but also to informal political systems (such as tribes, ethnic groups, and other centers of power). Long-term success in COIN is ultimately based on political efforts; all counterinsurgents must focus on the political impact of their actions. Therefore, tactical leaders may be expected to broker local political solutions.

Host Nation (HN), US and coalition political considerations drive the conduct of COIN operations. This is especially true concerning the involvement of the US Government and US public opinion. A major goal of most insurgencies is to influence US public opinion against US involvement as a counterinsurgent force. Successful counterinsurgents must therefore not only prevent insurgents from obtaining this goal, but also actively work to influence public opinion for the COIN mission.

Commanders must be prepared to operate within a broad range of political structures. The Host Nation's form of government may range from a despotic dictatorship to a struggling democracy. Commanders at all levels, including platoon leaders and company commanders, need to recognize the importance of establishing and reinforcing the HN as the lead authority for all operations. This reinforces the legitimacy of the HN government.

### Military

This variable includes the military capabilities of all armed forces. Most COIN units will need to analyze the insurgency's military forces (guerrillas), local militias, and the Host Nation security forces. Commanders should consider qualitative aspects, such as conscription or recruitment systems, economic basis (to include appropriations system), and position of forces in national and local government structure. Additional qualitative considerations are general organization, training and doctrine, efficiency, rapport with population, and the police role in the nation's internal security. For example, a typical US brigade in Iraq might have to analyze a Sunni guerrilla force, a Shia guerrilla force, an Iraqi National Police brigade, an Iraqi Army brigade, the Iraqi local police, and a Sons of Iraq militia unit.

### Economic

The economic variable consists of the general economic categories of an Area of Operations (AO), such as energy; raw materials; government development policy; distribution of labor and labor policies; income distribution; national food distribution; free market or socialist interface and functions; consumption patterns; external investment, taxation policy; port authorities; movement of goods; consumer issues; border controls; foreign trade; tariffs; and graft or corruption.

A low standard of living and a desire for economic reform may be a cause of resentment toward the government. Generally, the counterinsurgents plan their operations to minimize damage to the economic structure of an area to avoid causing adverse psychological and economic impacts and to support economic development.

### Social

The social variable describes societies within an operational environment. A society is a population whose members are subject to the same political authority, occupy a common territory, have a common culture, and share a sense of identity. Both insurgents and counterinsurgents need the support of the population to be successful. Most insurgencies attempt to increase friction between different groups in a society and to gain or increase support from any group that shares common elements with the insurgency. These groups may be aligned along racial, ethnic, religious, or social lines. Language similarities or tradition can also be a reason for alignment. Religious influences often play a major role in the sociological factors that affect the insurgent.

To be successful against insurgents in a particular area and to avoid alienating the populace, counterinsurgents must understand the local environment. This includes local social issues and national issues that effect the local environment. For instance, Afghanistan units may interact with groups of Pashtuns, Tajiks, Hazaras, Uzbeks, and Nuristani across their AO.

### Information

The information variable involves the collection, access, use, manipulation, rapid distribution, and reliance on data, media, and knowledge systems—both civilian and military—by the global and local communities. Insurgents seek to control and manipulate how the local, regional, national, and international community perceives its cause and events within their operational environment. To achieve this, they try to control, manipulate, and distribute information.

Understanding the existing communication system is important because it influences local, regional, national, and international audiences. Media coverage, in particular, influences US political decision-making, popular opinion, and the sensitivities of coalition members, while the local teahouse may control the community's opinion and the "word on the street." Commanders must use information engagements to fully achieve their tactical goals.

Insurgents observe the actions of both government and COIN forces. Insurgents often use propaganda to gain creditability and legitimacy with the population, while simultaneously undermining their opponents. Successful insurgents strive to seize the moral high ground on any counterinsurgent mistakes, both real and perceived. This includes political, military, economic, social, religious, cultural, or legal errors. They will use all available means, including the media, nongovernmental organizations, and religious and civic leaders, to get their information out to all audiences.

### Infrastructure

The infrastructure variable includes the basic facilities, services, and installations needed for a community or society to function. The state of the infrastructure determines the resources required for reconstruction. Typical key infrastructure includes sewers, water, electrical, academic, trash, medical facilities, safety, and other considerations (also known as SWEAT-MSO). The degradation or destruction of infrastructure will negatively affect both the Host Nation and its population. Thus, the degradation or destruction of infrastructure often helps the insurgency, especially with respect to propaganda and the population's perception of the HN.

### Physical Environment

The physical environment variable is often the most noticeable aspect of an operational environment. Terrain affects people, equipment, trafficability, visibility, and the employment of many weapons. The terrain aspects of each area of operations must be evaluated to determine the impact on both insurgent and counterinsurgent forces. For COIN operations, terrain is categorized as either rural or urban. Weather and climate influence insurgents, the population, and counterinsurgents, who analyze the weather to determine its effect on the population's well-being and operations. They pay particular attention to trafficability, visibility, and equipment. Despite weather extremes, most insurgents have an advantage, since they are usually native to the climate.

### Time

Time affects everything and influences all decisions. However, the population, the counterinsurgent, and the insurgent often view time differently. Insurgents may design operations with the intent to influence the American political process or elections. In contrast, counterinsurgents must understand that popular support for extended operations may diminish over time. Figure 1-1 shows the difference between a western counterinsurgent's and an insurgent's perspective of time.

### *Mission Variables*

While analysis, in terms of the operational variables, improves understanding of the operational environment, it does not lend itself directly to mission accomplishment. For operations at the tactical level, the Army uses the mission variables of METT-TC (mission, enemy, terrain and weather, troops available, time available, and civil considerations) to help a unit understand its mission within the context of its specific OE. For COIN, civil considerations are especially important. When commanders and staff receive a specific mission, or identify a particular problem, they can draw relevant information from their ongoing analysis of their OE (using operational variables) to further complement their analysis of mission variables. Use of the mission variables, combined with the knowledge of the operational variables, enables leaders to understand the threat, act effectively, and anticipate the consequences of their operations before and during mission execution.

### Mission

Mission is the task, together with the purpose, that clearly indicates the action to be taken and the reason therefore (FM 1-02). At the brigade, battalion, and company level, the COIN force conducts tactical operations, across seven COIN lines of effort. These incorporate the five stability tasks—establish civil security, establish civil control, support HN security forces, support to governance, restore essential services, support to economic development, and conduct information engagement. These tasks are described in detail in Chapter 3 and Chapter 4. Lethal efforts may include patrols, raids, and cordon and searches. Nonlethal efforts may include attending council meetings, engaging tribal leaders, or repairing damaged infrastructure.

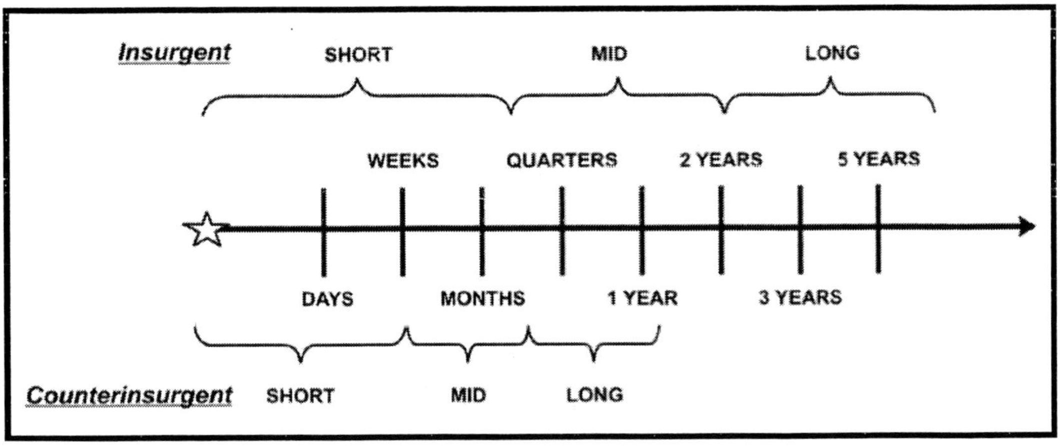


Figure 1-1. Comparison of insurgent's and counterinsurgent's perspectives of time.

### Enemy

COIN operations, by nature, involve a confusing enemy situation, since the enemy generally lacks a traditional task organization (FM 34-130). Moreover, the enemy (insurgents) can have a varying level of training, capability, commitment, involvement, and experience. In addition to analyzing the insurgent's disposition, composition, strengths, and weaknesses, counterinsurgents must identify and understand the five elements of the insurgency—leaders, guerrillas, auxiliary, underground, and mass base. Furthermore, it is important to understand the eight dynamics of the insurgency—its leadership, ideology, objectives, environment and geography, external support, internal support, phasing and timing, organizational and operational patterns. Finally, it is critical to identify which of the six strategies—urban, military focused, protracted popular war, conspiratorial, identity, and composite and coalition—the insurgent is employing. (For more on insurgency, see Chapter 2.)

### Terrain and Weather

Terrain includes natural features, such as rivers and mountains, and man-made features, such as cities, airfields, and bridges. Weather describes the conditions of temperature, wind velocity, precipitation, and visibility at a specific place and time. When evaluating the effects of terrain and weather on COIN operations, the commander should consider the effects of seasons of the year (to include planting and harvesting periods); phases of the moon; and coastal tides. In particular, he concentrates on—

- The *effects of the weather*—which mainly includes his Soldiers, equipment, and visibility, but also includes other factors such as mobility;
- The *suitability of terrain and road nets* for tactical and logistical operations.

He focuses on the effects of the terrain on Soldiers, equipment, visibility, and mobility. Units and staffs study the terrain in relation to the factors of OAKOC:

- Observation and fields of fire

- Avenues of approach
- Key and decisive terrain
- Obstacles
- Cover and concealment.

**Troops and Support Available**

Successful counterinsurgency operations depend upon the commander using his available assets to maximize force strengths and minimize vulnerabilities. To do this, the commander realistically appraises the capabilities and limitations of his assets, as well as joint, interagency, international, and multinational elements, to organize and employ them on suitable missions. In the COIN environment, the tactical unit must identify, account for, and leverage all HN security forces—police, army and paramilitary—to secure and control the population and disrupt the insurgency.

**Time Available**

For tactical operations, time available for planning and execution varies. Major operations need prolonged periods of time for detailed planning. Stability operations that address political, economic, and social issues usually take a considerable length of time to complete. As such, after the initial period of planning, the time available for modified or future planning is often quite long.

When planning short-term actions such as offensive operations against fleeting insurgent targets, planning time is usually short, and information is scarce. Commanders at all levels can use the time available to them more efficiently by planning contingency missions. One method to reduce planning time is to codify routine tasks common to similar missions in SOP. When the need to execute a contingency mission arises, the basic plan can be reviewed and planning expedited by making minor adjustment as required.

**Civil Considerations**

In COIN operations, the population is vital—since whoever the population supports has the advantage. Consequently, civil considerations are normally the most important mission variable for COIN. This variable comprises the influence of manmade infrastructure on the conduct of military operations.

## Section III—Analysis of Civil Considerations Using Ascope

An in-depth analysis of the civil considerations is vital for the long-term success of the counterinsurgent unit. There are six categories of civil considerations: areas, structures, capabilities, organizations, people, and events, represented by the useful memory aid, ASCOPE (Figure 1-2). During intelligence preparation of the battlefield (IPB), the commander and staff analyze civil considerations from several perspectives—the population, the insurgents, and the counterinsurgents—to determine the effects on friendly and enemy courses of action. Analyzing the six categories of civil considerations from multiple perspectives aids in understanding of the OE, and helps to isolate the insurgents from the population (FM 3-0 and FM 6-0).

| | |
|---|---|
| **A** | **Areas** |
| **S** | **Structures** |
| **C** | **Capabilities** |
| **O** | **Organizations** |
| **P** | **People** |
| **E** | **Events** |

Figure 1-2. ASCOPE.

**Table 1-1. Typical Civil Considerations within Each ASCOPE Category**

| Area | Structure | Capabilities | Organization | People | Events |
|---|---|---|---|---|---|
| Tribe | Cemeteries | Sewer | Tribal | Phones | Weddings |
| Families/ Clans | Religious shrines | Water | Family/clan | Speeches | Birthdays |
| Ethnicity | Houses of worship | Electrical | Religious | Face-to-face meetings | Religious gatherings |
| Religion | Bars/tea shops | Academic | Ethnic | Media/radio | Funerals |
| Economic districts | Social gathering places | Trash | US/coalition forces | Media/TV | Major religious events |
| Smuggling routes | Print shops | Medical | Governmental agencies | Media/print (newspaper) | Anniversaries of wars or battles |
| National | Internet cafes | Security | Farmers or Unions | Visual (graffiti, signs) | Holidays |
| Social classes | Television | Market (use and goods) | Community | Visual (videos, DVDs) | Harvests |
| Political districts | Radio station | Employment and commerce | Military or militia units | Audio (pirated or illegal radio) | Reconstructio n openings |
| Military districts | Hospitals | Crime and justice | Illicit organizations | Rallies or demonstrations | Town or council meetings |
| School districts | Banks | Basic needs | Insurgent groups | Restaurants | Elections |
| Road system | Dams | Public health | Gangs | Door-to-door | Sports events |
| Water sources | Bridges | Economic (jobs) | Businesses organizations | Internet | |
| Water coverage | Police stations | Religion | Police | Markets | |
| Water districts | Gas stations | Displaced persons and refugees | Nomads | Sports | |

**Table 1-1. (Continued)**

| Area | Structure | Capabilities | Organization | People | Events |
|---|---|---|---|---|---|
| Construction sites | Military barracks | Political voice | Displaced persons and refugees | Religious gatherings | |
| Gang territory | Jails | Civil rights, individual rights | Volunteer groups | Parks | |
| Safe areas/ sanctuary | Water pumping stations | | Intergovernmental organizations | Family gatherings | |
| Trade routes | Oil/gas pipelines | | Political | Gas lines | |
| Power grids | Water lines | | Contractors | Bars/tea shops | |
| | Power lines | | NGOs | Food lines | |
| | Storage facilities | | Labor unions | Job lines | |

### *Methodology*

While analyzing civil considerations, counterinsurgents should develop both ASCOPE matrixes and map overlays. Developing these products should be done in partnership with HN security forces and local government officials. Effective civil considerations analysis facilitates understanding. Table 1-1 lists typical examples in each of the ASCOPE categories.

### *Civil Considerations Overlay*

The civil considerations overlay provides insights to the area of operations by identifying sectarian fault lines, demographic groups, and their issues. Figure 1-3 shows an example of a civil considerations (ASCOPE) overlay.

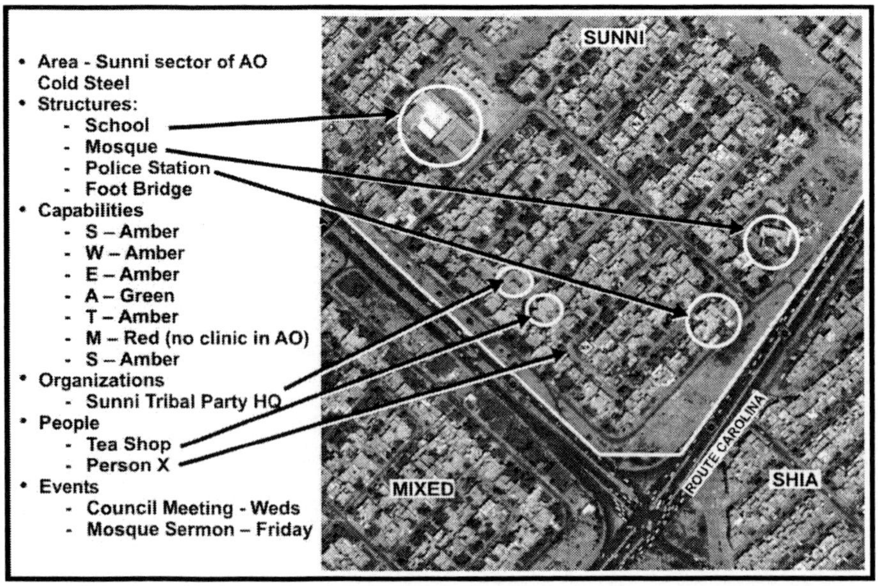


Figure 1-3. Civil considerations overlay.

*Civil Consideration Matrixes*

A separate matrix is a useful tool for each of the six ASCOPE categories for any demographic group. Also helpful is an ASCOPE map overlay that shows the boundaries of each identified demographic group. The government's districts (police, utilities, political) often fall along natural demographic lines. When they do not, the boundary conflicts with how the population defines its neighborhood. This can cause friction between the population and government. For each area, staffs might want to first identify the demographic group it contains. Then, they can work through the rest of the AS COPE categories. For example, a unit with an area that has two main tribes should develop a separate ASCOPE matrix for each tribe. Three key perspectives aid in understanding the categories of ASCOPE:

**Population or Populace**

This refers to the local community.

**Insurgent**

This refers to any group illegally impeding the functions of the government, such as insurgents, militias, gangs, criminals, foreign agents, and terrorists.

**Counterinsurgent**

This refers to the perspective of the forces and groups conducting the counterinsurgency and those supporting them. These elements can include HN security forces, local police, government leadership and employees, US military, government agencies, Provincial reconstruction teams, and coalition members.

**Areas**

This term refers to the specific localities within an AO, where a particular demographic group slives, neighborhood by neighborhood and block by block. Unless a unit occupies an ethnically homogenous area, it will have multiple "areas" within its AO. In addition, a single demographic area may cross several unit boundaries. Examples of specific areas include—

- Those defined by political boundaries such as city districts or regional municipalities.
- Social, political, religious, or criminal enclaves.

Once a unit defines the geographic area occupied by a demographic group, then it should complete the remainder of the ASCOPE for that area. They repeat this for other areas. Table 1-2 shows an example area matrix. Columns in this matrix include—

*Area*

This column names or describes each of the sub areas of a unit's AO, such as tribal, religious, economic, or political districts.

*Location*

This column describes the location or boundaries of each group. These boundaries will rarely be pure, since often groups overlap.

**Table 1-2. Example Area Matrix**

| Area | Location | From the Perspective of | | |
|---|---|---|---|---|
| | | Population | Insurgent | Counterinsurgent |
| Group/ Tribe A (specific name) | Grid coordinates/ boundaries, for example | Safe haven; Provides early warning from attack; community watches out for each other | Area provides freedom of movement, protection from government forces, cache sites, safe houses | Nonpermissive terrain. All roads in area are high threat. History of IEDs/SAFs. |
| Religious Group A | Boundary | Perspective is... | Perspective is... | Perspective is... |
| Political District A | Boundary | Perspective is... | Perspective is... | Perspective is... |

*Population*

This column describes how the population perceives and uses the area.

*Insurgent*

This column shows how the adversary perceives and uses the area.

*Counterinsurgent*

This column describes how counterinsurgents look at and use this area.

**Structures**

Existing structures can play many significant roles. Bridges, communications towers, power plants, and dams are important infrastructure. Others, such as churches, mosques, national libraries, and hospitals are cultural sites, play important roles in the community. Still others are facilities with practical applications such as jails, warehouses, television/radio stations, and print plants. Some aspects of the civilian infrastructure, such as the location of toxic industrial materials, may influence operations. Analyzing a structure involves determining how its location, functions, and capabilities support an operation. Commanders also consider the consequences of using a certain structure. Commanders must carefully weigh the expected military benefits against costs to the community that will have to be addressed in the future. Table 1-3 shows an example structures matrix. Considerations for each of the columns include—

*Structure*

This identifies, defines, and names the specific structures within the AO.

*Location*

This describes the specific location (grid coordinates) of each structure.

*Population*

This describes how the population perceives and uses the structure.

**Table 1-3. Example Structures Matrix**

| Area | Location | From the Perspective of | | |
|---|---|---|---|---|
| | | **Population** | **Insurgent** | **Counterinsurgent** |
| Hospital (specific name) | XM123456 | Needed medical care, however, shortage of staff and meds | Can provide black market meds to fill shortages | Funding and personnel shortfalls, needs expanding |
| Houses of worship | Grid coordinates | Perspective is... | Perspective is... | Perspective is... |
| Police stations | Grid coordinates | Perspective is... | Perspective is... | Perspective is... |

*Insurgent*

This describes how the insurgent perceives and uses the structure.

*Counterinsurgent*

This describes how counterinsurgents look at this structure.

## Capabilities

Capabilities refer to the ability of local authorities to provide citizens with key services such as public administration, public safety, emergency services, and food. Capabilities include areas in which the populace may need help after combat operations, such as public works and utilities, public health, economics, and commerce. Capabilities also refer to resources and services that can be contracted to support the military mission such as interpreters, laundry services, construction materials, and equipment. The Host Nation or other nations might provide these resources and services. Commanders and staffs analyze capabilities from different perspectives. They view capabilities in terms of those required to save, sustain, or enhance life, in that priority.

Within each demographic group, identify who is responsible overall for each item that is required to save, sustain, or enhance life. Include preexisting needs as well as the needs of the populace after combat operations or disaster. This will play a large part identifying root causes of the insurgency. These items are listed here as well but the focus is on who is responsible for each item. Table 1-4 shows an example capabilities matrix. Considerations for each of the columns include—

*Capabilities*

This, at a minimum, describes the SWEAT-MSO (sewer, water, electricity, academic, trash, medical, safety, and other considerations) items.

*Status*

1-61. This lists the status of each of the SWEAT-MSO items, for example—

***Red***

1-62. Nonexistent or nonfunctioning.

**Table 1-4. Example Capabilities Matrix**

| Area | Location | From the Perspective of | | |
|---|---|---|---|---|
| | | Population | Insurgent | Counterinsurgent |
| Sewer | Often red in slums | Blame tribal sheik; hurts economic development | Major issue to use against tribal sheiks; will destroy repairs | City manager places low priority on this tribal area |
| Water | Grid coordinates | Perspective is... | Perspective is... | Perspective is... |
| Electrical | Grid coordinates | Perspective is... | Perspective is... | Perspective is... |

***Yellow***

Present but not fulfilling the requirements of the population, needs labor/parts/fuel to maintain, expected to fail without support.

***Green***

Satisfactory to sustain population.

*Population*

This lists individuals that the population consider responsible for each specific SWEAT-MSO item.

*Insurgent*

This shows the perspective of the insurgency.

*Counterinsurgent*

This lists the individual the local government considers responsible for each SWEAT-MSO item.

## Organizations

Organizations are nonmilitary groups or institutions in the AO. They influence and interact with the populace, military units, and each other. Organizations generally have a hierarchical structure, defined goals, established operations, fixed facilities or meeting places, and a means of financial or logistic support. Some organizations may be indigenous to the area such as tribes and ethnic based groupings. Other organizations include church groups, fraternal, patriotic or service organizations, labor unions, criminal organizations, political parties, and community watch groups. Other organizations may come from outside the AO. Examples of these include multinational corporations, United Nations agencies, US governmental agencies, and nongovernmental organizations (NGOs). Table 1-5 shows an example organizations matrix. Considerations for each of the columns include—

*Organization*

This identifies, defines, and names the specific organizations within the AO. Some may be identified in the area matrix but this gives the details of the group where the area matrix identifies its location.

**Table 1-5. Example Organizations Matrix**

| Area | Location | From the Perspective of | | |
|---|---|---|---|---|
| | | **Population** | **Insurgent** | **Counterinsurgent** |
| Tribal | XM123456 | Tribal loyalties and interactions dominate life | Looks to increase intertribal strife, violence | Must include tribal leaders in government |
| Political | Grid coordinates | Perspective is... | Perspective is... | Perspective is... |
| Social | Grid coordinates | Perspective is... | Perspective is... | Perspective is... |

*Location*

This shows the specific location (grid coordinates) of each organization.

*Population*

This shows how the population perceives and uses the organization.

*Insurgent*

This shows how the insurgent perceives and uses the organization.

*Counterinsurgent*

This tackles how the counterinsurgents look at this organization.

**People (Means of Communication)**

An important aspect of people is how they communicate. The term people includes all civilians within the AO or AI whose actions or opinions can affect the mission. Both formal and informal means of passing information, actions, opinions and political influence, are critical to understanding the AO. All counterinsurgents should look for the obvious visual and audible signals as well as where people gather. Visual examples include graffiti, posters, signs, billboards, murals, videos and DVDs, and television. Audible examples include pirated radio broadcasts, loudspeakers from a Mosque, someone reading to a group, speeches, and religious teachings or services.

Most people who serve as the spokesmen in the community (community, labor, and religious leaders) should also appear in the people matrix. If the counterinsurgent's information dissemination techniques differ from the insurgents, this difference could explain why the enemy's propaganda is more credible, timely, and considered to be more legitimate by the target audience. Table 1-6 shows an example people matrix. Considerations for each of the columns include—

***People***

This column identifies, defines and names the specific methods people use to communicate in this area or key communicators.

***Location***

This column shows the locations where people communicate or where key communicators live and work.

**Table 1-6. Example People (Means of Communications) Matrix**

| Area | Location | From the Perspective of | | |
|---|---|---|---|---|
| | | Population | Insurgent | Counterinsurgent |
| Phones | Grid coordinates | Local store owner | Unknown | Tribal and community leader cell phones |
| Mass Media - Radio | Grid coordinates | Individual names... | Individual names... | Individual names... |
| Religious services | Grid coordinates | Individual names... | Individual names... | Individual names... |
| Individuals such as mayor, police chief, of store owner | Grid coordinates | Respected leader believed to be unbiased | Enemy of the people | Tolerated because not from same ethnic group |

*Population*

This column describes who the population perceives as being a key source of communication.

*Insurgent*

This column shows who the insurgents use to communicate with the population.

*Counterinsurgent*

This column shows who the HN uses to communicate with the population.

**Events**

Events, both public and private, are routine, cyclical, planned, or spontaneous activities that affect organizations, people, and military operations. Examples include national and religious holidays, agricultural crop/livestock and market cycles, elections, civil disturbances, and celebrations. Once tactical units determine significant events, they must template the events and analyze them for their political, economic, psychological, environmental, and legal implications. Table 1-7 shows an example event matrix.

*Event*

This column identifies and lists all events important to the populace. This includes annual events such as religious holidays; seasonal harvests or migration of insurgents; or more frequent events like council meetings, religious services, and special shopping days.

*Location*

This column shows the location, normally in military format and includes a date-time-group (DTG).

*Population*

This column describes the population's perception of the event.

*Insurgent*

This column comments on the insurgent's perception of the event.

**Table 1-7. Example Event Matrix**

| Area | Location | From the Perspective of | | |
|---|---|---|---|---|
| | | Population | Insurgent | Counterinsurgent |
| Religious Festival XX | Route XYZ | One religion approves; another disapproves | Gives an opportunity to promote religious strife | Religious freedom must be accepted |
| Fire Station XX Opening | Grid coordinates | Perspective is... | Perspective is... | Perspective is... |
| Funeral for XX | Grid coordinates | Perspective is... | Perspective is... | Perspective is... |

*Counterinsurgent*

This column comments on how the HN perceives the event.

## Section IV—Effects

Describing the effects of the operational environment is the second step in IPB. It involves taking the facts about an area of operations grouped by mission variables of terrain, weather, and civil considerations and analyzing them to arrive at a conclusion about their effects on enemy and friendly courses of action. In addition to the normal analytical tools, examining each of the prerequisites of an insurgency and the root causes that lead to the insurgency have proved to be useful in identifying long-term societal problems. These problems lie at the heart of the competition for the population's support between the insurgent and counterinsurgent.

***Prerequisites***

There are three prerequisites for an insurgency to be successful in an area—a vulnerable population, leadership available for direction, and lack of government control. When all three exist in an area, an insurgency can operate with some freedom of movement, gain the support of the people, and become entrenched over time.

### Vulnerable Population

A population is vulnerable if the people have real or perceived grievances that insurgents can exploit. The insurgents can exploit the population by offering hope for change as well as exploiting political, economic, or social dissatisfaction with the current government. A gap between population's expectations and the capability to meet these expectations may cause unrest within the population, including turning to insurgency. The larger the gap, the greater the population's perceived, or relative, sense of deprivation between what they have and what they perceive they should have. Similarly, the larger the gap, the more susceptible the population is to insurgent influence through promises to close the gap.

### Leadership Available for Direction

A vulnerable population alone will not support an insurgency. There must be a leadership element that can direct the frustrations of the population. If insurgents can recruit, co-opt, or coerce local leaders or the local leaders are part of the insurgency, these leaders can direct the frustrations of the populace.

### Lack of Government Control

Real or perceived lack of governmental control can allow insurgents to operate with little or no interference from security forces or other agencies. The greater the control the government has over the situation, the less likely are the chances for insurgent success. The opposite is also true. If the government is not providing what the people believe their government should, insurgents may provide an alternative government, or "shadow" government, or they may merely nullify governance to allow freedom of action and movement, depending on their end state. Host Nation failure to see or admit that there is an issue or outright refusal to change can further strengthen this prerequisite.

### *Root Causes*

There are five general categories of root causes for insurgencies. A root cause is the basis of a grievance among the population. Some or all of these grievances may fuel an insurgency to varying degrees. The importance of the root causes, or even their existence, can change over time. Additionally, insurgents may be adept at manipulating or creating root causes and grievances to serve their purpose.

### Identity

Many factors impact a person's sense of identity, but membership in a socio cultural group may have the deepest influence. Strong feelings based on identity can be in conflict with the group identity of the majority of the members of the Host Nation government, potentially leading to insurgencies with secession or political overthrow as goals. External nations with similar social identities as the insurgents may assist.

### Religion

While religion is often a primary identity, it can become important enough to be a considered a separate identity unto itself. In this way, religious fundamentalism or extremism can become a root cause of an insurgency in and of itself. External groups with similar extremist religious views as the insurgents may assist.

### Occupation or Exploitation

Popular perception of outsiders either occupying the HN, or excessive HN pandering to outsiders, can be a source of insurgency. For example, foreign businesses can dominate critical portions of the local economy. This can occur to the point that some may feel that they or their country are being exploited. An outside military presence or military treaty may offend national sentiment. The mere presence or specific actions of foreigners may offend religious or cultural sensibilities as well.

### Economic Failure

Pervasive and desperate poverty can often be a root cause of an insurgency. Starving young people without jobs or hope are ripe for insurgent recruitment. A large gap between the vast poor majority and a small extremely rich minority will exacerbate these issues.

**Corruption and Repression**

Corruption and repression can lead to popular dissatisfaction with the current government. Rampant corruption leads to the loss of HN legitimacy and possibly a desire to change or replace the Host Nation government.

## Section V—Cultural Competence and Situational Awareness

Culture can be defined as the set of a system of shared beliefs, values, customs, behaviors, and artifacts that members of a society use to cope with their world and one another. Since all wars are fought in and amongst a population, the Army seeks to develop an ability to understand and work with a culture for its Soldiers and leaders. Cultural capability is the blend of individual competence in understanding the general characteristics and the characteristics of specific cultures, derived from a cumulative set of cultural knowledge, skills, and attributes to help forecast and favorably influence the behavior of the target group or society. This section will define culture; identify the two major components of culture capability—cross-cultural competence and regional competence; and discuss the three cultural proficiency levels—cultural awareness, cultural understanding, and cultural expertise—within a culture capability.

***Culture***

Each society is composed of both a social structure and culture. Social structure refers to the relations among groups of persons within a system of groups. Social structures persist over time. That is, it is regular and continuous despite disturbances, and the relation between the parts holds steady even as groups export contract. In an army, for example, the structure consists of the arrangement into groups like divisions, battalions, and companies, and the hierarchy of ranks. In a society, the social structure includes groups, institutions, organizations, and networks. Social structure involves the arrangement of the parts that constitute society, the organization of social positions, and the distribution of people within those positions. Some elements of the social structure are considered here:

### Social Groups

- What are the major groups both inside and outside their AO?
- What are the formal relationships, such as treaties or alliances; and the informal relationships such as tolerance or friction between groups? What are the cleavages between groups and crosscutting ties, for example, religious alignments that cut across ethnic differences?
- Do the insurgent leadership and their rank and file belong to separate groups? Does the population belong to a different social group than the insurgents? Can seams among insurgents or between insurgents and the population be widened?
- How do people identify themselves (tribes, religions, ethnicity, provinces/regions, classes, occupations, and common language)?
- Are there a large number of homeless, refugees, squatters, internally displaced persons (IDPs)? What do the people think of them?

## Norms, Roles, and Statuses

- What are the expected behaviors *(roles) of* people in different social status? How should a parent, political leader, military figure, or religious leader behave? What are the appropriate treatments for women and children? What are the common courtesies, such as gift giving? What are the local business practices, such as bribes and haggling?
- What are the traditional roles of each family member?
- What are people in the society expected to do (norm)? Norms may be either moral (incest prohibition, homicide prohibition) or customary (prayer before a meal, removing shoes before entering a house). How important is being on time in the society (business, social gatherings)?
- What are the punishments (formal and informal) for role violations? What will the people disapprove of? What are the requirements for revenge if honor is lost?

### *Interrelated Nature of Culture*

Culture is learned, shared by members of a society, patterned, changeable, arbitrary, and internalized, in the sense that it is habitual, taken for granted, and perceived as "natural" by people in the society. Culture conditions the individual's range of action and ideas, including what to do and not do, how to do or not do it, and whom to do it with or not to do it with. Culture also includes under what circumstances the "rules" shift and change. Culture influences how people make judgments about what is right and wrong, assess what is important and unimportant, categorize things, and deal with things that do not fit into existing categories. Cultural rules are flexible in practice.

### *Taxonomy*

One simple way to show a culture is to build a chart that systematically distinguishes, orders, and names groups—a cultural taxonomy. In order to do this, leaders and staffs must define a culture's influences, variations, and manifestations. Cultural influences and cultural variations explain *why* the culture is the way it is. Cultural manifestations refer to *what* one may encounter in a culture. Figure 1-4 shows an example of a culture's taxonomy.

| ***Cultural Influences*** | ***Cultural Variations*** | ***Cultural Manifestations*** |
|---|---|---|
| History | *Values* | Planning Style |
| Language | Individualism vs Collectivism | View of Authority |
| Geography | Power Distance | Negotiation Styles |
| Religion | Formality vs Informality | Willingness to Compromise |
| Communications | Uncertainity Avoidance | Risk Avoidance |
| Political Science | Relationship vs Deal Focus | Time to Decision |
| Military Arts and Science | Long vs Short Term Orientation | |
| Sociology | Time Orientation | |
| Cultural Anthropology | *Cognition* | |
| Economics | Reasoning Styles | |
| Education | | |
| Art, Music, Entertainment | | |
| Literature | | |
| Food and Drink | | |
| Psychology | | |
| Law and Criminal Justice | | |
| Science and Technology | | |
| Philosophy | | |

Figure 1-4. Taxonomy of culture.

**Elements of Culture**

The size of a nation, its diverse subcultures, different educational levels and geographic backgrounds contribute to a great range of cultural variances amongst individuals and groups. Members of the population view cultural influences differently depending on their geographic location or identifying group. Some elements of culture should be identified and evaluated in a counterinsurgency operation. The following questions can aid units in defining the different elements of culture:

*History*

- What are the major wars, massacres, and conflicts that shaped the culture?
- What are some of the great leaders, heroes, or legends in the nation's history?
- What are some of the villains (infamous) people in the nation's history?
- Who founded the country? Who brought it to its modern form?
- What are some of the significant eras, generations or major shifts in a nation that are significant?

*Language*

- What are the common languages or dialects spoken?
- Standard words and phrases universal in all cultures (formal and informal):
    - Hello.
    - Goodbye.
    - Please.
    - Thank you.
    - You're welcome.
    - How are you?
    - May I help you?
- Common sayings in a culture, for example, "God bless you" or "God save the Queen."
- "Excuse me" or "Pardon me."
- Toasts with appropriate beverages, such as coffee or beer, are sociable.
- Grace or well wishes (for food)—meals are sociable.
- What are the common sayings, clichés, or slang?
- What is customary during greeting and departing (shake hands, kiss, and bowing)?

*Geography*

- What typically defines a community or neighborhoods, for example, economic, ethnic, tribal, religious, or political traits? Where are the neighborhood boundaries?
- Where do new arrivals, immigrants, workers, and IDPs typically come from? Why did they migrate here? Is the migration seasonal, temporary, or permanent?
- What are the most significant local natural and man-made landmarks and structures (for example, religious, historical, cultural)?

*Religion*

- What are the main religions? Is there an official religion?
- What are important religious events and holidays?
- What is the role of religious leaders within the society?
- What is traditional for funerals and mourning?
- Are there any tensions in the nation due to religious differences?
- What are the tenets of the main religions?

*Communications*

- How do people communicate?
- How do the people receive information? Radio, TV, newspaper, meetings, word of mouth)?
- Where do the people usually gather? Bars, tea or coffee shops, cafes, or markets?
- Where do people socialize or congregate randomly in previously unrecognized manners, for example, wait in long lines, for day labor, in traffic, or at sporting events and tournaments)?
- Who are the principal communicators within the local community?

*Political Science*

- How do people view the role of the Host Nation government?
- What are the roles of and how important are civilian (nongovernmental) community leaders?
- What are the major political parties?
- Is the local government effective? Why or why not?
- What civil and human rights do the populace hold most sacred?
- Is the government secular or religious?
- Does the country have a constitution, document, guideline that lays out the role of government, rights of the people, and laws?

*Military Arts and Science*

- How respected is military service in the culture? How are veterans treated?
- Who are some of the famous military leaders or revolutionaries in the country?
- Is there an NCO corps? How are NCOs selected? What are their duties and responsibilities? What is their relationship with officers and Soldiers?
- What oath of allegiance, if any, do members of the military swear upon enrollment?
- What colors, banners, symbols, or uniforms do antigovernment forces use?

*Sociology*

- Do the people identify themselves with organizations or affiliations (tribes, religions, ethnicity, provinces/regions, classes, occupations, and common language)?
- What are the major problems and underlying issues with the people (root causes)?
- Are there a large number of homeless, refugees, squatters, internally displaced persons (IDPs)? What do the people think of them?
- What is customary in dealing with guests or strangers? Are people friendly or guarded with strangers?
- What are some of the core values of the people that define who they are?
- What is the daily or weekly schedule of most citizens (wakeup, meals, work, social time, sleep)?

*Cultural Anthropology*

- Have current warring groups/factions ever lived side-by-side in peace? What changed?
- What are some of the key cultural aspects of the local tribes or nomadic groups?
- What are the traditional roles of each family member?
- Whom do the people look to for leadership (governmental and nongovernmental)?
- How important is being on time in the society (business, social gatherings)?
- How do they correct social mistakes?

*Economics*

- What export or local product are the people known for and the most proud of?
- Are bribes or "gifts" normal in dealing with businesses, government officials, or police? What is acceptable (levels of corruption)?
- What infrastructure is required to support economic growth (electricity for factories, roads to move produce, security to minimize extortion/black market)?
- What is the daily wage of an average worker/laborer? Which jobs are considered honorable?
- What economic organizations are important and influential in the society (labor unions, merchant guilds)?
- Is there a local black market? Who is involved, what products, and how tied to the community and local government?
- Are prices fixed or negotiated in normal commerce?

*Education*

- What is the literacy rate?
- Who goes to school (males, females, all, optional)? What is the last year of general public education?
- Are public schools secular or religious? How?

- What influence do local universities have, for example, do the professors promote radicalism and do the schools serve as recruiting centers?

*Art, Music, and Entertainment*

- How important is the national anthem to the populace? What do the lyrics mean?
- What types of music do the most people listen like?
- What types of movies do they like?
- What are their favorite holidays, and how do they celebrate them? What types of food do they eat during holidays and special occasions?
- What are some of the most popular hobbies and recreations?

*Literature*

- What types of stories do children read? What are the morals of these stories?
- What stories, fables, and epics, oral or written, pass down through families or communities? Do these help define the culture?
- Who are the most famous—or infamous—characters in popular literature?
- What are some the legends of the nation's past?
- What are the popular books, and who are the controversial authors (past and present)?

*Food and Drink*

- What is the local cuisine?
- What are some typical or traditional foods and drinks?
- How are they prepared?
- How important is sharing a meal?
- Is there any food or drink culturally forbidden?

*Psychology*

- Who or what do people fear?
- Rank the following from 1 (most important) to 7 (least important):
    - God
    - Family
    - Tribe
    - Neighborhood
    - Country
    - Political party
    - Ethnic group
- Rank the following from 1 to 7 (most to least important):
    - Esteem needs (self-esteem and respect of others).
    - Safety needs (security and stability).
    - Self-actualization (meet ones potential).

- Love needs (belonging).
- Physiological needs (basics necessities—water, food, shelter).

*Law and Criminal Justice*

- Who makes and enforces the local laws? What justice can the victims or their families exercise?
- What are the basic rules of the road, traffic laws, and right of way? Are they followed?
- What types of organized crime exists? What symbols, colors, graffiti, or uniforms do local gangs or organized crime use? What does each mean (marks territory, identifies targets, intimidates populace)?
- How do the people feel about corporal punishment and capital punishment?
- What could dishonor an individual, family, or group? How do you correct serious situations between individuals, families, or groups or families?

*Science and Technology*

- Does the country or area have Internet service? Satellite or hardwired?
- Does the enemy use the Internet? How and why?
- What Internet sites are forbidden or blocked?
- What is the country recognized for inventing or discovering? What do the people generally believe their country invented or discovered?

## Cultural Variations

Cultural variations are the behaviors, values, and interests common to a culture. Understanding these variations allows US and HN security forces to interact and thus to operate more effectively in negotiations, advisory roles, population control, and daily interaction with the populace to gain better cultural competence.

*Behaviors*

A culture's behavior consists of actions which can be sensed; specifically a group's language, social mores, customs, structures, and institutions. Culturally competent units understand and train to recognize these behaviors as a means to identify insurgent actions, anticipate the population actions, and detect subtle changes within the population. Actions inconsistent with the population's behavioral norms could be indicators of guerrilla activity, internal conflict, or the confirmation or denial of intelligence. Living and operating among the population is essential to understanding population behavior.

*Values*

Values are the principles the population uses to evaluate alternatives or consequences of decisions and actions. A value is a concept that describes the beliefs of an *individual* or *culture* and is identity based, for example, Army Values. Values are how people understand what they are and what they will and will not tolerate. Values define their sense of honor and respect. Values are often unchangeable. Soldiers never attempt to change the population's values, confuse its interests with its values, or use its interests in an attempt to alter its values. During tactical operations, counterinsurgent forces prioritize the population's values over its interests to demonstrate the Host Nation government's legitimacy in supporting the population.

*Interests*

An interest is what the population wants or desires for a group's benefit or advantage (it is often perceived as a right or legitimate claim). An interest may be flexible and can change. Interests are linked to the situation, such as what people want at present. US perceptions should not dictate what the population needs and wants. The insurgency likely understands these needs and wants and eagerly exploits them to gain support, as well as to turn the population away from the HN government. A vulnerable population gravitates towards who it feels best understands and satisfies its needs and wants. During tactical COIN operations, the counterinsurgent must know the local population's difference between a need and a want and not operate with a US cultural bias.

## Cultural Manifestations

Cultural manifestations are the concrete displays of a culture's thought and behavior measured by the senses. It is how a population demonstrates its views on authority, legitimacy, negotiation style, compromise, and other similar thoughts and behaviors.

### *Cultural Capability*

Cultural capability has two major components:

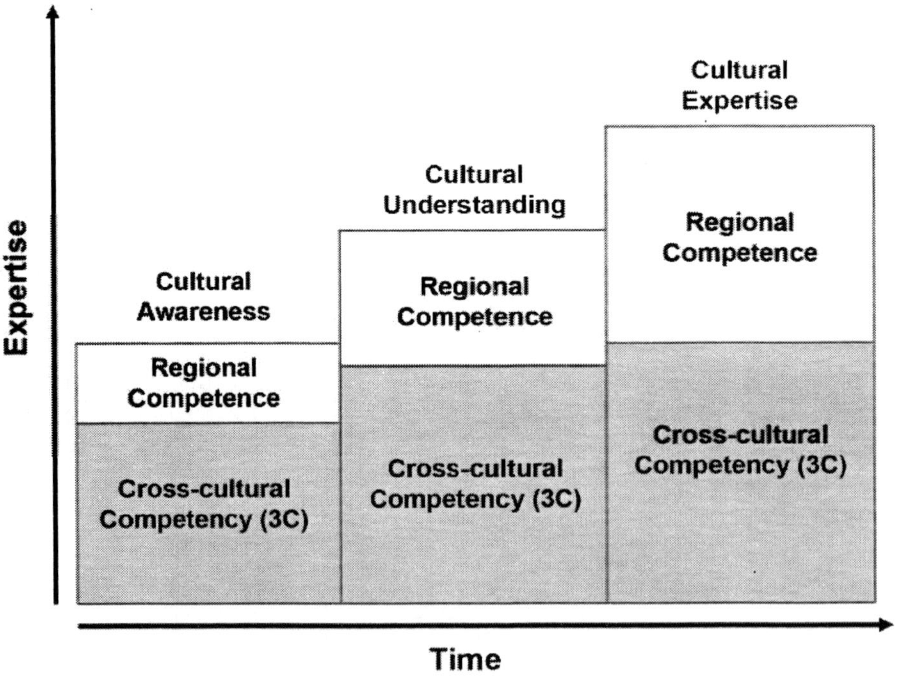


Figure 1-5. Changes in cultural capability over time.

## Cross-Cultural Competency

Cross-cultural competency (3C) includes general cultural knowledge, skills, and attributes. All Soldiers must devote time to developing cross-cultural competency. It forms the foundation for understanding any culture, and is developed by studying the humanities, including movies and other media; traveling to other countries; and personally interacting with people from countries outside the US.

## Regional Competence

Regional competence includes culture-specific knowledge, skills, and attributes that pertain to a given country or region. Regional competence is developed by lifelong study of a region and tailored training during preparation for a deployment.

### *Cultural Proficiency Levels*

As Soldiers develop cross-cultural competence and regional competencies over time, broad descriptions of their proficiency levels show the depth of their knowledge, skills, and attributes in those competencies. These descriptions represent a standard that culture and foreign language education and training are designed to achieve. The following paragraphs define the cultural proficiency levels, and Figure 1-5 shows how they change over time:

## Cultural Awareness

This proficiency level describes Soldiers who demonstrate basic cross-cultural competence in all three subcomponent areas: culture fundamentals, cultural self-awareness, and culture skills. They will have a minimal level of regional competence necessary to perform assigned tasks in a specific geographic area. These Soldiers will be able to describe key culture terms, factors, and concepts. Additional characteristics of cultural awareness are—

- Cultural awareness sets the conditions to learn about foreign cultures and people.
- Cultural awareness includes Soldiers who have an appropriate mind-set and a basic culture capability.

## Cultural Understanding

This proficiency level describes Soldiers and leaders with well-developed cross-cultural competence. They will have a comprehensive level of regional competence that allows them to accomplish the mission in a specific geographic area. These Soldiers will be able to apply relevant terms, factors, concepts, and regional information to their tasks and mission. Additional characteristics of cultural understanding are—

- The proficiency category of cultural understanding includes Soldiers who are familiar with a specific region and have the ability to identify economic, religious, legal, governmental, political, and infrastructural features of a specific region.
- Cultural understanding also includes Soldiers who are aware of regional sensitivities regarding gender, race, local observances and local perception of the US and its allies.

### Cultural Expertise

Cultural expertise is a proficiency level that describes culture professionals and leaders who possess an advanced level of cross-cultural competence. They will have an advanced and sophisticated level of regional competence pertaining to a specific geographic area. These Soldiers will be able to integrate and synthesize terms, factors, concepts, and regional information into plans, programs, and advice to commanders. In addition—

- In most cases, cultural expertise entails some degree of proficiency in a language or a few relevant languages; proficiency in the skills that enable effective cross-cultural persuasion, negotiation, conflict resolution, influence, or leadership; and an understanding of the most salient historic and present-day regional structural and cultural factors of a specific geographic area.
- Cultural expertise also describes Soldiers and leaders with the ability to advise commanders of the region on military operations.

### *Culturally Influenced Situational Awareness*

Situational awareness, the goal of every leader and Soldier, is the immediate knowledge of the conditions of the operation, constrained geographically and in time (FM 3-0). Culturally influenced situational awareness allows counterinsurgents to detect subtle indicators of change or threat in the operational environment and understand how this will affect insurgent decisions and planning. When conducting counterinsurgency operations, cultural capability is a key part of achieving culturally influenced situational awareness for a Soldier, leader, or tactical unit. Within small units, superb cultural capability and improved situational awareness results in a greater chance of mission accomplishment, tactical effectiveness, and protection.

### Assessment

A leader or Soldier has begun to achieve culturally influenced situational awareness when he/she can ask and answer such questions accurately:

- What is my adversary thinking and why?
- What are my Host Nation security forces thinking and why?
- What are groups of people thinking and why?
- What will my adversaries, groups of people, adjacent units, and coalition partners, and Host Nation security forces do if I take action X, and why?
- How are cultural factors influencing my operations?
- How can I make groups of people and Host Nation security forces do what I want them to do?

## SUMMARY

Counterinsurgency can be extremely complex. At its core, COIN is a struggle for the population's support. Understanding that struggle or becoming "the world expert on your district" (28 Articles, Kilcullen) is the foundation for any unit. A unit that uses the four tools described in this chapter dramatically increases its likelihood of success against an insurgency. These four tools are:

- Studies carefully its operational environment (OE) using the operational variables of PMESII-PT (political, military, economic, social, information, infrastructure, physical environment and time).
- Defines its situation using METT-TC.
- Determines the root causes of the insurgency and analyzes the three prerequisites.
- Develops cultural capability to increase their ability to understand and interact with the population.

## 2. Foundations of Insurgency

*"It is certainly easier to launch an insurgency than it is to repress it."*
David Galula, Counterinsurgency Warfare.

An insurgency and the operational environment must be understood before the insurgency can be defeated. Tools that assist the counterinsurgent in understanding and predicting the insurgent's actions are the five elements of an insurgency, the eight dynamics of an insurgency, the six insurgent strategies, insurgent tactics, and the strengths and vulnerabilities of insurgents. Together these tools, known as the components and manifestations of an insurgency, provide leaders at all levels a means to comprehend and defeat the insurgent.

### Section I—Overview

This section introduces the relationship between the components and the manifestations of an insurgency. If a counterinsurgent understands both the components and manifestations of the insurgency, then the unit can correctly apply pressure along the seven counterinsurgency lines of effort (Chapter 3) to defeat it.

#### *Components*

The components of an insurgency are comprised of the five elements, the eight dynamics, and six strategies. The elements are the five groups of people—leaders, guerrillas, underground, auxiliary, and mass base—that form the insurgency's organization. The dynamics are the eight categories that define an insurgency—leadership, ideology, objectives, environment and geography, external support, internal support, phasing and timing, organizational and operational patterns. Finally, the six insurgent strategies are the urban, military focused, protracted popular war, identity focused, conspiratorial, and the composite and coalition. Together, the components—the five elements, the eight dynamics, and the six strategies—are tools of analysis that allow the counterinsurgent to fully grasp the nature of the insurgency.

### *Manifestations*

The manifestations are the visible outputs of the insurgency. Made up of the insurgent's tactics, strengths and vulnerabilities, the counterinsurgent will be able to track, categorize, and develop the insurgency's pattern, and a means to defeat it. Figure 2-1 shows the relationship between the components and manifestations of an insurgency.

### *Historical Insurgency*

With an understanding of the components of an insurgency, a counterinsurgent unit can identify, describe, and categorize any insurgency. Figure 2-2 shows the components of an historical insurgency.

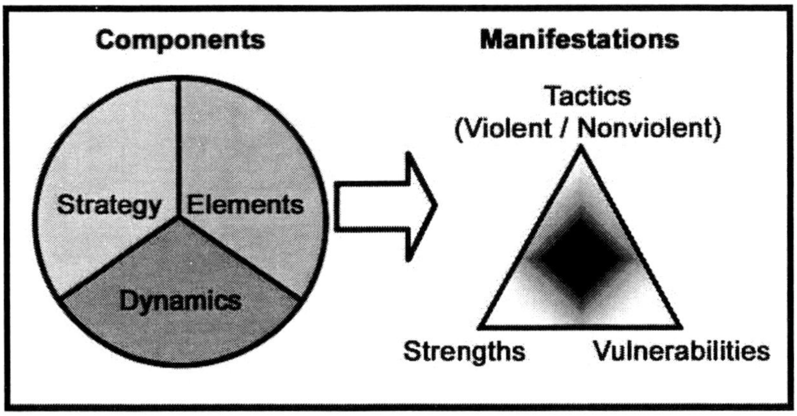


Figure 2-1. Insurgency.

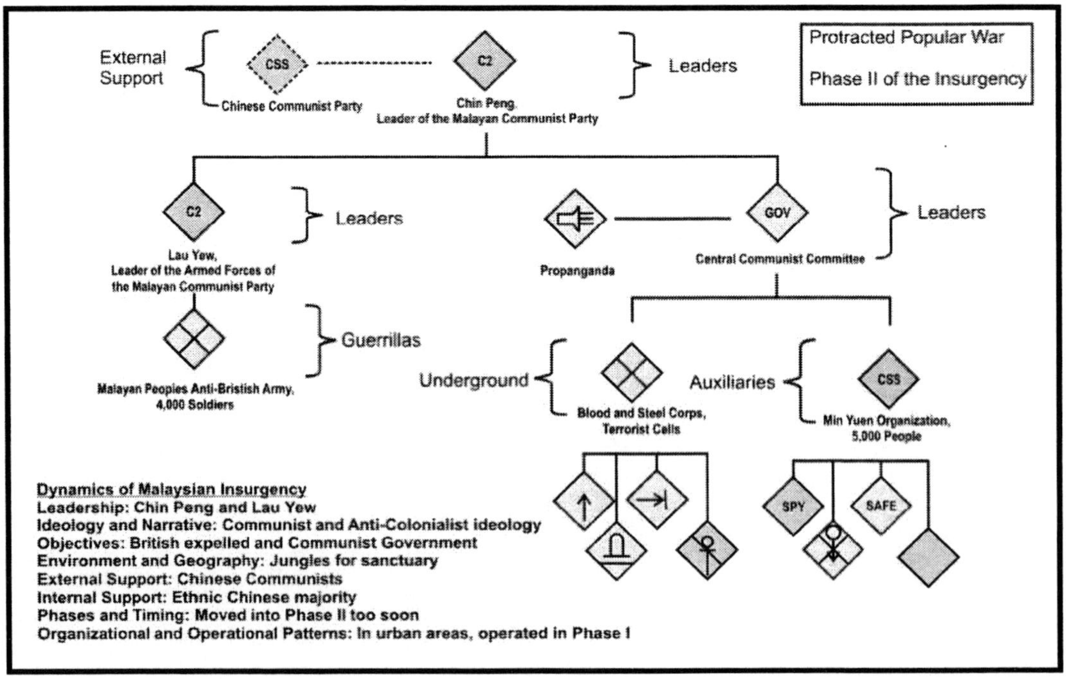


Figure 2-2. Components of Malaysian Insurgency (circa 1950).

## Section II—Elements

Insurgent organizations vary considerably, but are typically made up of five elements supported by a military wing and a political wing. The proportions of each element depend upon insurgent strategy and the degree of active support obtained from the populace. If the existing government presence is eliminated in any particular area, these elements can exist openly. If the HN government presence is strong in a particular area, the elements of an insurgency will maintain a clandestine existence. The five elements of an insurgency are—leaders, guerrillas, underground, auxiliaries, and mass base (Figure 2-3).

### *Leaders*

Leaders provide direction to the insurgency. They are the "idea people" and the planners. They usually exercise leadership through force of personality, the power of revolutionary ideas, and personal charisma. Generally, they convey the ideology of the insurgency into objectives and direct the military efforts of the guerrillas. In some insurgencies, they may hold their position through religious, clan, or tribal authority.

Leaders who form the political core of the insurgency are often called the political leaders. They are actively engaged politically in the struggle to accomplish the goals of the insurgency. They may also make up a formal political party to signify their political importance. These political leaders are the driving force behind propaganda. Insurgencies based on religious extremism usually include religious and spiritual advisors among their political cadre.

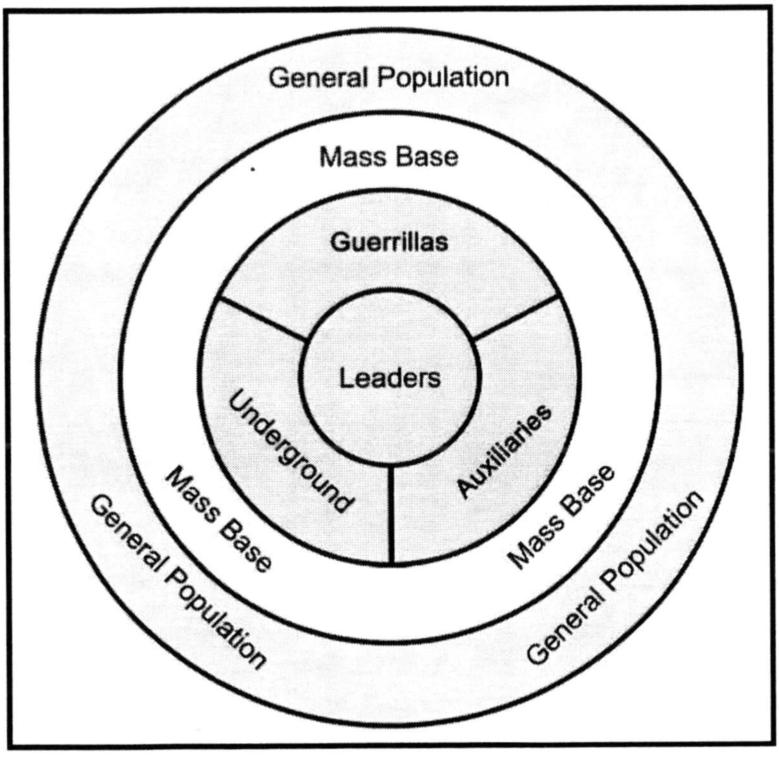


Figure 2-3. Organizational elements of an insurgency.

At a tactical level, units may identify leaders as IED cell leaders; political, religious, or social leaders who direct the propaganda and nonviolent efforts of the insurgency; or as business leaders who provide extensive resource support to the insurgency, and other roles.

### *Guerrillas*

A guerrilla is any insurgent who uses a weapon of any sort and does the actual fighting for the insurgency. They may conduct acts of terror, guerrilla warfare, criminal activities, or conventional operations. They are often mistaken for the movement or insurgency itself; but they are merely the foot soldiers of the movement or insurgency. Guerrillas vary widely in size, make-up, tactics, and methods from one insurgency to another. They even vary widely within an insurgency, especially in each of the three phases of an insurgency.

Guerrillas may continue in their normal positions in society and lead clandestine lives for the insurgent movement. Guerrillas tend to organize themselves based upon the activity they will be conducting. Those focused on using terrorism usually operate individually or in small cells and are often armed with explosives instead of weapons. Guerrilla bands, historically, have lived in remote areas and conducted raids on HN government infrastructure. Historical examples of guerrillas include the *Maquis* in World War II France, the *Viet Cong* in the Vietnam War and the *Mahdi* Army in Iraq.

### *Underground*

The underground is a cellular organization of active supporters of the insurgency, which may contain an element that works in the HN government. Keeping the nature of their work for the insurgency secret is often paramount to them. They are more engaged than the auxiliaries are and may at times be guerrillas, if they use weapons or conduct combat operations. They operate in all areas; especially in areas denied to any established guerrilla force and where operations are not suitable for guerrilla forces. They conduct clandestine, covert, and overt operations, sometimes infiltrating the HN government. Members of the underground often continue in their normal positions in society, but lead second, clandestine lives for the insurgent movement. Some insurgencies are unique in that they conduct most of their political activities inside the underground while a different section trains recruits, maintains propaganda, and helps in population control. The underground may—

- Spread propaganda.
- Support sabotage, assassination and subversion.
- Support intelligence and counterintelligence operations.
- Run safe houses.
- Provide transportation.
- Manufacture and maintain arms and explosives.

### *Auxiliaries*

An auxiliary is the support element of the insurgency. Auxiliaries are active sympathizers who provide important logistical services but do not directly participate in combat operations. If they participate in guerrilla activities, they become guerrillas. Auxiliaries may work full time or part time for the insurgency and generally conduct safer activities than the underground. They often include women, children and other individuals that tend to be less scrutinized by counterinsurgent forces. Examples of auxiliaries include shepherds or street merchants that may openly operate near a counterinsurgent base and provide intelligence on that site. Examples of support that auxiliaries provide include—

- Store weapons and supplies.
- Perform courier operations.
- Provide passive intelligence collection.
- Give early warning of counterinsurgent movements.
- Acquire funds from lawful and unlawful sources.
- Provide forged or stolen documents.
- Promote and facilitate desertion of security forces.
- Recruit and screen new members.
- Create and spread propaganda.
- Provide medical support.
- Manufacture and maintain equipment.

### *Mass Base*

The mass base consists of the population of the state who are sympathetic to the insurgent movement. This sympathy varies between the specific elements within the population such as religious and ethnic groups and within those specific elements themselves. This mass base, by default, passively supports the insurgency. As occasions arise, they may provide active support. Leaders often recruit members of the mass base, who are more actively oriented, to serve as auxiliaries, underground or guerrillas. Mass base members are the true silent supporters of the insurgency and are often the most available for the HN government to positively influence.

### *Example*

Although no two insurgencies or insurgent organizations are alike, they still have elements that can be identified and some form of hierarchy. Figure 2-4 shows an example insurgent organization with a developed structure. In this insurgency, the underground finances the insurgency.

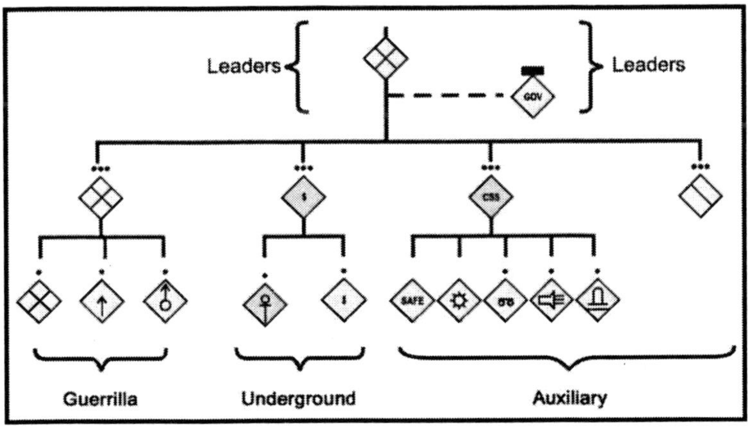


Figure 2-4. Example insurgent organization.

## Section III—Dynamics

Insurgencies are political movements that result from real or perceived grievances, or neglect that leads to alienation from an established government. Eight dynamics are common to an insurgency. Knowing and understanding these dynamics helps to understanding the insurgency. The eight dynamics are leadership, objectives, ideology, environment and geography, external support, internal support, phasing and timing, organizational and operational patterns.

### *Leadership*

Insurgent leaders provide vision, direction, guidance, coordination, organization for an insurgent movement. Successful insurgent leaders use, interpret and shape the frustrations of a dissatisfied populace into the insurgent strategy. They often provide solutions to grievances by advancing alternatives to existing conditions that support the insurgency. Effective insurgent leaders make their cause and solutions known to the people to gain popular support. Individual leaders of an insurgency are often members of the elite of society who have been somehow alienated from the power structure. Their education, background, family connections, social standing, and experiences contribute to their ability to organize and inspire the people who makeup the insurgency. To be successful, they must break the ties between the people and the government and establish credibility for their movement.

### Structure

In general, there are three categories of leadership found in insurgencies. They are—

#### *Single person*

One person may be the overall leader of an insurgency. This leadership structure has one person who provides cohesion, motivation, and direction for the insurgency. Cuba's Fidel Castro is an excellent example of a single person leadership structure. However, this single leader may centralize power or decentralize decision-making and execution, leaving decision-making and execution to subordinates. At the local level, most units will see organizations controlled by a single person.

#### *Single group or party*

The insurgency may be headed by a ruling council that makes and executes policy. This leadership group or party may also control other groups involved in the insurgency. China's Communist Party, before the ascendancy of Mao, is an example of a group leading an insurgency.

#### *Group of groups*

Different groups that have different concepts of how the country should be governed make up the leadership. Under this kind of leadership, there will be many leaders, possibly unified only by their opposition to the government. They compete with each other and the government. Example: The partisan forces in Yugoslavia and Greece during World War II were united in their fight against the German occupation, but ranged from monarchist to democratic to communist. Sometimes, they fought each other more violently than they did the Germans.

As a group, insurgent leaders operate in either a decentralized or centralized manner.

*Decentralized*

The power base of some insurgencies is collective and does not depend on specific leaders or personalities to be effective. These insurgencies are easier to penetrate, but recover rapidly when they lose key personnel. Decentralization restricts an insurgency in its ability to function as a coherent body and to create a viable counter state. However, decentralized insurgencies are very hard to destroy and can continue to sow disorder, even when degraded. For example, Al Qaeda in 2008 is a loose, decentralized organization held together by an ideal of re-establishing the Caliphate.

*Centralized*

Other organizations depend on a single, often charismatic personality to provide cohesion, motivation, and direction. Centralized insurgencies make decisions and initiate new actions rapidly. However, they are vulnerable to disruptions if key personalities are removed, co-opted, discredited, or eliminated. These insurgencies are often led by traditional authority figures such as tribal sheikhs, local warlords, and religious leaders. For example, Tecumseh was the major factor in the creation of Shawnee confederation. After he was killed in battle, the confederation fell apart.

**Identification of Key Leaders**

Identifying leaders can be critical in determining an insurgency's organizational structure, strategy, and tactics. Understanding the basic beliefs, intentions, capabilities, and vulnerabilities of known key leaders is extremely important to countering insurgency. Removing these key leaders will normally degrade an organization's capabilities. Che Guevara was a well-known, centralized leader that counterinsurgents effectively targeted, when he was killed, his insurgency fell apart. Alternately, in the case of the decentralized leader Abu Musab al-Zarqawi, his organization, Al-Qaeda in Iraq, although severely disrupted, continued as a functioning organization after his death. The following information is vital to understand and target an insurgent leader:

- Role.
- Activities.
- Associates.
- Personal background.
- Personal beliefs, motivations, and ideology.
- Education and training.
- Temperament.
- Position within the organization.
- Public popularity.

### *Objective*

Insurgencies normally seek to achieve one of three objectives: to overthrow the existing government in order to reallocate power, to expel whom they perceive to be "outsiders" or "occupiers," or to seek to create or maintain a region where there is little or no governmental control that they can exploit. Insurgents' objectives include struggles for independence against colonial powers; the rising up of political, ethnic, or religious groups against their rivals; and resistance to foreign invaders. The central issue in an insurgency is the reallocation of power. Usually, an insurgency mounts a political challenge to the existing state through the formation of a counter state, which is promoted as an alternative to the existing state.

Understanding the root causes of the insurgency is essential to analyzing the insurgents' objectives. Effective analysis of an insurgency requires knowing its strategic, operational, and tactical objectives. These objectives can be psychological in nature, physical in nature or a combination of the two.

### Strategic

The insurgent's overall political goals are their strategic objectives. The set of required conditions that define achievement of the insurgents' objectives are their desired end state. Examples of strategic objectives include—

- Overthrow an established government.
- Establish an autonomous national territory.
- Cause the withdrawal of a foreign occupying power.
- Extract political concessions.

### Operational

Insurgent operational objectives are those that insurgents pursue to destroy government legitimacy and progressively achieve their political end state. These are the means they use to link tactical goals with strategic end states. They often target the government's inability to address the root causes that lead to the insurgency. The insurgent's operational objectives define their overall plan, although most insurgents will not have a formal plan. These objectives are based on the insurgent's strategy across political, military, economic, and social objectives. Examples of operational objectives include—

#### *Political*

Disrupt elections; develop or strengthen an insurgency's political wing, attack government legitimacy; gain recognition of their political party by the government; attack the legitimacy of the government; or gain ability to run candidates for elected office

#### *Military*

Disrupt operational lines of communication and supply routes; force US or HN units out of an area; keep government forces in their bases; draw US or HN forces into fight; or provoke over-reaction from US or HN forces, which result in media-reported civilian casualties.

*Economic*

Disrupt reconstruction and attack infrastructure; thereby preventing the government from addressing root causes; prevent government from addressing the populace's issues.

*Social*

Spark sectarian violence; subvert education system; cause population to question the government; highlight corrupt or oppressive police; or establish a regular means for mass communication (radio or paper).

**Tactical**

Tactical objectives are the immediate aims of insurgent acts such as disseminating propaganda such as posters, compact discs and handbills, killing individuals, or the attack and seizure of key facilities. Tactical objectives can be both physical and psychological aspects. Counterinsurgents can often gain insight into tactical goals by analyzing the insurgent propaganda. Examples of tactical objectives include—

*Political*

Intimidation of a local official or the dissemination of propaganda products.

*Military*

Attack a US or HN government convoy or checkpoint or random indirect fire attacks on bases.

*Economic*

Attack manufacturing centers, stores and markets; sabotage roads, bridges, electrical lines and pipelines. Threaten violence against storeowners, employees and customers.

*Social*

Attack a police station or directly interact with the population in order to communicate directly to the populace.

**General Insurgent Goals for Terrorism and Guerrilla Warfare**

Insurgents typically work toward achieving ten general objectives as they gain the support of the population. Insurgent activity, like all other aspect of the insurgency, develops and evolves over time.

*Obtaining Popular Support*

Insurgent operations are normally focused directly or indirectly at decreasing support to the existing government. Simultaneously, competent insurgents try to gain support for the insurgent movement through propaganda, coercion, or by causing terror. If they cannot gain active support, they will seek passive support such as silence.

*Undermining Host Nation Legitimacy*

Insurgent efforts can damage or destroy both real and perceived economic and political targets. The loss of government services, overreaction by government forces, or the belief that the government is powerless against the insurgency contributes to the population's dissatisfaction with the government.

*Lessening Government Control*

By defeating small government forces, remaining government forces can be forced to consolidate in larger size elements or bases. This can reduce the overall presence of the counterinsurgent forces and allow an insurgent force to operate openly where counterinsurgent forces are not, if even for a short amount of time. This can create the perception that the insurgency is increasing its control or the government is losing control and legitimacy.

*Providing Psychological Victories*

The guerrilla seeks to gain small psychological victories. These victories do not need to be significant in terms of material damage to the government or its armed forces. These tactical victories show that a small guerrilla force can defeat the much larger government force.

*Tying up or Blocking Host Nation Government Resources*

By forcing the HN government to expend resources on military operations against guerrillas, the insurgency seeks to tie up resources that could best be used by the government to provide services and development programs to the populace.

*Weakening Host Nation Government Resolve*

By defeating small elements of the HN's security forces and attacking government agencies, the guerrilla weakens the resolve of government employees and forces. Small government forces and agencies become aware of their vulnerability. Desertion, absence from work, willingness to compromise, difficulty in recruiting, or limitations on services to dangerous areas all benefit the insurgency's purpose.

*Intimidating the Population*

By attacking vocal opponents and certain types of individuals, such as teachers, the insurgents seek to frighten the population.

*Acquiring Supplies and Equipment*

Guerrillas seek to acquire government weapons, uniforms, equipment, supplies, or vehicles.

*Infiltrating Host Nation Government and Government Forces*

Insurgents will often attempt to infiltrate various government ministries and security forces by emplacing sympathizers and by converting existing members of those agencies. These infiltrators are used to gather intelligence and to subvert operations.

*Causing COIN Security Force Overreaction*

Since a COIN fight is the fight for the population, often the intent of insurgents using terrorism or guerrilla warfare is to cause a heavy-handed response to attacks on the part of the

COIN force or HN security force. These responses will often drive the population to support the insurgency.

### *Ideology*

Insurgents often use their ideology to show the population how they can address the root causes that the government cannot provide or is not providing. Insurgent ideology attempts to provide a vision of how a society, including a political and economic system, should be structured. Ideology should not be confused with the insurgent strategy, which is the way that the insurgents intend to achieve their end state. Two of the most identifiable insurgent ideologies have been communism and religious extremism.

Ideology is a motivating factor in insurgent activities. Insurgencies can gather recruits and amass popular support through ideological appeal, which includes religious or other cultural factors. The insurgency's ideology explains its followers' difficulties and provides a means to remedy those ills. The most powerful ideologies tap latent, emotional concerns of the populace. Tactical units could see ideology expressed in the propaganda and recruitment techniques of local insurgents.

#### Communism

Communism is a political system where private property is eliminated and controlled by the state. Historically, it has been advocated in countries where wealth is unevenly distributed among the classes. Communism was once the most typical form of insurgent ideology and often experienced success. With the fall of Soviet Union, communism has not been a successful motivation for insurgents, especially since there is currently little or no external support for communist-based insurgency.

#### Religious Extremism

Religious extremism, often a byproduct of religious fundamentalism, can be defined by strict adherence to a set of religious principles and the rejection of compromise. These ideologies are often energized by inequities in social, political or economic development and further helped by counterinsurgent attempts to marginalize religious issues. Globalization creates opportunities for an increase in religious extremism based on both real and perceived inequities. Insurgencies based on religious extremism want their values incorporated into the nation's governmental structure. This fulfills the frustration and dissatisfaction of the religious extremists. Adherents often receive formal instruction on the religion's fundamentals and use adherence to these fundamentals as a recruiting tool. Religious leaders are often leaders of the insurgency and nonextremist leaders are often replaced. Characteristics of insurgencies motivated by religious extremism that differ from traditional insurgencies are—

- Individual duty with an indifference to popular support.
- Use of violence that maximizes shock, awe, and casualties.
- Disinterest with governance and lack of a practical political objective in organizations with global reach or establishing a theocracy in local cases.
- Ability to regenerate guerrillas without popular support.

## Narrative

The central mechanism through which ideologies are expressed and absorbed is a narrative. A narrative is when a story is used to display the benefits of a certain ideology. Narratives are often central to representing identity, particularly the collective identity of religious sects, ethnic groupings, and tribal elements. Stories about a community's history provide models of how actions and consequences are linked. Stories are often the basis for strategies and actions, as well as for interpreting others' intentions. Insurgent organizations use narratives and religious-based concepts very effectively in developing, spreading, and mobilizing followers.

In the Al Qaeda narrative, Osama bin Laden shows himself as a man purified in the mountains of Afghanistan who is gathering and inspiring followers and punishing infidels. In the collective imagination of Bin Laden and his followers, they are agents of Islamic history who will reverse the decline of the *umma* [Muslim community], reestablish the Caliphate, and bring about its inevitable triumph over Western imperialism. This image mobilizes support for Al Qaeda among some of the most traditional Muslims.

### *Environment and Geography*

Environment and geography, including cultural, religious, tribal affiliation, and other demographic factors along with terrain and weather, affect all participants in an operational environment. How insurgents and counterinsurgents adapt to these realities creates advantages and disadvantages for each. In Chapter 1, this manual stressed the importance of understanding an AO's civil considerations (ASCOPE). Considerations for environment and geography include—

- Population density and distribution, especially degree of urbanization
- Root causes of the insurgency within an urban population, such as lack of basic services, security, markets, governance, municipal council representation, or schools.
- Root causes that lead to the insurgency within a rural population, such as lack of land ownership, grazing rights, water rights, isolation or inclusion in political process, access to markets, or schools.
- Growing seasons, rainy/dry seasons, planting/harvest time (key events). These may influence level and type of insurgent activity.
- Use of structures and infrastructures by both the insurgent and population.
- Tribal, religious, or other affiliations, which may have tremendous effect on the local populace's willingness to support an insurgency, or where people and insurgents can expect sanctuary or will avoid.
- Proximity to international borders that may provide sanctuary or support.
- Rugged, inaccessible terrain with often hostile populations to outsiders that may provide sanctuary or support.
- Economic enclaves, such as slums, market areas, middle class areas, and wealthy areas.
- Geographic divisions along ethnic, tribal, religious, political or other factors.

***External Support***

External support includes moral support, political support, resource support, or sanctuary support. External support can come from any entity outside of the Host Nation—not just neighboring states. Countries from outside the region seeking political or economic influence can also support insurgencies. Insurgencies may turn to transnational criminal elements for funding or use the Internet to create a support network. Ethnic or religious communities in other states may also provide a form of external support and sanctuary, particularly for transnational insurgencies. Access to external support influences the effectiveness of insurgencies.

Accepting external support can affect the legitimacy of both insurgents and counterinsurgents. The act of acceptance implies the inability to sustain oneself. In addition, the country or group providing support attaches its legitimacy along with the insurgent group it supports. The consequences can affect programs in the supporting nation wholly unrelated to the insurgent situation.

**Moral Support**

Moral support is the acknowledgement that the insurgent or their cause is just and admirable. It starts as outside popular approval and can manifest with negative media attention focused towards the counterinsurgent or Host Nation. Moral support often leads to political, resource, and sanctuary support.

**Political Support**

Political support is the active promotion of insurgents' strategic goals in international forums. International forums such as the United Nations, trade sanctions and embargoes, and the creation of political discussion in the Unites States can all provide political support, which negatively influences the counterinsurgent's effort. Another form of political support can come when a legitimate state actually recognizes an insurgent group as a legitimate authority. Political support is the most dangerous form of support as it can result in an insurgency gaining international legitimacy, forcing the counterinsurgent to stop actively targeting them.

**Resource Support**

Resource support is typically guerrillas, money, weapons, equipment, food, advisors, and training. Resource support is often the most important form of support, such as, during the French experience in Algeria. Although insurgents were numerous, weapons were not, because the French closed international borders, preventing arms smuggling.

**Sanctuary Support**

Sanctuary support is a secure site to train, obtain sustainment, rest and refit. Historically, sanctuaries provided insurgents a place to rebuild and reorganize without fear of Host Nation or counterinsurgent interference. Often these were in neighboring countries or remote areas difficult to access. Sanctuaries may also include areas within a state, including neighborhoods where HN security forces cannot or will not conduct operations. The meaning of the term sanctuary is evolving. Today, insurgents can draw on "virtual" sanctuaries in the Internet, global financial systems, and the international media. These virtual sanctuaries can be used to present insurgent actions as acceptable activities worthy of internal and external support. Effective COIN operations work to eliminate all sanctuaries.

### *Internal Support*

Internal support is any support provided from inside the country. It is normally broken down into the two general categories: popular and logistical support. Together, these two form the mass base. For the purposes of this manual, one other category is added—insurgent bases.

## Popular Support

An insurgent movement requires popular support (Figure 2-5) to survive, and popular support is even more essential for an insurgency to succeed. Typically, there is also a relationship between the amount of popular support and the size of the insurgency. To grow, an insurgency needs an adequately sized mass base that will support this growth. One of the best means of defeating insurgencies is to shrink this mass bass by causing the local population to become hostile or at least apathetic toward the insurgents.

#### *Types of Support*

Popular support can be either active or passive and may come from only a small segment of the population or from a broad base of the population. Supporters of an insurgency may also be overt or clandestine.

##### *Active Support*

Active supporters provide open sympathy to the movement, participate in legal insurgent activities, such as strikes, find new recruits, and may transition to being an active element of the insurgency such as guerrillas, underground, or auxiliary. Active supporters are usually central to the insurgency's propaganda efforts.

##### *Passive Support*

Passive supporters vary from those who are sympathetic yet inactive to those who are not sympathetic, but who choose to remain silent about insurgent activities. Silence on the part of the populace concerning insurgent activities provides passive insurgent support.

#### *Recruitment*

Local insurgent representatives address local grievances and conduct recruiting. Elements of the population who are dissatisfied with existing conditions or those who have been marginalized through psychological alienation are prone to insurgent recruitment. The cadre often gives credit to the insurgent movement for all local successes and blames government forces for all failures and problems. Every promise and appeal made by cadre members is associated with tangible solutions and deeds. Competent insurgents and counterinsurgents both seek to mobilize and sustain popular support for their cause while discouraging popular support for their adversaries. There are five common methods, used individually or in various combinations, to mobilize popular support. Knowing these five means provides an opportunity for the counterinsurgent to identify when they are being used and then counter them. The five common methods are—

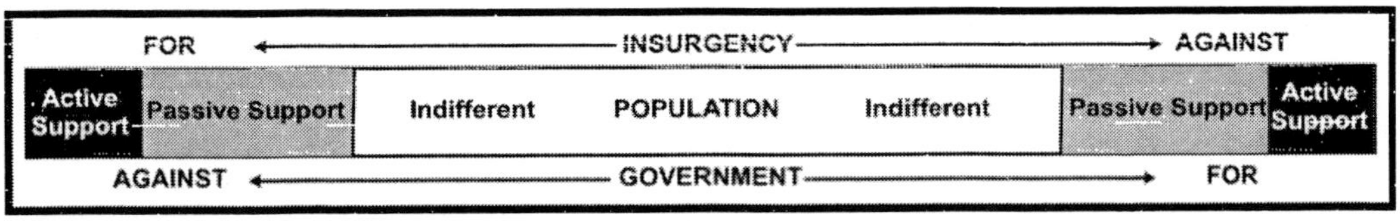


Figure 2-5. Range of popular support.

***Persuasion***

Political, social, religious, security, and economic promises can often entice people to support one side or the other.

***Coercion***

Through threat of violence or abuse, insurgents can force people to support them. Citizens seek to ally with groups that can guarantee their safety.

***Reaction to Abuses***

Though firmness by security forces is often necessary to establish a secure operational environment, a tyrannical government can generate resistance to its rule.

***Foreign Support***

Foreign governments can provide the expertise, international legitimacy, and money needed to start or intensify a conflict.

***Apolitical Motivations***

Insurgencies attract foreign volunteers, criminals, and mercenaries who are often motivated by money or extremism.

*Measurement*

The reality of insurgencies is that the support of the population fluctuates between the government and the insurgency due to many factors, but no one side will ever possess the support of the total population. Trying to quantify why the populace favors the government or the insurgency is difficult, but evaluating the issue is important. Gauging aspects such as the reaction of a local populace to the presence of troops or government leaders can help estimate popular support at the tactical level. Asking the population directly or using surveys can produce valuable insight into popular support and attitudes.

## Logistical Support

Modern warfare is inherently resource intensive. The insurgent needs to ensure continual access to supplies, weapons, ammunition, and money, and this requires insurgents to maintain open supply lines. While logistical support may be an insurgent's greatest vulnerability, it can be difficult to interdict. As an insurgency develops and expands, logistical support increasingly relies on external sources. In a conventional war, insurgent forces often receive significant amounts of supplies from external sources— usually from one of the opposing nations involved in the conflict. The protracted popular war strategy emphasizes mobilization of the masses and requires considerable resources to build and maintain a counter state. The urban strategy requires significantly less support.

*Supplies*

Once the insurgent's on-hand supplies are exhausted, supplies must come from the populace or an external source. In most insurgencies, initial support comes from the populace (either voluntarily or coerced). Identifying types of supplies the insurgency needs and then discovering how the insurgent is obtaining these supplies are significant steps in the process of identifying insurgent supporters and insurgent bases. Medical supplies are often the most critical supply that the insurgency requires.

*Weapons and Ammunition*

External sources are often required to sustain specialized arms and equipment. Insurgents can obtain these weapons through legal or illegal purchases, or from foreign sources. Another common tactic is to capture weapons from government forces.

*Money*

Money is essential to an insurgency for the purchase of critical supplies, especially high-tech weapons and ammunition or bomb making materials. Money is also essential for the payment of guerrillas and the bribery of corrupt officials. Money can be obtained through many sources and in today's electronic world, it crosses all boundaries. While money is often shipped to insurgents just like other supplies, it may also be moved and held by local financial institutions. In some cases, insurgencies develop an "underground banking" system, sometimes known as *hawala*, which can be used to launder money. Just like supply lines, the money path must be discovered, tracked, and disrupted. Funding greatly influences an insurgency's character and vulnerabilities. Local supporters or international front organizations may provide donations. Sometimes legitimate businesses are established to furnish funding. In areas controlled by insurgents, confiscation or taxation might be used, especially in cases when insurgencies provide their own essential services to the population. Another common source of funding is criminal activity, especially the illegal trade of drugs.

*Supply Lines*

In all cases, either the insurgents must go to their suppliers, or the suppliers must come to the insurgents. Discovering and tracking these supply lines can provide key information on insurgent forces and support. Although rarely overt, all insurgent organizations will have some system of supply lines, means of transportation, and storage facilities. Skillful counterinsurgents attempt to cut off the flow of supplies, especially weapons and ammunition. Often, a nation neighboring an insurgent AO is used as a depot.

**Insurgent Bases**

In counterinsurgencies, there are two general types of insurgent bases: safe houses and guerrilla base camps. Insurgents will normally establish multiple safe houses of guerrilla camps based upon functioning groups or cells. Urban insurgents tend to use safe houses, while rural insurgents tend use guerrilla base camps. Additionally, training camps may be established. These training camps may be established within a safe house; however, they normally will be established either in a rural guerrilla base camp or outside the territory controlled by the existing government, often in a foreign country.

*Safe Houses*

Many legitimate and illegitimate organizations use safe houses. A safe house is typically occupied by a member of the auxiliary and temporarily used to hide insurgents. Normally a system of "safe houses" have been carefully selected. The occupant of the safe house has procedures to move insurgents along selected routes at the best times to avoid detection.

*Guerrilla Base Camps*

Like any other armed force, guerrillas have requirements for command and control, rest, resupply, refit, and training. The larger the guerrilla force and more active they are, the more they will need established bases, both semipermanent and temporary. Base camps must be relatively safe and secure in areas where insurgents can rest, eat, and plan. More sophisticated guerrilla base camps have command posts, training areas, communications facilities, medical stations, and sustainment centers. These base camps, however, are not the same as conventional bases. They are usually small in overall scope, spread out, and sometimes underground. Insurgents try to locate base camps within insurgent-controlled areas where cover and concealment provide security against detection. In rural areas, base camps tend to be in remote areas characterized by rough, inaccessible terrain. In urban areas, base camps tend to be located in areas where the insurgent has popular support. Urban insurgents may rent houses for use as temporary base camps. Insurgents will normally avoid battling over their base camp. Once detected, they will move to an alternate location. Routes into a base camp will be constantly observed for security. Mines, booby traps, special-purpose munitions, expedient devices, and ambushes are used as standard security enhancements. If surprised and cornered, they will vigorously defend themselves with a delaying action while evacuating key personnel and equipment.

*Insurgent Training Camps*

Insurgent training camps are established both in urban and rural environments. While some training is accomplished at urban safe houses or rural guerrilla base camps, most training is accomplished at locations focused on training. This may be a special urban safe house, a remote guerrilla base camp or an insurgent training center in another country.

***Phasing and Timing***

Insurgencies often progress through three phases in their efforts. While the use of these three phases is common in most writings concerning insurgencies, the titles used for these three phases vary considerably. However, what makes up each phase remains nearly identical.

**Phase I—Latent and Incipient**

The first phase of an insurgency tends to begin with the government having stronger forces than the insurgents do. As a result, insurgents often must concentrate on survival and building support. Insurgent efforts may include—

- Establishing and expanding the organization
- Spreading its ideology through information operations
- Starting or supporting antigovernment activity such as demonstrations and strikes
- Raising funds through illegal taxation and crime

- Organizing small, local guerrilla forces that conduct small-scale intermittent operations
- Using terrorism and sabotage to intimidate uncooperative government officials and members of the population

**Phase II—Guerrilla Warfare**

The second phase of an insurgency starts when force correlations approach equilibrium and guerrilla warfare becomes the most important, pervasive activity. Insurgent efforts may include—

- Increased scale of guerrilla attacks; attempt to force government forces into the defense.
- Increased use of sabotage and terrorism.
- Intensified propaganda.
- Attempts to gain control of isolated geographic areas and develop bases for further operations.
- Government officials are being driven out of areas with strong insurgent support.
- Establishment of local shadow governments.
- Increased efforts to gain international recognition and support.

**Phase III—War of Movement**

The third phase of an insurgency normally begins when insurgents believe they have superior strength and their military forces attempt conventional operations to destroy the government's military capability. Insurgent efforts may include—

- Combining guerrilla forces and training them to fight as conventional forces. Over time, these conventional forces form multiple echelons. These forces confront the counterinsurgents in conventional battle.
- Acquiring more powerful and sophisticated weapons through foreign assistance or capture.
- Obtaining support from external combat forces, such as special or conventional forces.
- Expanding areas of insurgent control and increasing political activity.

Not all insurgencies progress through all three phases, and linear progression through all three phases is certainly not a requirement for success. Insurgent success can occur in any phase. Also, insurgencies can revert to an earlier phase and resume development when favorable conditions return. Insurgent success can occur in any phase.

Movement from one phase to another phase does not end the operational and tactical activities of earlier phases; it incorporates them. Therefore, it is difficult to determine when an insurgency moves from one phase to another. In addition, a single insurgency may be in different phases in different parts of the country. Advanced insurgencies can rapidly shift, split, combine, or reorganize—they are dynamic and adaptive.

### *Organizational and Operational Patterns*

Insurgencies develop organizational and operational patterns from the interaction of many factors. Although each insurgency organization is unique, there are often similarities among them and knowing the commonly accepted general patterns or strategies of insurgency helps in predicting the tactics and techniques they may employ against the government. Other considerations include the operational environment, level of success of the insurgency, and the length of time an insurgency has been operating. A counterinsurgent must learn about the insurgency and adapt to it based off its operational organizational patterns.

Insurgent organizational and operational patterns vary widely between one province or urban area and another. Different insurgent groups using different methods may form loose coalitions when it serves their interests. These groups may also fight among themselves. The result is more than just a "three-block war"—it is a shifting, "mosaic war" that is difficult for counterinsurgents to envision as a whole. In such situations, an effective COIN strategy must be multifaceted and flexible. Insurgents gain ground by sowing chaos and disorder anywhere. Counterinsurgents lose ground by failing to maintain stability order everywhere. Insurgents normally begin substantially weaker than the established government. This sets the tone for how they operate and how they fight

## Section IV—Strategies

Even if modern insurgencies use more than one doctrinal model or theory for their strategy, aspects of these strategies and recognizable characteristics do exist. The six common insurgent strategies are urban, military-focused, protracted popular war, identity-focused, conspiratorial, and composite and coalition. At the tactical level, a counterinsurgent will deal usually only with the urban, military-focused, and protracted popular war strategies. These insurgent strategies provide a common frame of reference for the counterinsurgent. The savvy counterinsurgent can identify if an insurgency is using one or a combination of the strategies. Knowing what strategy the insurgents are using facilitates the anticipation of insurgent courses of action (COAs).

### *Urban Strategy*

In the urban strategy, the insurgents attack government targets with the intention of causing government forces to overreact against the population. The insurgents want the government's repressive measures to enrage the people so they will rise up, support the insurgency and overthrow the government. This strategy can be initiated without popular support. Its success relies almost exclusively on a spontaneous uprising sparked by rage at government oppression. However, an insurgency occurring in an urban area does not necessarily mean that it uses the urban strategy. On November 1, 1954, the National Liberation Front in Algeria used a form of the urban strategy when they launched a series of bombings and attacks, causing significant civilian casualties, in order to shock the French into negotiations. The urban strategy actions are often predictable and possess these characteristics—

- Insurgents often use terrorist attacks, which they hope are highly visible and produce high casualties. Their true intention may not necessarily be to cause fear or terror, but to provoke the government into overreaction.

- Insurgent propaganda tends to focus on government brutality, calling attention to specific harsh government actions such as massacres, torture of political prisoners, disappearances of individuals, and brutal responses to peaceful demonstrations.
- Insurgent political organization is minimal with no sustained effort to indoctrinate political cadre or the masses.
- Insurgents make little or no effort to subvert the government from within (however, infiltration of HN government and security forces still possible.
- Insurgents require only a small amount of popular support.

### *Military-Focused Strategy*

The military-focused strategy believes that military action can create the conditions needed for success. Military-focused insurgents often believe that a small group of guerrillas operating in an area where grievances exist can eventually gather enough support to achieve their aims. The success of this small group depends upon successful military action and popular uprising. The most iconic examples of military-focused strategy are Che Guevara and Fidel Castro; both proposed attacks on military and government targets until they gathered the support necessary to seize power. Military-focused strategy actions include—

- Attacks on Host Nation targets to gain popular support.
- Propaganda that incites people to join the insurgency and rise up against the government and that focuses on demonstrating the Host Nation Government's weakness and illegitimacy.
- Little evidence of long-term efforts at building a political base.
- Little effort to building the political wing or infiltrating legitimate organizations.

### *Protracted Popular War Strategy*

The protracted popular war strategy is based on Mao Zedong's theory of protracted popular war. This strategy is broken down into three distinct phases—latent or incipient, guerrilla warfare, and war of movement. Each phase builds upon the previous phase, and continues activities from the previous phases. The protracted popular war strategy has both a political wing and a military wing. This strategy requires a high level of organization and indoctrination, actions along multiple lines of effort, and leadership to direct the shifting of phases according to circumstances. In all the variations of this strategy, certain characteristics tend to stand out, such as—

- Continuous, long-term efforts to build popular support, infiltrate legitimate government organizations, and establish and maintain a clandestine organization.
- Highly-indoctrinated leadership, political cadre, and guerrilla fighters.
- Extensive, well-organized, unarmed auxiliary.
- Leadership that is able to exert control over the insurgency.
- Ability to shift phases at the direction of its leadership; including return to previous phase if necessary.

- Attacks on infrastructure and attacks designed to wear down the government and counterinsurgents.
- Continuous operations along multiple lines of effort, although some phases will emphasize different lines of effort.

### *Identity-Focused Strategy*

The identity-focused strategy mobilizes support based on the common identity of religious affiliation, clan, tribe, or ethnic group. In this strategy, legitimacy and popular support are tied to their identity and, often, no effort is made to garner popular support outside their identity. Rather, communities often join the insurgent movement as a whole, bringing with them their existing social or military hierarchy. External support is garnered from international elements of the same identity. In Sri Lanka, the Tamil Tigers have fought an insurgency against the government for decades in order to establish an ethnic Tamil state and, at times, have received support from India. Contemporary characteristics of the identity-focused strategy include—

- Attacks on those who threaten the traditions and social structure of the identity.
- Little or no need to establish a shadow government; already established.
- Protects what it considers the interest of the identity.
- Willingness to use tactics of other strategies.
- Mass base easily aligns with insurgency objectives.

### *Conspiratorial Strategy*

The conspiratorial strategy attempts to subvert the government from within and often involves a few leaders and a militant cadre. Although subversive activities may take place in other strategies, particularly in the protracted popular war or urban strategies, conspiratorial strategies often attempt to have its illegal political party become a legitimate political party, enter the government legitimately and then take control of the government. Insurgents using the conspiratorial strategy do not intend to integrate into the national government, but to overthrow the government. Once the insurgency succeeds in gaining legitimate political representation, the newly legitimized politicians, who may have been previously targeted by counterinsurgents, must be re-evaluated to see if they are still legitimate targets. In the Russian Revolution in 1917, the Bolshevik Party actively worked to break existing government, while being elected to serve in the administration. Distinguishing characteristics of this strategy include—

- Insurgents seeking meetings with HN government to discuss ceasefires.
- Attacks on infrastructure designed to wear down and reduce the credibility of government.
- Political cadre distancing itself from the insurgency by making public statements denouncing violence, yet the insurgent leadership still controls the cadre.
- A public breach between militant and political elements of the insurgency, although this is often a deception.
- Formation of new alliances, often with groups that seem to have little in common with the insurgency or its ideology apart from the desire for governmental change.

- An end or reduction in guerrilla activity with an increase in political activity.
- Intensive efforts to gain international moral and political support.
- Using sophisticated propaganda, aimed at specific target audiences.
- Insurgent political wing seeks recognition and entry into politics, including election to local, district, departmental, regional, or national offices.

### Composite and Coalition Strategy

The composite and coalition strategy applies when different insurgent groups using different strategies combine to form loose coalitions that serve the purposes of the different groups. However, the composite and coalition strategy is usually united in opposition to something, for example an occupier or specific grievance, rather than for a positive objective. Within a single AO, there may be multiple competing entities, each seeking to maximize its survivability and influence. Recently, Al—Qaeda in Iraq provides the best example of a composite and coalition strategy; they founded the Islamic State of Iraq and became an umbrella group for many other established insurgent groups. Contemporary actions of the composite and coalition strategy include—

- Unclear or vague objective.
- Multiple or disjointed strategies within a single area of operations.
- More likely to ally with criminal actors.
- Attacks on other members of the coalition.

### Shifts between Strategies and Phases

Insurgencies often operate using different strategies, in different phases, in different geographical areas. These decisions are based on the operational environment and insurgent objectives. Most insurgencies eventually move to the protracted popular war or subversive strategy after another strategy proves unsuccessful.

Insurgencies can also shift both phases and strategy. Pressure from the counterinsurgent can force an insurgent to move laterally to a new strategy or return to an earlier phase (Figure 2-6). For example, an insurgency in Phase I, Latent and Incipient, using an urban strategy, builds sufficient strength to progress to Phase II, Guerrilla Warfare. Once in Phase II, the insurgents may believe they are strong enough to initiate a series of attacks. Then, if counterinsurgents successfully drive out the insurgents, the insurgents will normally consolidate and reorganize elsewhere. However, this failure may force the insurgency back to Phase I in that area. Additionally, the insurgent leadership may transition from the urban strategy to a protracted popular war strategy. When the insurgent leaders believe they are ready, the insurgency will return to Phase II, Guerrilla Warfare.

Insurgencies are often vulnerable when they shift between strategy or phases. These shifts may be due to fractures within the insurgent leadership or setbacks. Shifts may also occur due to time, changes in external support, changes in leadership, or counterinsurgent action. These shifts are often rapid so counterinsurgents must be able to recognize and exploit them. Indicators of a shift in strategies may be:

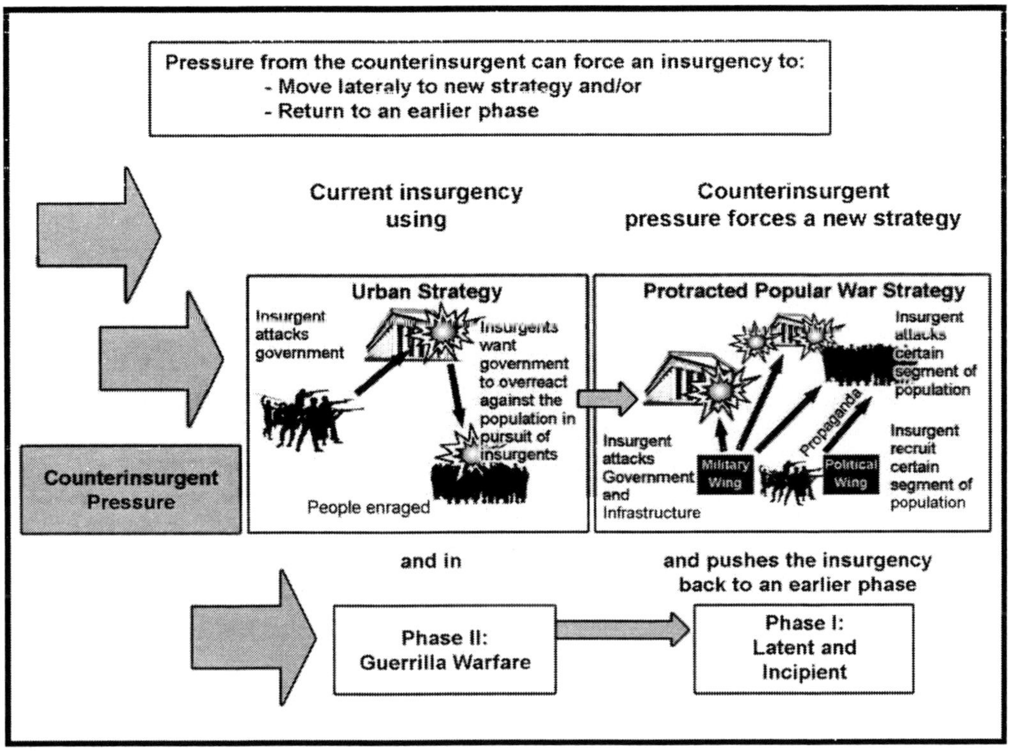


Figure 2-6. Shifts between strategies and phases.

- Changes in propaganda message content
- Uncharacteristic increase in communications
- Unexplained pauses or sudden increases in guerrilla attacks
- Shift of effort between urban and rural efforts
- Displacement of insurgents from one location to another
- Unanticipated statements of support from external actors for an insurgency
- Increased organization, indoctrination, and secure means of communications
- Increased efforts to infiltrate legitimate organizations such as trade unions, professional or business organizations, universities, and so on.
- New advocacy for rights of peasants, farmers, or other groups
- Change of in focus of attacks, such as the targeting of a specific sector

## Section V—Tactics

Insurgencies employ both violent and nonviolent tactics to achieve their objectives. Nonviolent tactics attempt to achieve political goals without the use of force. Insurgent violent tactics are often accompanied by a variety of nonviolent tactics. Used together, these tactics, along with supporting propaganda, can assist in recruiting and gathering popular support. Historically insurgencies have fielded dedicated personnel to foment nonviolent action such as strikes and protests to supplement violent action. Insurgents are by nature an asymmetric threat. *Asymmetric Warfare* is a conflict in which a weaker opponent uses unorthodox or surprise tactics to attack weak points of a stronger opponent, especially if the tactics include terrorism, guerrilla warfare, criminal activity, subversion, or propaganda. Violent tactics by insurgents can include *terrorism*, *guerrilla warfare*, *sabotage*, or *conventional operations*. Insurgents often use terrorism and guerrilla tactics to achieve their goals, because they do not have the capability to contest the government or counterinsurgency forces in conventional operations.

### *Violent Tactics*

Violent insurgent tactics are normally characterized by elusiveness, surprise, and brief, violent action. These tactics are often divided between terrorism and guerrilla warfare early in the insurgency. In Phase III of the insurgency, it is common to see more conventional operations. There are three general tactics available to insurgents that use violence—terrorism, guerrilla tactics, conventional tactics and criminal activity. The challenge is that at any given time the insurgent could use any of them. For clarity, the term 'guerrilla' applies to any insurgent forces performing any of these four types of violence.

Although violence can alienate the population when not linked to a vision of a better life or the violence is indiscriminant, the commitment to use violence is a potent insurgent weapon. Targets of violence can be anything insurgents deem to be obstructions to their cause. Host nation security forces, foreign forces, aid workers, members of the population who do not accept insurgent claims, and infrastructure are typical insurgent targets.

Normally, organized groups of insurgents using terrorism or guerrilla warfare use violent tactics. The differences between these two can become blurred, especially within an urban environment or where the government exerts strong control. Although potentially nonviolent, criminal activity provides a ready source of income for an insurgency and, for the purposes of this FM, has a violent nature. Violent tactics include, but are not limited to (for more see FM 3-24):

- Ambushes.
- Assassination.
- Arson.
- Bombing and high explosives.
- Chemical, biological, radiological, or nuclear weapons.
- Hijacking and skyjacking.
- Hostage taking.
- Indirect fire.
- Kidnapping.
- Raids or attacks on facilities.
- Sabotage.
- Seizure.

## Terrorism

A terrorist is an individual who uses violence, terror, and intimidation to achieve a result (DOD). Insurgents may use terrorism. Terrorist attacks employ violence primarily against noncombatants as a way to increase the population's vulnerability and decrease their perception of security. Insurgent terrorism techniques include assassination, arson, blackmail, bombings, hijacking, kidnapping, threats, murder, mutilation, and torture. The insurgent using terrorism often targets economic and political symbols to undermine the legitimacy of the government. Any overreaction by government forces or other authorities adds to the population's resentment toward the government and turns its support to the insurgency.

Insurgents using terrorism generally require fewer personnel than guerrilla warfare or conventional operations. Inherently, these activities have greater security and lower support requirements. Insurgents using terrorism often select targets for their political and psychological impact. Their attacks can be effective in generating popular support within one faction when used against a competing faction. They can also be effective in forcing government reaction that alters government policies to benefit insurgent objectives.

## Guerrilla Tactics

Guerrilla tactics are typified by hit-and-run attacks by lightly armed, small groups. The guerrilla is the combat element of the insurgency. Guerrilla tactics emphasize ambushes, raids, snipers, rocket and mortar attacks, and the use of explosive devices. Guerrilla tactics may also include assassination, coercion, and kidnapping to achieve support or eliminate opposition.

### *Principles*

The principles of guerrilla tactics as stated by Mao Zedong are still valid: "Enemy advances, we retreat. Enemy halts, we harass. Enemy tires, we attack. Enemy retreats, we pursue."

Guerrilla operations are generally offensive, not defensive, and are often harassing in nature. Guerrillas seldom attempt to seize and defend physical objectives and, in general, avoid decisive engagement. Their overall aim is often to cause confusion, to destroy infrastructure or security forces, and to lower public morale. Guerrilla harassment attempts to keep government forces on the defensive and weaken the Host Nation, which can include destroying resources and disrupting lines of communication. One advantage of harassment is that it may create the perception that the guerrilla can strike anywhere and that the Host Nation cannot prevent it. Because of this, counterinsurgents the must not only win small battles, but also win the battle of public perception. In rural areas, guerrillas may seize a remote area or conduct raids and small-scale attacks on remote targets and lines of communications.

### *Numerical Superiority*

While government forces outnumber the guerrilla, the guerrilla seeks to attain local numerical superiority. If guerrillas can successfully concentrate against counterinsurgents, they can attain victory over small elements of government forces. Guerrillas often use simple techniques of speed, surprise, maneuver, and especially infiltration. Near the target area, small guerrilla elements will often mass in order to conduct a specific, larger-scale operation. The baited ambush is a favorite guerrilla technique. Guerrillas often create incidents, such as arsons, bombings, and hoaxes, as the bait. These baited ambushes can attempt to lure small government forces of all types into a mechanical or manned ambush.

### Conventional Tactics

Conventional operations are not always necessary for success; however, guerrillas may engage in conventional operations after an insurgency develops extensive popular and logistical support. When they feel the conditions are set, insurgents may generate a conventional military force that can directly confront HN security forces. These conventional operations may vary from a small regional operation to general conventional warfare. Large conventional operations are usually an attempt to obtain the strategic or operational objectives.

### Criminal Activity

Sustainment requirements, especially funding, often bring insurgents into relationships with organized crime or insurgents may turn to criminal activity themselves. Reaping windfall profits and avoiding the costs and difficulties involved in securing external support makes illegal activity attractive to insurgents. Taxing a mass base usually yields low returns and alienates the population, especially in cases where the insurgency provides essential services to the population. In contrast, kidnapping or hostage taking, extortion, armed robbery, and trafficking (drug, human, black market goods, and so on)—four possible insurgent criminal activities—are very lucrative, although they also alienate the population. The activities of the Fuerzas Armadas Revolucionarias de Colombia (FARC) are a perfect example. The FARC often receives millions of US dollars from a single high-profile kidnapping. Similarly, failed and failing states with rich natural resources like oil or poppies are particularly lucrative areas for criminal activity.

Devoting exceptional amounts of time and effort to fund-raising requires insurgents to shortchange ideological or armed action. Indeed, the method of raising funds is often at the heart of internal debates within the insurgency. For example, the FARCs involvement in the drug trade has made it the richest self-sustaining insurgent group in history; yet it continues to claim to pursue "Bolivarian" and "socialist" or "Marxist-Leninist" ends. FARC activities have increasingly been labeled "narco-terrorist" or, simply "criminal," by a variety of critics.

Many insurgencies have degenerated into criminality. This occurred as the primary movements disintegrated and the remaining elements were cast adrift. Such disintegration is desirable for the counterinsurgent. It replaces a dangerous, ideologically inspired body of disaffiliated individuals with a less dangerous but more diverse body. This transition would mean the counterinsurgency would also transition to more of a law-and-order approach. Successful counterinsurgents must recognize that the ideal approach eliminates both the insurgency and any criminal threats.

#### *Nonviolent Tactics*

Successful insurgents use nonviolent tactics in conjunction with violent tactics. Subversion and propaganda are the two most prevalent forms of nonviolent warfare. Although some subversive activities can bleed over to violent activities, for the purposes of this FM, subversion will emphasize the nonviolent activities that define the nonviolent nature of subversion. Nonviolent tactics include, but are not limited to (for more see FM 3-24):

- Demonstrations
- Denial and Deception

- Hoaxes
- Infiltration
- Strikes

**Subversion**

*Subversion* is action designed to undermine the military, economic, psychological, or political strength or morale of a regime (DOD). In addition, Subversive activity is anyone lending aid, comfort, and moral support to individuals, groups, or organizations that advocate the overthrow of incumbent governments by force and violence (DOD). All willful acts that are intended to be detrimental to the best interests of the government that do not fall into the categories of treason, sedition, sabotage, or espionage are subversive activity.

Insurgents use various subversive techniques in their attempt to convince the populace to resist the government and COIN forces and support their insurgency. These techniques include demonstrations, boycotts, clandestine radio broadcasts, newspapers, and pamphlets. In addition, movement leaders organize or develop cooperative relationships with legitimate political action groups, youth groups, and trade unions. This approach develops popular support for later political and military activities. Subversive activities often openly challenge, in an organized pattern and just short of violence, the control and legitimacy of the established government and COIN forces.

One of the most visual forms of subversion is civil unrest. The fomenting of riots, organizing of strikes, and staging of demonstrations can drain the power, presence, and capabilities of the government and conversely increase the power and prestige of the insurgency. Another means of subversion is infiltration of government organizations, political parties, labor unions, community groups, universities, and charitable organizations.

To increase public credibility, attract new supporters, generate revenue, and acquire other resources, insurgent groups may establish their own *front groups*. Front groups are organizations that purport to be independent, but are in fact created and controlled by the leaders of the insurgency. A historic example of a front group is the Sinn Fein. The Irish Republican Army (IRA) used Sinn Fein, their "political wing" made up of political cadre, to serve as the IRA's respectable public face.

**Propaganda**

Insurgent groups commonly use propaganda to increase their base of support or reduce support for COIN forces. The joint definition of propaganda *is* any form of communication in support of national objectives designed to influence the opinions, emotions, attitudes, or behavior of any group in order to benefit the sponsor, either directly or indirectly (JP 1-02). In this case, the insurgents use propaganda to further their own ends.

The insurgent's propaganda efforts use activities such as clandestine radio broadcasts, the Internet, newspapers, graffiti and pamphlets that openly challenge the control and legitimacy of the established government. Insurgents will search for any leverage they can use in their propaganda. This includes seeking support based on the common identity of religious affiliation, clan, tribe, or ethnic group. Additionally, insurgents often create new problems and reinforce existing problems they then can exploit. Insurgents will arrange for the "COINcidental" presence of photographers or cameras where planned events occur. If the government is unwilling or unable to address these problems successfully, the insurgency can claim they will solve them. Common insurgent propaganda efforts include the following:

- Encouraging the HN populace or specific neutral parties to avoid supporting HN or other friendly government's forces.
- Increasing insurgent will to resist by fanning hatreds, biases, and predispositions.
- Inciting riots or organizing rallies, which may include honoring "martyred" insurgents.
- Causing or exacerbating a dislocated civilian crises.
- Creating or fostering public distrust of the HN security forces.
- Undermining the support of specific HN local leaders or businessmen.
- Creating or intensifying general ethnic or religious unrest or friction.
- Supporting or revitalizing dissident or opposition organizations.
- Linking local groups with similar groups in neighboring countries or regions.
- Discrediting or ridiculing specific HN or counterinsurgent officials.
- Characterizing government leaders as puppets and tools of foreign COIN forces.
- Spreading hostile coverage of COIN personnel, especially counterinsurgent mistakes.

Effective counterinsurgents must counter insurgent propaganda. This can be accomplished by conducting information engagement (IE) to exploit inconsistencies in the insurgents' propaganda and their excessive use of force or intimidation. Additionally, counterinsurgents must have a coherent and unified information engagement plan. This IE plan must be planned, prepared, and executed with input from all appropriate civil and military agencies, especially the HN.

## Section VI—Strengths and Vulnerabilities

Understanding insurgents' typical strengths and vulnerabilities allows counterinsurgents at the tactical level to work toward neutralizing or avoiding their capabilities and exploiting their weaknesses. The typical strengths and vulnerabilities of insurgents are explained below.

### *Strengths*

The recognized strengths of insurgent warfare provide a base to analyze the specific insurgent threat. No two insurgent forces are identical. Known strengths are applied against a specific situation the COIN force encounters, and are refined because of local analysis. Regardless, identified insurgent strengths must be reduced or circumvented. Table 2-1 presents insurgent strengths and countermeasures for analysis during COIN operations.

### *Vulnerabilities*

The vulnerabilities of insurgents also provide a base for analysis. To gain the advantage over the insurgent and enhance effectiveness, these weaknesses must be targeted and exploited. Table 2-2 displays insurgent vulnerabilities and considerations for analysis during COIN operations.

**Table 2-1. Insurgent Strengths and Countermeasures**

| | *Insurgent Strengths* | *Countermeasures* |
|---|---|---|
| **INDIGENOUS** | Insurgents are usually indigenous to the local area and have the support of at least some of the populace. Therefore, they have the ability to blend with the local populace. In many cases they have two roles—a local resident one moment, an insurgent the next moment. This enhances their capability to operate without discovery in a given area. | The counterinsurgent force must separate the insurgent from the rest of the populace. This is best accomplished through the effective use of populace and resources control. Care must be taken to ensure that civilians are not injured or mistreated as a result of counterinsurgent operations |
| **KNOWLEDGE** | Since most insurgents are indigenous, their knowledge of the local populace, customs, issues, language and terrain are first hand. The insurgents use this understanding to develop working relationships with the populace. The insurgent can apply this knowledge to the effective use of propaganda. If insurgents can get some of the local populace to identify to some degree with their cause, they can win their support and gain new recruits. If insurgents cannot persuade locals to help or refrain from hindering, they may resort to coercion. | The counterinsurgent force must overcome the insurgent's advantage of local knowledge by fostering a strong relationship with government security forces, other counterinsurgent forces, and the populace. If possible, counterinsurgent forces should include members of the local populace and reside within the local area. Maintaining continuous counterinsurgent operations in a given area through a permanently stationed counterinsurgent force is important. The skillful use of local assets or creating a local civilian defense force that has the support and backing of the government are some ways to accomplish this. |
| **INTELLIGENCE OPERATIONS** | The insurgent's inherent advantage with the population normally allows them to develop intelligence networks and infrastructure within the government and population. These networks can provide insurgents with continuous and current information on government or counterinsurgent force dispositions, strengths, weaknesses, and capabilities | The counterinsurgent force must place counterintelligence operations, intelligence collecting, and intelligence analyses as a high priority. The use of compartmentalization, deception, operations security, and communications security must be constantly emphasized. Since insurgents are indigenous, their intelligence networks and infrastructures can be infiltrated to gather intelligence and turn insurgent operatives into double agents. |
| **MOTIVATION** | Some insurgents may be devoted to their cause to the point of fanaticism. On the other hand, insurgents who wish to abandon the movement face major challenges. their ability to cease being an insurgent is difficult. The government and the local populace may not welcome them back, and the remaining insurgents will view them as a turncoat and threat. Motivation within the insurgency is usually kept high | The counterinsurgent force must therefore promote the belief that remaining an insurgent leads only to death and defeat while at the same time creating a viable method for "former" insurgents to return to normal life. Host government reintegration, reconciliation, and amnesty programs remain the important elements of this success. |

**Table 2-1. (Continued)**

| | *Insurgent Strengths* | *Countermeasures* |
|---|---|---|
| | through intimidation and threats of violence on them and their families. | |
| **FOCUSED RESPONSIBILITY** | Insurgents do not have the responsibility to maintain normal governmental obligations toward society. This frees their efforts to conduct focused operations in support of their goals. However, they often provide some aid and services to the local community, especially where government services fall short and they highly advertise this fact. | Counterinsurgent forces can use the insurgent's lack of provisions for the society, as a tool to increase government support of counterinsurgent operations and to decrease populace's insurgent support. In addition, they can show that the insurgents have acted irresponsibly. Increasing the HN governments ability to provide services to the population may marginalize insurgent efforts. |
| **INSURGENT TACTICS** | The insurgent can use a broad range of tactics, from conventional warfare to terrorism. They can escalate or deescalate their activity in reaction to government or counterinsurgent activity almost at will. | The counterinsurgents must remain flexible and adaptive to engage and, if possible, disband, defeat or destroy the insurgent force while at the same time preventing the insurgent force from having tactical successes. They must also learn and adapt quickly. |
| **ENDURING HARDSHIP** | Insurgents often come from impoverished backgrounds, are young, and are in good shape. They can make do with less by both design and background. Successful insurgents are innovative in their tactics, techniques and procedures. They learn and adapt to changes in the operational environment. | The counterinsurgent force must therefore establish controls and eliminate resource support. Tight security and control on arms and ammunition must be maintained, to include thorough destruct-tion of unused, abandoned, discarded equipment. Hardships are still hardships, and the harder life is for the insurgent, the more likely they are to quit. |

**Table 2-2. Insurgent Vulnerabilities and Considerations**

| | *Insurgent Vulnerabilities* | *Considerations* |
|---|---|---|
| **LIMITED PERSONNEL** | Insurgents operate in small bands to avoid detection by government and counterinsurgent forces. Due to the challenges of recruiting new insurgents, insurgent personnel losses are not easily replaced. | Insurgent methods to recruit replacements can be identified and exploited. |
| **LIMITED RESOURCES** | Because of their covert nature, insurgents must rely on resources that are stolen or clandestinely delivered from friendly entities such as internal or external supporters. This is especially true for sophisticated equipment, cash, replacement parts and expendable supplies. | Insurgent equipment losses are not easily replaced, and methods of re-supply can be discovered, severed, or tracked to sources and destinations. |

| | *Insurgent Vulnerabilities* | *Considerations* |
|---|---|---|
| **COMBAT POWER** | The insurgents normally lack the combat power for a sustained fight. As a result, insurgents typically avoid decisive engagements with government forces. By their own design, insurgent operations are usually limited in scope and are driven by both engagement and disengagement plans. | Insurgents will usually break off engagements when they become too intense, as the potential for higher losses is not usually worth the risk. However, they may conduct a stand up fight if they have a large base of expendable guerrillas; they think they can manipulate a large political gain; or in some cases a rites of passage event. |
| **INDIVIDUAL FACTORS** | The insurgent endures a life of physical danger, privation and many types of stress. Stresses include combat and the fear of combat, the need to live covertly (constantly fearing discovery by the government) recognizing the numeric superiority of the government forces they face; fear of criminal treatment if captured by the government; and fear of violence to self and family (often imposed by the insurgent organization to ensure cooperation). Besides a belief in the cause, the insurgent may remain an insurgent due to fear of government reprisal or reprisals from the insurgent organization. | Counterinsurgent forces can exploit these stresses. Offering insurgents, a pardon, to include food, shelter and protections from both the government and insurgent, is sometimes enough incentive to induce insurgent desertions. |
| **POPULAR SUPPORT** | The dependence of the insurgent on popular support is a major weakness. This popular support includes direct aid and active intelligence reporting. | If the popular support is withdrawn, the insurgent will not be able to operate effectively. If the populace turns against the insurgent, the government can reap significant benefits. Willing locals can help locate, capture, or kill once "popular" insurgents. |
| **OPERATIONAL FACTORS** | Insurgent operational weaknesses can include security (requiring extensive resources that may slow down responsiveness); bases and safe houses (that are difficult to acquire and operate); a lack of sophisticated communications (requiring insurgents to spend excessive amounts of time preparing to launch operations); and a lack of technology (including the ability to maintain captured high-technology items). | The counterinsurgent force can exploit these insurgent weaknesses by interdicting supply routes and facilities, following the supply trail, forcing desertion because of hardships, and inflicting combat losses that are hard to replace. |

## SUMMARY

The central struggle of the majority of insurgencies is to gain and maintain the support of the population. However, the five elements of each insurgency, the eight dynamics, the strategy, the tactics, and the specific strengths and vulnerabilities are each unique to an individual insurgent group. For the tactical leader, from the platoon to the brigade, it is imperative that they first identify and understand the insurgent group or groups that they are fighting, before determining potential insurgent COAs and friendly COAs.

# 3. Foundations of Counterinsurgency

*"The best way to attain peace is to combine force with politics, We must remember that destruction must be used as a last recourse, and even then only in order to build something better in the end... Each time an officer is required to act against a village in a war, he needs to remember that his first duty, after securing submission of the local population, is to rebuild the village, reorganize the local market and establish a school."*

General Joseph Gallieni, Fundamental Instructions, 1898

US and Host Nation participation in combating an insurgency includes simultaneous military and nonmilitary efforts that normally build upon the Host Nation government's efforts, its institutions and plans. This chapter discusses the foundations of a successful counterinsurgency through sections on the seven of counterinsurgency lines of effort, tactical force consideration, *clear-hold-build* operations, other major COIN operations, and the phases of COIN operations.

## Section I—Overview

A counterinsurgency is a complex subset of warfare. This section defines counterinsurgency, explains the concepts of foreign internal defense (FID) and internal defense and development (IDAD), and describes the scores of units, agencies, organizations that units could work with in their area of operations. This discussion also covers full-spectrum operations and their applications in counterinsurgency. Finally, it introduces the seven counterinsurgency lines of effort (LOEs) to help units achieve unity of effort.

### *Definition of Counterinsurgency*

Counterinsurgency is those military, paramilitary, economic, psychological and civil actions taken by a government to defeat an insurgency (JP 1-02). In a counterinsurgency, Host Nation forces and partners operate to defeat armed resistance, reduce passive opposition, and establish or reestablish the legitimacy of the Host Nation's government (FM 3-0). Counterinsurgency is a proactive approach involving all elements of national power; even down to the tactical level. COIN operations strive to achieve unity of effort amongst many joint, interagency, intergovernmental, and multinational organizations. COIN includes tactical planning; intelligence development and analysis; training; materiel, technical, organizational assistance; advice; infrastructure development; tactical-level operations; and information engagement. US forces often lead the US government's counterinsurgency efforts because the US military can quickly project a counterinsurgent force and sustain not only its force but also other agencies.

### *Goal of Counterinsurgency Operations*

The end state of counterinsurgency operations is a legitimate Host Nation government that can provide effective governance. This includes providing for their populace, eliminating the root causes of the insurgency and preventing those root causes from returning. Counterinsurgent operations can successfully defeat an insurgency; achieve unity of effort along multiple lines of effort; isolate the insurgent from the people; and increase the legitimacy of the Host Nation government. The five requirements for successful COIN operations at the tactical level include:

- Together, US and HN military commanders devise the plan for attacking the insurgent strategy, and focus on bolstering governmental legitimacy.
- HN and US forces establish control of area and secure the population continuously.
- Operations should be initiated from the HN government's area of strength against areas under insurgent control.
- Regaining control of the insurgent's areas requires the HN government to expand operations to secure and support the population.
- Information engagements favorably influence perceptions of HN legitimacy, obtain local support for COIN operations, publicize insurgent violence, and discredit insurgent propaganda.

### *Foreign Internal Defense*

*Foreign internal defense* (FID) is the participation by civilian and military agencies of a government in any of the action programs taken by another government or other designated organization to free and protect its society from subversion, lawlessness, and insurgency (JP 3-07.1). The FID programs are used to support friendly nations operating in or threatened with potential hostilities. As a tool of US foreign policy, FID is a national-level effort that involves numerous US Government agencies all working together to eliminate the root causes of an insurgency.

US military involvement in FID has traditionally been focused toward counterinsurgency. Although much of the FID effort remains focused on this important area, US FID programs may aim at other threats to an HN's internal stability, such as civil disorder, illicit drug trafficking, or terrorism. These threats may, in fact, predominate in the future as traditional power centers shift, suppressed cultural and ethnic rivalries surface, and the economic incentives of illegal drug trafficking continue. Typical tactical involvement in FID entails conducting combat operations against guerrillas, conducting actions across all seven lines of effort to defeat insurgency, and training Host Nation security forces.

### *Internal Defense and Development*

*Internal defense and development (IDAD)* is the full range of measures taken by a nation to promote its growth and protect itself from subversion, lawlessness, and insurgency. It focuses on building viable institutions (political, economic, social, and military) that respond to the need of the society (FM 1-02). IDAD is not only a single master plan, it is a compilation of all internal defense and development plans at the strategic, operational, and tactical level that a nation possesses. The fundamental goal of IDAD is to prevent subversion, lawlessness and insurgency by forestalling or defeating the threat and by working to correct the conditions that prompted the violence. Tactical units may only see the local government strategy, economic development plan, or the long-range military plan of their partner HN security force unit, which are part of the IDAD plan.

### *Joint, Interagency, Intergovernmental, Multinational*

Successfully conducting a counterinsurgency requires a host of organizations. Brigade, battalion, and company commanders must integrate and synchronize their operations directly with the activities and operations of other military forces and nonmilitary organizations in their area of operations.

### Terms

Leaders must understand the terminology between joint, interagency, intergovernmental, and multinational operations.

#### *Joint*

Between military services (Army, Navy, USAF, and USMC).

#### *Interagency*

Between other government agencies, for example, Department of State, Central Intelligence Agency, Federal Bureau of Investigation, National Security Agency, USAID.

#### *Intergovernmental*

Between international government organizations, for example, United Nations, European Union, NATO, African Union.

#### *Multinational*

Between foreign government organizations, for example, Great Britain, Saudi Arabia, Poland.

### Command Responsibilities

When working with interagency, intergovernmental, nongovernmental and multinational groups, Army commanders have inherent responsibilities that include clarifying the military's mission; determining controlling legal and policy authorities; and sustaining and caring for these organizations and individuals. Information sharing between the elements is essential to establish ground truth. Not all agencies may agree on the nature or scope of support required or on the operation's progress.

Gaining and maintaining popular support presents a formidable challenge that the military cannot accomplish alone. Achieving these aims requires synchronizing the efforts of many nonmilitary and HN agencies in a coordinated approach. Coordination at the lowest level is essential for unity of effort. Likely participants in COIN operations include the following:

- US military forces.
- Multinational (including HN) military forces.
- US governmental organizations.

- Intergovernmental organizations (IGOs).
- Nongovernmental Organizations (NGOs).
- Multinational corporations and contractors.
- Indigenous population and institutions (IPI).

### US Military Forces

The military's contribution is vital for COIN efforts. Demanding and complex, COIN draws heavily on a broad range of the force's capabilities and requires a different mix of offensive, defensive, and stability operations from that expected in major combat operations. Air, land, and maritime components all contribute to successful operations and to the vital effort to separate insurgents from the people. The Army and Marine Corps usually furnish the principal US military contributions to COIN forces.

The most important military assets in COIN are disciplined Soldiers and Marines with adaptive, self-aware, and intelligent leaders. Tactical units may have specially trained or attached personnel who bring certain capabilities such as:

- Civil affairs.
- PSYOP.
- Language specialists.
- Human intelligence.
- Logistic support.
- Contractors.
- Medical units.
- Military police.
- Engineers.
- Legal affairs.

### Multinational Military Forces

Soldiers and Marines may function as part of a multinational force. In COIN operations, US forces usually operate with the Host Nation security forces. Each multinational participant provides capabilities and strengths that US forces may not have. Other countries' military forces bring different cultural backgrounds, historical experiences, languages and other capabilities that can be particularly valuable to COIN efforts.

### US Governmental Organizations

Commanders' situational awareness includes being familiar with other US governmental organizations participating in the COIN effort and their capabilities. Commanders and leaders of other US governmental organizations should collaboratively plan and coordinate actions to avoid duplication or conflicting purposes. Within the US Government, key organizations that tactical units may work with are shown in Table 3-1.

### Inter-Governmental Organizations

An *intergovernmental organization* is an organization created by a formal agreement, for example, a treaty, between two or more governments. It may be established on a global, regional, or functional basis for wide-ranging or narrowly defined purposes. IGOs are formed to protect and promote national interests shared by member states (JP 1-02). The most notable IGO is the United Nations. Depending on the situation and HN needs, tactical units can expect to encounter any number of UN organizations in their AOs, such as the following:

- Department of Peacekeeping Operations.
- World Food Program.
- UN Refugee Agency (the UN High Commissioner for Refugees).
- UN High Commissioner for Human Rights.

### Non-Governmental Organizations

Joint doctrine defines a nongovernmental organization as a private, self-governing, not-for-profit organization dedicated to alleviating human suffering; or promoting education, health care, economic development, environmental protection, human rights, and conflict resolution; or encouraging the establishment of democratic institutions and civil society. (JP 1-02). There are several thousand NGOs of many different types. NGO activities are governed by their organizing charters and their members' motivations. Typical NGOs that tactical units may encounter include—

- Médecins sans Frontières (Doctors without Borders).
- Cooperative for Assistance and Relief Everywhere (CARE).
- Oxford Committee for Famine Relief (OXFAM).
- Save the Children.

### Indigenous Population and Institutions

IPI is the civilian construct of an area of operations to include its population, governmental, tribal, commercial, and private organizations and entities. The population includes legal citizens, legal and illegal immigrants, and all categories of dislocated civilians. As stated earlier, this is the most important group for counterinsurgent units to protect, engage, and synchronize actions.

**Table 3-1. US Governmental Organizations**

| |
|---|
| Department of State |
| • Office of the Coordinator for Stabilization and Reconstruction |
| |
| US Agency for International Development (USAID) |
| Central Intelligence Agency |
| Department of Justice |
| • Drug Enforcement Administration |
| • Federal Bureau of Investigation |
| Department of the Treasury |
| Department of Homeland Security |
| • US Coast Guard |
| • Immigration Customs Enforcement |
| Department of Agriculture |

## Multinational Corporations and Contractors

Multinational corporations often engage in reconstruction, economic development, security and governance activities. At a minimum, commanders should know which companies are present in their AO and where those companies are conducting business. Such information can prevent fratricide and destruction of private property.

### *Interagency Coordination in COIN*

A tactical unit conducting COIN operations can expect to work with a wide range of agencies and will be expected to track, manage, support, and coordinate with all of these agencies, some of whose agendas or desired outcomes may or may not COINcide with the overall COIN effort. Commanders at all levels will have to overcome many challenges and work to integrate all the agencies organizations present in the AO, so that everyone works towards a common end state.

Challenges to the integration of military and civilian agencies include different organizational cultures, capabilities and structures. Some civilian organizations may not be trained and equipped to operate in austere, unsecured environments or they may not be able to sustain themselves in remote areas. Some organizations may be averse to assuming risk, which may hinder the overall COIN effort. Military and governmental organizations are heavily dependent on establishing and implementing their internal operating procedures and they are often reluctant to adapt or change those procedures to accommodate the addition of new actors. Different expectations amongst the organizations can also serve to further heighten tensions and create friction among agencies.

US government and civilian agencies provide capabilities critical to the successful accomplishment of the counterinsurgency mission. Most agencies have their own budgets, and often will have more robust financial capabilities than military units. An effective interagency effort can use all of the organizations' budgets in a complementary way to facilitate stability and development efforts. Civilian agencies can also provide links to strategic resources that may not be available to tactical units. They may also have access to information, resources and enablers that can help military units develop a comprehensive understanding of the operational environment.

Military forces have several complementary characteristics that can be of value to other agencies. The most obvious is the ability to provide security and transportation assets. Military forces also have the ability to provide an accurate assessment of whether an area is ready for or capable of supporting essential services and economic development projects. Once these indicators become apparent, counterinsurgent forces have the ability to provide logistical support, security, and expertise to a project.

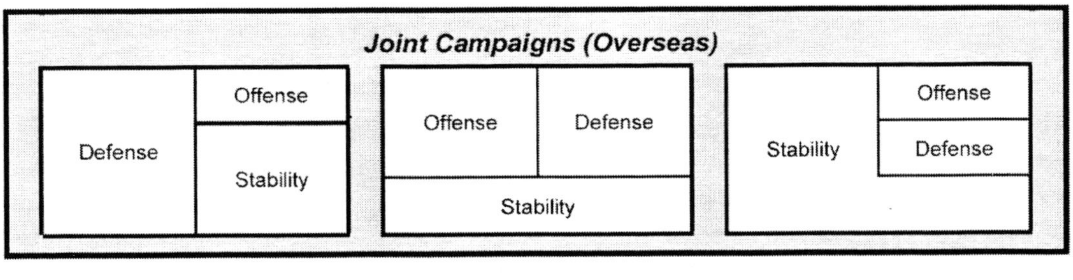


Figure 3-1. Full-spectrum operations.

The formation of effective civil-military teams creates complementary capabilities that mitigate the inherent weaknesses of both the counterinsurgent force and civilian agencies. Effective interagency teams can conduct concurrent operations to capitalize on gains made through security operations. In Afghanistan on 2002, USAID representatives embedded with units were able to render immediate aid to returning families whose homes were damaged during combat operations.

## Full Spectrum Operations

FM 3-0 states that Army forces conduct full-spectrum operations outside the United States by executing offensive, defensive, and stability operations as part of integrated joint, interagency, and multinational teams. Full spectrum operations entail simultaneous and continuous combinations of offensive, defensive, and stability or civil support operations. Based on the mission, one type of operation may predominate. Commanders shift the predominant type of operation based on the current situation and their assessment as they shape the operational environment and set the conditions to achieve the end-state (Figure 3-1).

Like any Army operation, at the tactical level, counterinsurgency operations are a combination of offensive, defensive, and stability operations. Counterinsurgent offensive operations focus on eliminating the insurgents. Counterinsurgent defensive operations focus on protecting the populace and infrastructure from insurgent attacks. Stability operations focus on addressing the root causes that allowed to insurgency to come into existence. Determining the combination is not easy, since it varies depending on the situation, the mission, and the commander's desired end-state. All three of these operations may be ongoing within different parts of the area of operations at the same time.

## Offensive Operations

These are combat operations conducted to defeat and destroy enemy forces and seize terrain, resources and population centers. They impose the commander's will on the enemy. This active imposition of land power potentially makes the offense the decisive type of military operation, whether undertaken against irregular forces or the armed forces of a nation-state supporting the insurgency. The physical presence of land forces also enhances stability operations through the threat of offensive action in areas they occupy. During an insurgency, offensive operations aim at destroying the guerrilla, underground or leader in order to establish a secure environment for the establishment or the re-establishment of the rule of law, legitimate government, and economic development. The successful counterinsurgent unit designs their offensive operations to complement their other defensive and stability operations. Chapter 5 further discusses offensive operations.

## Defensive Operations

These are combat operations conducted to defeat an enemy attack, gain time, economize forces, and develop conditions favorable for offensive or stability operations. They defeat attacks, destroying as many attackers as necessary. The defense preserves physical dominance over land, resources, and populations. Defensive operations retain terrain, as well as protecting the HN population and key resources. Defensive operations during an insurgency aim at securing the population, protecting counterinsurgent forces, securing key sites, and securing key personnel. The successful counterinsurgent unit designs its defensive operations to complement its other offensive and stability operations. Chapter 6 further discusses defensive operations.

### Stability Operations

These encompass various military missions, tasks, and activities conducted outside the United States in coordination with other instruments of national power to maintain or reestablish a safe and secure environment, provide essential government, services, emergency infrastructure, reconstruction, and humanitarian relief. Most stability operations are both multiagency and multinational. Forces engaged in stability operations may have to conduct offensive and defensive operations to defend themselves or destroy forces seeking to undermine the effectiveness or credibility of the stability mission.

Stability operations consist of five primary tasks—establish civil security, establish civil control, support to governance, restore essential services, and support to economic and infrastructure development. At the tactical level, the primary stability tasks may serve as lines of effort or simply as guideposts to ensure broader unity of effort. In this manual, they become the nucleus for the seven COIN lines of effort.

The degree to which Army forces engage in stability operations is dependent on the specific circumstances of any given operation. In some operations, the Host Nation can carry out most security operations and Army forces are engaged in stability operations to offset any negative impact of military presence on the populace. In other operations, Army forces within a failed state may be responsible for the well-being of the local population, to include providing basic civil functions, while working with other agencies to restore essential services to the area or region. An example of this was Operation Restore Hope in Somalia in 1992 and 1993. Chapter 7 further discusses stability operations.

#### *Lines of Effort in Counterinsurgency*

Commanders use LOEs to visualize, describe, and direct operations when positional reference to enemy forces has little relevance, such as an insurgency. FM 3-0 defines a *line of effort* as a line that links multiple tasks and missions using the logic of purpose—cause and effect—to focus efforts toward establishing operational and strategic conditions. A plan based on LOEs unifies the efforts of all actors participating in a counterinsurgency toward a common purpose. Each LOE represents a conceptual category along which the HN government and COIN force commander intend to attack the insurgent strategy and tactics and establish HN government legitimacy. LOEs are closely related and are not sequential in nature. Successful achievement of the end state requires careful coordination of actions undertaken along all LOEs. Figure 3-2 shows example COIN LOEs. The figure also shows how the LOEs try to gain the support of the population to reach the end state.

Success in one LOE reinforces successes in the others. Progress along each LOE contributes to attaining a stable and secure environment for the Host Nation. Once a measure of stability is achieved, achievements in other LOEs, like popular recognition of the HN government's legitimacy, improved governance, and progressive, substantive reduction of the root causes that lead to the insurgency, follow. No single list of LOEs applies to all insurgencies. Commanders select LOEs based on their understanding of the nature of the insurgency and what the COIN force must do to counter it. Commanders designate LOEs that best focus counterinsurgent efforts against the insurgents' strategy.

Commanders at all echelons can use LOEs. Lower echelon operations are nested within the higher echelon's LOEs; however, lower echelon operations are conducted based on each unit's AO. Commanders and staffs synchronize activities along all LOEs to gain unity of effort. This approach ensures the LOEs converge on a well-defined, commonly understood end state.

Commanders at all levels should select the LOEs that relate best to achieving the desired end state. The following list of possible LOEs is not all-inclusive. However, it gives commanders a place to start:

- Establish civil security.
- Establish civil control.
- Support HN security forces.
- Support to governance.
- Restore essential services.
- Support to economic and infrastructure development.
- Conduct information engagement.

These lines can be customized, renamed, changed altogether, or simply not used. Commanders may combine two or more of the listed LOEs or split one LOE into several. For example, some commanders may combine the LOEs, restore essential services, and support to economic and infrastructure development into one LOE. Likewise, other commanders may split out rule of law from the civil control LOE.

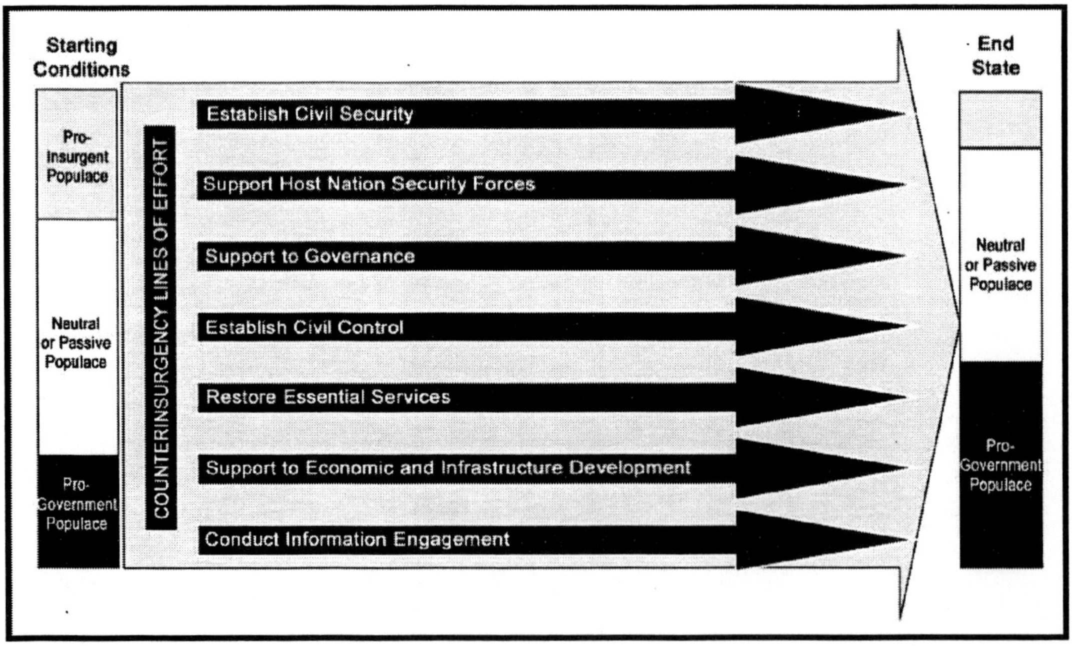


Figure 3-2. Example lines of effort for a counterinsurgency.

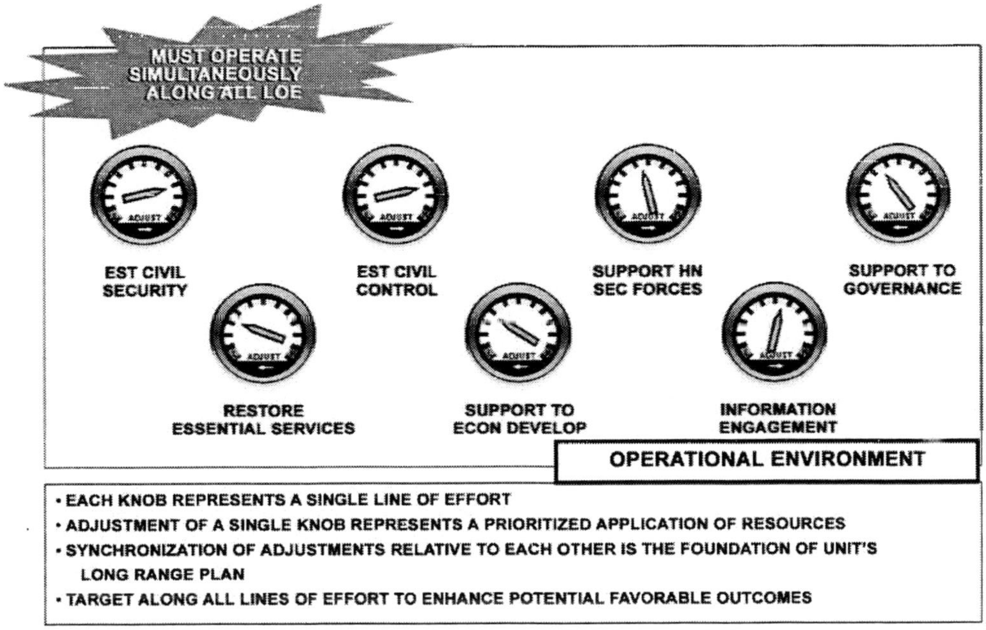


Figure 3-3. Rheostat approach to the lines of effort.

Tactical units, HN security forces and the HN's government can only accomplish a limited number of tasks at any one time. The seven COIN LOEs help prioritize and synchronize efforts along all of the LOEs. One useful construct is the rheostat approach with LOEs—increased effort along the establish civil security LOE usually means less effort along the other LOEs. Units may see this during operations against a powerful insurgency or during the *clear* phase of a *clear-hold-build* operation. In most cases, restoring security to an AO enables units and the HN government to rapidly meet objectives along the other lines of effort. In other AOs, units may increase effort along the support to governance and support to economic and infrastructure development LOEs, while seeing a reduction in effort along the establish civil control LOE. This occurs during operations against a weak insurgency or during the build phase of a *clear-hold-build* operation. Figure 3-3 shows the rheostat approach to the LOEs.

Operations designed using LOEs typically employ an extended, event-driven timeline with short-, mid-, and long-term goals. These operations combine the effects of long-term operations, such as neutralizing the insurgent infrastructure, with cyclic and short-term events, like regular trash collection and attacks against insurgent bases. Chapter 4 discusses considerations for planning LOEs and horizons.

## Section II—Historical Theories

This section compares three historic theories of counterinsurgency. Together, they provide a reference for forces engaged in COIN operations. COIN operations are complicated. Even considering these theories will not guarantee success. However, understanding these aspects will help illuminate the challenges inherent in defeating an insurgency. Three counterinsurgency experts, Robert Thompson, David Galula and Charles Callwell, had very specific theories concerning the conduct of counterinsurgency based on their experience.

### *Robert Thompson's Principles for Successful COIN*

Robert Thompson's *Five Principles for Successful Counterinsurgency* based on his experience in the Malayan emergency follow:

- The need for government to have a clear political aim.
- To function within the law.
- To establish an overall plan, whereby all political, socio-economic and military responses were coordinated.
- To give priority to the elimination of political subversion.
- To secure the government's base area before conducting a military campaign.

### *David Galula's Four Laws for Successful COIN*

David Galula's *Four Laws for Successful Counterinsurgency* based on his experience in Indochina and Algeria follow:

- The support of the population is necessary for the counterinsurgent as it is the insurgent.
- Support is gained through the active minority.
- Support from the population is conditional.
- Intensity of efforts and vastness of means are essential.

### *Charles Callwell's Ideas for Successful COIN*

Charles Callwell's *Ideas for Successful Counterinsurgency*, based on his experience in the Boer War and other British Imperial operations, follow:

- [Following up] successes...; a single blow will often achieve results, but a succession of blows paralyzes the enemy.
- Matching the enemy in mobility and inventiveness.
- Collecting actionable intelligence.
- Seizing what the enemy prizes most.

## Section III—Tactical Considerations

This section contains basic tactical considerations common to counterinsurgency operations, categorized first by terrain, and then by troops. As for any type of operation, the commander and staff must consider all relevant operational and mission variables, and their effects on operations. These additional considerations for terrain and troops allow the commander to better visualize the conduct of counterinsurgency operations as he accomplishes the assigned mission.

### *Terrain*

As with any type military operation, terrain plays a key role in counterinsurgency operations. Insurgencies are fought in either urban or a rural terrain, each with its own characteristics.

## Urban Areas

The urbanization of the world population continues to increase. The concealment and anonymity that was once only provided by remote rural areas to the insurgent is now available in urban areas. The transient nature and size of urban populations increasingly hinder a counterinsurgent's ability to detect and identify insurgents. Insurgents use urban centers for freedom of movement, easy access to their mass base and proximity to targets. Popular support at the outset of this kind of insurgency is not necessary, but can be gained through intimidation and attacking basic services provided by the government. With a greater population density, urban areas need more government functions and services than rural areas. This requires more government organizations for operations and a balanced approach using all LOEs.

Insurgencies thrive in an urban environment. Operations against urban insurgents vary from operations designed to control the population to operations that involve seeking out and killing or capturing the insurgent. In these urban areas, counterinsurgency forces may have to emphasize intelligence and police operations to counter clandestine organizational, intelligence, logistic, and terrorist activities. Counterinsurgency forces may be required to reinforce HN police forces in combating riots and disorders provoked by the insurgents as well as conducting raids and cordon and searches. Military counterinsurgency forces must be able to communicate with HN police forces and other agencies involved in operations.

### *Considerations during Counterinsurgencies*

When military forces must reinforce police or defeat insurgent forces inside the urban area, leaders must closely control and coordinate operations. However, the local government determines the level of intensity at which urban operations are conducted. Military forces should be withdrawn as soon as police forces can manage the situation. Basic urban counterinsurgency considerations include—

- Maintaining a constant, forward presence with the population.
- Acquiring and disseminating accurate and timely intelligence.
- Avoiding overreaction to insurgent activity.
- Ensuring the population has the basic level of essential services.
- Developing relationships with competent Host Nation officials.
- Using countersniper operations, especially at roadblocks, outposts, and sentry posts.
- Reacting to the ambush of patrols and firing on helicopters.
- Emphasizing countermeasures against explosive hazards (to include improvised explosive devices and mines) and booby traps of all types.
- Controlling access to weapons, uniforms and other supplies the insurgents may use.
- Protecting industry and public services from attack and sabotage.
- Preventing riots, protests and other large population incidents.

*Lines of Effort*

All urban operations require careful planning and coordination, particularly those operations involving application of force. The most vital requirement is accurate and timely intelligence. Military forces must be able to communicate with police and other agencies involved in the operations. During urban operations, actions must be taken across the all LOEs. Examples of actions along each LOE are—

***Establish Civil Security***

Conduct targeted raids with HN security forces to attack key insurgent cells.

***Establish Civil Control***

Conduct investigative training for HN police forces to increase arrests and convictions; establish a block committee to control the population.

***Support HN Security Forces***

Maximize the number, effectiveness, and use of HN security forces to secure and control the population and to prevent the insurgent's freedom of movement.

***Support to Governance***

Establish or convene local or tribal council meetings to identify and solve the underlying issues of the insurgency.

***Restore Essential Services***

Provide projects, such as the restoration of electrical power and sewer systems to blocks that support the government. Projects should provide positive reinforcement of desired behavior.

***Support to Economic and Infrastructure Development***

Provide funds to city managers to hire additional personnel and initiate economic development projects. Hire from blocks that are neutral to increase their support for the government.

***Conduct Information Engagement***

Conduct information engagements to highlight government successes and expose insurgent defeats and excesses.

**Rural Areas**

Much of the guerrilla warfare in the 20th Century occurred in a rural setting—the mountains of Yugoslavia in WWII, the jungles of South Vietnam in the 1950s and 1 960s, and the jungles of Columbia in the 1980s and 1990s. The reduced potential for collateral damage, limited infrastructure, and the lack of urban development, and the characteristics of the rural setting itself influence both insurgent and counterinsurgent operations.

*Considerations during Counterinsurgencies*

Counterinsurgency operations in a rural environment may allow for the potential application of full military combat power, conventional fire and maneuver, and less restrictive rules of engagement due to a lower population density. However, leaders, still plan for the application of the minimum-essential force required to accomplish the mission.

In the rural environment, insurgents rely on friendly elements within the population to provide supplies and intelligence. Rural insurgents prefer to operate in regions providing plenty of cover and concealment such as heavily forested and mountainous areas that hinder the counterinsurgency force in gaining access and intelligence. Often operating from their own home or village, rural insurgents will often move to camps if security does not permit them to remain at home. Insurgent camps are also chosen with a view toward easy access to the target population, access to a friendly or neutral border, prepared escape routes, and good observation of counterinsurgency force approach routes. When counterinsurgency operations force the insurgents out of his preferred base camps, he tends to establish camps in rugged inhospitable areas not easily penetrated. Like COIN in urban areas, rural counterinsurgency operations must focus on both locating and killing the guerrilla and on severing the supportive element of the population, such as the mass base and auxiliary, from providing supplies and intelligence.

Offensive operations are usually conducted in areas that either are under insurgent control or are contested. In these areas, ground or water modes of entry are often used, but air assault or parachute operations can also be employed. These operations use offensive tactics such as a raid, a reconnaissance in force, a cordon and search, a hasty or deliberate attack, a pursuit, or a combination of these.

*Purpose*

Their purpose is to—

- Destroy the insurgent force and its base complexes.
- Expand controlled areas.
- Isolate guerrillas from their support.
- Demonstrate support for the government and for the populace in the local area.
- Harass the insurgent to prevent the buildup of personnel and logistical resources.

*Lines of Effort*

All rural operations must address all LOEs. Examples along each LOE to consider include—

***Establish Civil Security***

Conduct raids with HN security forces to attack key guerrilla groups.

***Establish Civil Control***

Reinforce police operations, especially police stations and checkpoints, as a means to increase control of the population; conduct census.

***Support HN Security Forces***

Use HN security forces to increase combat power, expand the AO, increase the number of villages secured, and increase the legitimacy of the operation.

***Support to Governance***

Establish or convene a village, district, provincial or tribal council to identify and solve underlying issues.

***Restore Essential Services***

Provide projects such as wells to villages that support the government.

***Support to Economic and Infrastructure Development***

Provide economic stimuli such as the donation of a tractor or gasoline to villages that are neutral to increase their support for the government.

***Conduct Information Engagement***

Conduct information engagements to generate support for the HN government, highlight government victories and expose insurgent defeats and excesses.

## Border Areas

In addition to the typical external support, usually supplies, that an insurgent group may receive from across a border, insurgents may also establish sanctuary base camps and conduct cross-border operations from adjacent countries.

Host nation police, customs, or paramilitary border forces should be responsible for border security. However, the guerrilla threat may force the military to conduct border operations, particularly in rural areas. US forces advise and assist Host Nation security forces to interdict the infiltration of insurgent personnel and materiel across international boundaries with the intent of isolating the insurgent forces from their external support and sanctuaries. Border operations normally require restrictive measures for tribal and ethnic groups who do not recognize the international boundary.

Border operations require close coordination and cooperation between the armed forces, paramilitary forces, and government agencies involved. Physically sealing the border may be impossible, since doing so could increase the requirement for forces and materiel beyond available resources. Placing forces or barriers at every crossing and entry site may also be impossible. Commanders should prioritize where to place their forces or barriers.

*Routes*

Based on detailed terrain analysis and intelligence, commanders can determine infiltration and exfiltration routes, support sites, frequency and volume of traffic, type of transportation, number and type of personnel, amount and type of materiel, terrain and traffic conditions, and the probable location of base areas and sanctuaries. Continuous and detailed surveillance is required.

*Zones*

Restricted zones or friendly population buffer zones can be established if needed. Either of these operations, which could require relocating many persons, must be carefully planned. Although armed forces may assist, civil authorities normally are responsible for planning and carrying out a relocation program. Forced relocation is held to a minimum. The 1949 Geneva Conventions prohibit forced population resettlement unless there is clear military necessity.

***Restricted Zone***

This is a carefully selected area, varied in width and contiguous to the border. Authorities normally relocate all persons living in this zone. Authorities give public notice that they will regard all unauthorized individuals or groups encountered in the restricted zone as infiltrators or insurgents.

***Friendly Population Buffer Zone***

This is an area where only civilians believed to be loyal to the government live in the AO. The government relocates all persons whose loyalty it cannot establish. The government may use this operation to establish information nets and employ loyal citizens in paramilitary units. The operation denies insurgents potential civilian contacts and base areas for border-crossing activities.

*Lines of Effort*

Border operations must use actions across all LOEs. One or two examples along each LOE are—

***Establish Civil Security***

Conduct ambushes with and without HN security forces at likely insurgent crossing sites.

***Establish Civil Control***

Conduct training of HN police and border forces to increase the speed and results of searches.

***Support HN Security Forces***

Integrate and maximize the number and effectiveness of HN border, police and Army forces.

***Support to Governance***

Establish or convene a village or tribal council from both sides of the border to identify and solve issues.

***Restore Essential Services***

Provide projects, such as restoration of irrigation systems to villages that assist the government's efforts to close the border. Projects should provide positive reinforcement of desired behavior.

***Support to Economic and Infrastructure Development***

Provide economic stimuli such as hiring village personnel to build border fences or to act as guards. Hire from villages that are neutral to increase their support for the government.

***Conduct Information Engagement***

Create information engagements to canalize movement through official points of entry and establish the legitimacy of the border in the eyes of the populace.

***Troops***

Counterinsurgency operations typically involve actions that combine joint, interagency, multinational, and nongovernmental organizations efforts. The increased number of military and nonmilitary participants and their divergent missions and methods are a coordination and unity of effort challenge. Achieving unity of effort requires the greatest possible common purpose and direction among all agencies. One means of achieving this is using a long-range plan designed around the seven COIN lines of effort.

**Task Organization**

The organization for, and conduct of, counterinsurgency operations depends on the mission variables and the OE. However, COIN operations place a premium on boots on the ground. *Task organization* is the temporary grouping of forces designed to accomplish a particular mission (FM 3-0). Task organization for counterinsurgency operations is often substantially different from how units are task organized in conventional operations. During counterinsurgency operations, many units do not perform their traditional role. For example, in Iraq and Afghanistan, many artillerymen have served in infantry, civil affairs, military police, or intelligence roles.

**Troops-to-Task Analysis**

Given the extended nature of COIN, tactical units must maximize all available assets to accomplish missions in each of the COIN LOEs and sustain Soldier proficiency, physical fitness, and emotional and psychological well-being. A means to maximize the employment of all assets is referred to as troops-to-task analysis. This process enables units to correctly assign tasks to units of appropriate size and capabilities.

Often conducted during MDMP, staffs and tactical units begin troops-to-task analysis by determining a standard size element to use as a baseline (Infantry platoon, Armor platoon, Cavalry platoon, company, or others). The staff determines the forces available (often including HN security forces). Then, the staff or tactical unit lists and prioritizes each task and determines the forces required to accomplish each task using the baseline unit as a measure,. Finally, the tactical unit and staff uses the commander's planning guidance to array forces and assign tasks.

Figure 3-4 shows an example format for a troops-to-task analysis worksheet. It identifies specified and implied tasks, a baseline number of unit or units, including HN security forces, required to accomplish a task, and assigned them higher headquarters. The troops-to-task process is also useful for establishing unit AOs.

| | | AO Platoon Requirements | | | | |
|---|---|---|---|---|---|---|
| | | AO 1 | AO 2 | AO 3 | AO 4 | AO 5 |
| Specified or Implied Tasks | Secure key infrastructure | 3 | 2 | 3 | 1 | |
| | Establish and secure CMOC | | | | | 2 |
| | Conduct population control | 3 | 3 | 4 | 3 | |
| | Conduct route security | | | | | 3 |
| | Conduct base security (COPs) | 1 | 1 | 1 | 0 | 1 |
| | Secure PRT | | | | | 1 |
| | Totals: | 7 | 6 | 8 | 4 | 6 |

**Steps:**

1. Subdivide urban areas into AOs based on demographics.
2. List specified and implied tasks.
3. Determine a combat power metric such as 1 x PLT per key infrastructure, or 2 x PLT to secure the CMOC.
4. Fill out table with number of platoons, and tally at bottom to determine how many platoons are required for each AO.
5. Consolidate AOs if possible, and assign HQ.

Figure 3-4. Example format for AO platoon requirements worksheet.

## Brigade Combat Teams

In larger COIN efforts, BCTs will usually provide the required command and control apparatus to conduct sustained tactical operations over an AO. Typically, a BCT's boundary is aligned with a Host Nation governmental boundary. BCTs also allocate resources to their subordinate battalions and coordinate with higher military headquarters, Host Nation government officials, HN security forces, NGOs, or US agencies.

Once a BCT is given an AO, they, along with the Host Nation, should be the controlling headquarters for all other elements in their AO. This should include the temporary attachment for control, if not command, of any element that is physically within their AO. Examples would include the United States Agency for Internal Development (USAID), Corps of Engineers, Military Police, advisor teams, reconstruction teams, Host Nation security forces, or private contracting security firms, since these elements may not completely understand the intricacies in the BCT's assigned AO.

The military force conducting counterinsurgency operations will not always consist of maneuver forces. Counterinsurgency operations are manpower intensive and therefore infantry, armor, artillery, reconnaissance, and military police battalions are the primary tactical building blocks for combat in a counterinsurgency environment. The composition of the tactical force in counterinsurgency operations depends upon these available forces and the threat that is faced. Aviation, engineers, military working dog teams, special operations forces, and reconstruction teams are key force multipliers for the counterinsurgency force. In deciding how to use these forces, leaders assess the factors of METT-TC. Normally, most ground combat elements are organized to fight as maneuver forces. However, the proper use of other forces can provide the counterinsurgency force with many advantages.

### Host Nation Security Forces

The Host Nation security forces must be included or, in some cases, drive the unity of effort of the counterinsurgent effort. HN security forces can be integrated in planning cells, share bases with US forces and, at a tactical level, conduct parallel planning in corresponding staff sections. Additionally, HN security forces give many benefits to counterinsurgency efforts. For more information, see Chapter 8.

### Capabilities

Units conducting counterinsurgency operations have identified the need for additional capabilities beyond their standard task organization. Human terrain teams, document and media exploitation teams, personal security detachments, detainee holding areas, Host Nation security force advisor teams, base commanders and base defense commanders, explosive ordnance disposal teams, company intelligence support teams, and tactical site exploitation teams have all proved useful during recent operations.

A *human terrain team* is a group of civilian anthropologists attached to brigades and battalions. This team helps the unit understand local cultures. These social scientists aid leaders in better understanding relevant cultural history, engaging locals in a positive way, and incorporating knowledge of tribal traditions to help resolve conflicts.

*Document and media exploitation teams (DOMEX)* process, translate, analyze, exploit, and share hard copy documents and electronic media collected during operations. This capability increases in importance as the rule of law is re-established and insurgents go to trial, rather than long term detention.

Due to the organization of brigade and battalion headquarters, commanders and sergeant majors lack the combat power needed to move freely around the battlefield. In recent operations, units have either built ad hoc organizations or tasked platoons to serve as escorts so that they can reconnoiter the AO, attend meetings, engage locals, check on Soldiers, and better visualize the fight. These are often called personal security detachments. Most theaters have implemented a three or four vehicle rule for convoys, which make this unit's strength at least twelve Soldiers strong.

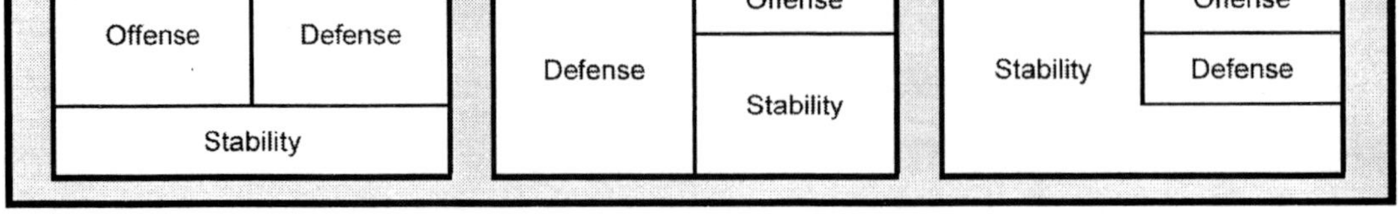


Figure 3-5. Clear-hold-build framework.

A *detainee holding area* (DHA) is a temporary location used to field process and house any person captured or otherwise detained by an armed force, and provide resources for intelligence exploitation. Detainees are kept here for a short period of time before being released or being sent to a theater internment facility. The DHA generally consists of a semipermanent structure designed to house detainees. Basic infrastructure includes shelter, latrines, basic hygiene facilities, medical care, interrogation facilities, and evidence holding areas. For more see FM 3-19.40.

Host nation security force *advisor teams* conduct operations to train HN military individuals and units in tactical employment, sustainment and integration of land, air, and maritime skills; provide advice and assistance to military leaders; and provide training on tactics, techniques, and procedures. These teams can be resourced by the BCT or battalion, or provided by DA. Their size and capability varies on the size of the HN security force being advised. (For more information, see Chapter 8.)

Base commanders and base defense commanders are typically internally resourced personnel and units of the brigade, battalion, and company level who command and control the base and supervise the defense of the base. (For more information, see Chapter 6.)

Explosive ordnance disposal (EOD) support provides the capability to neutralize domestic of foreign conventional explosive hazards, which include unexploded ordnance (UXO), booby traps, improvised explosive devices (IEDs), captured enemy ammunition, and bulk explosive. EOD units detect, mark, identify, render safe, and dispose of explosive hazards. Also, EOD specialists work with intelligence personnel to conduct explosive forensics to help identify the makers of the devices, as well as their ever evolving methods of construction, placement, concealment, and detonation.

Human intelligence collection teams (HCTs) are teams with trained HUMINT collectors that collect information for people and their associated documents and media sources to identify elements, intentions, capability, strength, disposition, tactics, and equipment. The team uses human sources as tools and a variety of collection methods to satisfy the commander's intelligence requirements. They can conduct source operations and interrogate detainees.

A PSYOP team is a team that reinforces the effects of tactical and nontactical operations, as well as discrediting and demoralizing the insurgency. Their actions enhance the probability of accomplishing the unit's mission.

A civil affairs team (CAT) is a team that helps a unit establish, maintain, influence, or exploit relations between the unit and civilian organizations, governments, authorities, and populace in an area of operations. Civil affairs core tasks include populace and resource control (PRC), foreign humanitarian assistance, civil information management, nation assistance, and support to civil administration.

Company operations teams sometimes referred to as company intelligence support teams are a group of two to six individuals at the company level who enhance the company commanders' situational awareness of their area of operations by producing intelligence at the company level. They collect and analyze patrol reports, human intelligence reports, and battalion intelligence reports. As a result, they are able to conduct link analysis, conduct pattern analysis, create target folders, and enemy situation templates.

Site exploitations teams are teams at the company or battalion level that execute systematic actions with the appropriate equipment, to ensure that personnel, document, electronic data, and other material at any site are identified, evaluated, collected, and protected to gather information to be developed into intelligence and facilitate future operations. They may collect biometric, physical, digital, and spoken data. (See also Chapter 5 of this manual or CALL product 07-26.)

## Section IV—Clear-Hold-Build Operations

A *clear-hold-build* operation is a full spectrum operation that combines offense (finding and eliminating the insurgent), defense (protecting the local populace) and stability (rebuilding the infrastructure, increasing the legitimacy of the local government and bringing the rule of law to the area) operations. Each phase—clear, hold, and build—combines offensive, defensive, and stability operations in varying degrees. In the clear phase, offensive operations usually dominate; in the *hold* phase, defensive operations are emphasized; and in the *hold* phase stability operations are preeminent. It is usually a relatively long-term operation and requires the commitment of a large number of forces. Figure 3-5 shows the typical combination of offense, defense and stability operations.

### *Overview*

This pattern of operation is to *clear*, *hold*, and *build* one village, area, or city at a time—and then expand into another area. In previous conflicts, this was known as the "*Oil Spot Strategy*." This type of operation was used successfully in Algeria and Indochina by the French, who called it *tache d'huile*; in Malaysia by the British, who referred to it as "The Briggs Plan"; and in Tal Afar by the Americans, who named it "*clear-hold-build*." This operation aims to develop a long-term, effective Host Nation government framework and presence in the area, which secures the people and facilitates meeting their basic needs, and provides legitimate governance.

The purpose of America's ground forces is to fight and win the Nation's wars. Throughout history, however, the Army has been called on to perform many tasks beyond pure combat; this has been particularly true during the conduct of COIN operations. COIN operations will continue to require Soldiers to be ready both to fight and to build—depending on the security situation and a variety of other factors.

### Resources

*Clear-hold-build* operations require a substantial commitment of resources, time and a clear unity of effort by civil authorities, other agencies, and security forces. Counterinsurgent commanders must first plan and prepare for a long-term effort. Before conducting a *clear-hold-build* operation, units must ensure that they have identified or coordinated for—

- Adequate troops, US and HN, to clear the insurgents from a designated area.
- Interagency cooperation and unity of effort.
- Sufficient resources, expertise, and labor to restore essential services.
- A local government that will support the operation.
- Means for US and HN security (military, police, and paramilitary) forces to share intelligence.

### Objectives

Actual operations begin by controlling access to the area and then by controlling key points within the area. Security and influence can then spread out from these areas. Often, the steps of *clear-hold-build* operations will overlap, especially between hold and build, where activities are often conducted simultaneously. *Clear-hold-build* operations have the following objectives:

- Create a secure physical and psychological environment.
- Provide continuous security for the local populace.
- Eliminate the insurgent presence.
- Reinforce political primacy.
- Enforce the rule of law.
- Rebuild local Host Nation institutions.
- Gain the populace's support.

#### *Clear*

*Clear* is a tactical mission task that requires the commander to remove all enemy forces and eliminate organized resistance in an assigned area (FM 3-90). The force does this by destroying, capturing, or forcing the withdrawal of insurgent combatants and leaders. This task is most effectively initiated by a clear-in-zone or cordon-and-search operation, as well as patrolling, ambushes, and targeted raids.

### Offensive Operations

Offensive operations predominate during the clear phase; however, defensive and stability operations are still conducted. Eliminating guerrilla forces does not remove the entrenched insurgent infrastructure. While leaders and the underground exist, insurgents will continue to recruit among the population, undermine the Host Nation government, receive supplies, and coerce the populace through intimidation and violence. After guerrilla forces have been eliminated, removing the insurgent infrastructure such as leaders, underground, and auxiliaries begins. This should be done in a way that minimizes the impact on the local populace.

If insurgent forces are not eliminated, but are expelled or have broken into smaller groups instead, they must be prevented from reentering the area. Once counterinsurgent units have established their bases, platoons and companies cannot become static. They should be mobile and should patrol throughout the area. The local populace should be compensated for damages that occur while clearing the area of insurgents.

### Defensive and Stability Operations

Defensive and stability operations are continued to maintain gains and set the conditions for future activities. These include—

- Isolating the area to cut off external support and to kill or capture escaping insurgents.
- Conducting periodic patrols to identify, disrupt, eliminate, or expel insurgents, as well as secure the population.

- Employing security forces and government representatives throughout the area to secure the populace and facilitate follow-on stages of development.

**Lines of Effort**

Although the clear phase emphasizes establishing civil security, operations across the other LOEs must happen. Examples of complimentary actions across all LOEs include—

*Establish Civil Security*

Conduct targeting of insurgent leaders and guerrilla bands, cordon and searches of insurgent controlled areas, and raids on safehouses.

*Establish Civil Control*

Train and support police forces or paramilitary forces to implement curfews to interdict insurgent movement.

*Support HN Security Forces*

Train and use HN security forces to increase combat power during clearing operations.

*Support to Governance*

Identify key government and local leaders that can support the re-establishment of local government that can administer the area. Identifying the underlying issues of the population.

*Restore Essential Services*

Identify essential services that need immediate attention.

*Support to Economic and Infrastructure Development*

Identify potential "quick win" projects to stimulate the local economy and create additional support for the government.

*Conduct Information Engagement*

The message to the populace focuses on gaining and maintaining their overt support for the counterinsurgency effort, as well as informing them that active support for the insurgency would prolong combat operations, creating a risk to themselves and their neighbors. The message to the insurgent force focuses on convincing them that they cannot win, and that the most constructive alternatives are to surrender or cease their activities.

***Hold***

After clearing the area of guerrillas, the counterinsurgent force must then assign sufficient troops to the cleared area to prevent their return, to defeat any remnants, and to secure the population. This is the hold task. Ideally, Host Nation security forces execute this part of the clear-hold-build operation. Success or failure depends on effectively and continuously securing the populace and on reestablishing an HN local government. Although offensive and stability operations continue, in this phase, uses defensive operations to secure the population.

## Bases

Cleared areas are best held by establishing counterinsurgent forces in bases among or adjacent to the area. From these bases, the counterinsurgent forces can then concentrate on two primary purposes: to disrupt, identify, and ultimately eliminate the insurgents, especially their leadership and infrastructure; and to end popular support for the insurgency and to gain popular support for the government

Exactly where these bases are established and their actual force composition is a key counterinsurgent decision. The main consideration should be towards the desired effect on the population, especially increasing their security. If a area is supportive of the insurgency, then a base established in known insurgent strongholds may sever the relationships between insurgents and the populace. However, if the area is neutral or has pockets of support for the government, then the base should be established where it can best help in developing the area.

## Securing the Populace

If adequate HN security forces are not available, units should consider hiring and training local paramilitary forces to secure the cleared village or neighborhood. Not only do the members of the paramilitary have a stake in their area's security, they also receive a wage. Providing jobs stimulates the economy. Having a job improves morale and allows locals to become a potential member of the local governmental process.

Contact with the population should be continuous, and both planned and unplanned. Being based within the population forces information engagement and provides opportunities for contact with the population. From the first day, Soldier's actions in these areas, if culturally astute, can build bonds with the local populace and in many cases change the attitudes of the people. Ultimately, the goal of this contact is twofold: to gain a better picture of the actual situation and to turn the population's support toward the government.

Another consideration is to secure key physical infrastructure. Because resources are always limited, parts of the infrastructure vital for stability and vulnerable to attack must receive the priority of protection. This can be accomplished by analyzing the risk and likelihood of attack on various sites in the AO. (Chapter 7 discusses one technique for this.)

## Lines of Effort

Although the *hold* phase emphasizes defensive actions, operations across all LOEs must be employed. Examples of complimentary actions along multiple LOEs include—

### *Establish Civil Security*

Continuously secure the people and separate them from the insurgents. Establish a firm government presence and control over the area and populace by recruiting, organizing, arming, and training local paramilitary forces and integrate them into operations against the insurgents. Continue to conduct raids on insurgent leaders or members of the underground.

### *Establish Civil Control*

Train and support police force to combat crime and enforce curfews to curtail insurgent movement.

*Support HN Security Forces*

Conduct combined patrols, checkpoints, cordon and searches, and raids with HN security forces.

*Support to Governance*

Establish or reestablish a government political apparatus to replace the insurgent apparatus.

*Restore Essential Services*

Establish contracts, empower or finance local governments to initiate SWEAT-MSO repairs.

*Support to Economic and Infrastructure Development*

Conduct local improvements designed to convince the populace to support the Host Nation government, participate in securing their area, and contribute to the reconstruction effort.

*Conduct Information Engagement*

Information engagements should affirm that security forces supporting the Host Nation government are in the area for the long term and are securing the population from insurgent intimidation, coercion, and reprisals.

***Build***

The *build* phase of *clear-hold-build* operations consists of carrying out programs designed to remove the root causes that led to the insurgency, improve the lives of the inhabitants, and strengthen the Host Nation's ability to provide effective governance. Stability operations predominate in this phase, with many important activities being conducted by nonmilitary agencies. During this phase, the Host Nation security forces should have primary responsibility for security. Progress in building support for the Host Nation government requires protecting the local populace. People who do not believe they are secure from insurgent intimidation, coercion, and reprisals will not risk overtly supporting counterinsurgent efforts.

**Securing the Populace**

To secure the populace, security forces should continuously conduct patrols and use measured force against insurgent targets of opportunity. Just like in the *hold* phase, if the HN security forces are inadequate, units should consider hiring a paramilitary force to secure the village or neighborhood. Contact with the people is critical to the local counterinsurgency effort's success. Actions to eliminate the remaining covert insurgent support infrastructure, such as the underground, auxiliary, and mass base, must be continued, because any insurgent presence will continue to threaten and influence people.

**End State**

During the build phase, Host Nation government representatives reestablish government offices and normal administrative procedures. National and international development agencies rebuild infrastructure and key facilities. Local leaders are developed and given authority. Life for the area's inhabitants begins the return to normal.

**Lines of Effort**

Although the build phase continues to secure the population and separate them from the insurgents, the focus will shift to the other LOEs. Examples of complimentary actions along multiple LOEs include—

*Establish Civil Security*

Conduct targeted raids on insurgent leaders and the underground, led by HN security forces. Provide a US Quick Reaction Force (QRF) to HN security forces.

*Establish Civil Control*

Continue to patrol the area and control the population while improving HN police training and equipment. Police forces may continue to expand their role.

*Support HN Security Forces*

Increase the number of patrols by HN security forces; US forces begin reducing their roles.

*Support to Governance*

Continue to support and enhance the local government.

*Restore Essential Services*

Continue projects to restore SWEAT-MSO services such as building roads, digging wells, building schools and establishing emergency services.

*Support to Economic and Infrastructure Development*

Continue to stimulate the local economy through projects such as market repairs.

*Conduct Information Engagement*

Information engagements should affirm that security forces supporting the Host Nation government are in the area for the long term and are eliminating insurgent leaders, organizations and infrastructure and improving essential services.

## The Strategic Hamlet Program

Successful *clear-hold build* operations (CORDS program and Tal Afar) are discussed in FM 3-24. This vignette highlights the challenges of executing a *clear-hold-build* operation properly.

In 1962, the Military Assistance Command Vietnam (MACV) launched the Strategic Hamlet Program to compliment its advisory efforts with the Republic of Vietnam's military. The Strategic Hamlet Program was an extremely ambitious program designed to build fortified hamlets, relocate the population to the hamlets, and train paramilitary forces, known as the Regional or Provincial Forces, across South Vietnam. Together this would increase security and quality of life for the population. With the goal of fortifying half the country's hamlets in only 18 months, the program struggled with providing governance and solving bureaucratic issues, ensuring security for all the villages, and limiting corruption. Additionally, the program met resistance from locals, who felt an ancestral connection to their original hamlets.

Although the program was abandoned in 1964, after the war, the North Vietnamese Army acknowledged that the well-run hamlets forced their guerrilla forces to relocate to other, insurgent controlled areas. The South Vietnamese Strategic Hamlet program failed to meet all five prerequisites for a *clear-hold-build* operation, especially obtaining an adequate number of troops, securing sufficient resources, and providing a local government up to the task.

## Section V—Other Major COIN Tactical Operations

Major counterinsurgency tactical operations include Strike operations and Populace and Resource Control operations. Strike operations are operations to find, fix and finish insurgent forces in areas under insurgent control where the counterinsurgent does not want to maintain a permanent presence afterwards. Population and resources control operations are government operations to control the populace, deny insurgents access to the populace and resources, and reestablish law order.

### *Strike Operations*

Strike operations are short duration (generally one day to several weeks) offensive, tactical operations conducted in contested or insurgent controlled urban or rural areas to find, fix and destroy insurgent forces. Small, highly mobile combat forces operate in dispersed formations to locate and fix the insurgents. Upon locating the insurgents, commanders direct their forces to attack, pursue, and destroy them. If contact is lost, the units resume aggressive patrolling to reestablish contact and destroy insurgent forces before they can rest, reorganize, and resume operations. Strike operations seek to destroy insurgent forces and base areas, isolate insurgent forces from their support, and interdict insurgent infiltration routes and lines of communications (LOCs). Strike forces are organized as self-sufficient task forces capable of operating in areas remote from logistical bases. Ground or water borne means of entry may be used, as well as air assault or parachute deliveries. Strike operations use offensive tactics such as raids, reconnaissance in force, cordons and attacks, hasty or deliberate attacks, and pursuits. It is often a complementary operation to a *clear-hold-build* operation.

Speed and surprise are important in strike operations. The sudden and unexpected delivery of combat forces into an insurgent-held or contested area provides significant advantages to the forces conducting these operations. Speed and surprise can be achieved by using air assaults to insert the first forces into the area of operations. Subsequent forces can be delivered on later airlifts or by other modes of transportation. Fires can also be used to block escape routes or areas that are not secured by ground forces.

A strike on an insurgent force normally requires superior combat power. COIN forces attempt to immediately engage and destroy insurgents before they can disperse. Depending on the situation, hasty or deliberate attacks are made on bases that contain fortifications. After a successful attack on insurgent forces, troops thoroughly search the area for insurgent personnel, supplies, equipment, and documents. All captured enemy documents must be placed into intelligence channels as soon as possible to ensure that it is properly exploited by DOMEX personnel. Pursuit operations are undertaken to destroy or capture forces attempting to flee. Artillery, air support, and air assault forces support ground pursuit.

**Purpose**

Strike operations may also be used as a means to encourage reconcilable insurgents to the negotiation table. Strike operations are conducted to—

- Harass the insurgent to prevent the buildup of personnel and logistical resources.
- Destroy the insurgent force and its base complexes.
- Demonstrate government resolve and garner support from the populace in the local area.
- Set the conditions to expand *clear-hold-build* operations.

**Lines of Effort**

Along the LOEs in a strike operation, tactical units should examine—

*Establish Civil Security*

Conduct attacks, raids, and cordon and searches with HN security forces on insurgent forces and bases.

*Establish Civil Control*

Conduct PRC operations, such as a census, a checkpoint or a search operation in villages or blocks to control the populace.

*Support HN Security Forces*

Increase combat power by using HN security forces.

*Support to Governance*

Establish or convene local or tribal council meetings to identify and solve the underlying issues of the insurgency.

*Restore Essential Services*

Ensure projects provide positive reinforcement of desired behavior, such as digging a well for a tribal sheik who provides intelligence.

*Support to Economic and Infrastructure Development*

Hire people from villages or blocks that are neutral to increase support for the government.

*Conduct Information Engagements*

Use information engagements to highlight government success and expose insurgent defeats and excesses.

***Populace and Resource Control Operations***

Populace and resource control (PRC) operations are government actions to protect the populace and its materiel resources from insurgents, to deny insurgents access to the populace and material resources and to identify and eliminate the insurgents, their organization, their activities, and influence while doing so. The objective of populace and resources control is to assist in preserving or reestablishing a state of law order within an area or entire nation. PRC operations are normally nontactical, police-type operations and a responsibility of HN governments. However, US forces may be required to conduct PRC operations until HN security forces possess the will and capability. PRC operations may be conducted independently of *clear-hold-build* operations or Strike operations or as an integrated part in each of these operations.

In peacetime, police forces protect the population and resources of a state from criminal activity by both armed and unarmed criminals. During an insurgency, however, the number of armed insurgents, the willingness of insurgents to use violence against the police and the populace, and the potential for civilians to become displaced or refugees, often creates a situation where police forces and their operations cannot successfully curtail the insurgents nor protect the population and resources.

## Objectives

Typical objectives for populace and resources control operations include—

- Establish and maintain a secure physical and psychological environment for the population.
- Limit insurgent freedom of movement and initiative.
- Sever relationships between the population and insurgents.
- Identify and neutralize insurgent support activities.
- Establish and maintain security of resources.

## Categories

Populace and resources control measures can be classified into four general categories:

- Surveillance and intelligence measures.
- Establish control measures.
- Enforce control measures.
- Protection measures.

*Surveillance and Intelligence Measures*

Surveillance and intelligence measures include both overt and covert surveillance of known, likely or potential insurgents, their targets, and the creation of a QRF to exploit intelligence or conduct greater surveillance of insurgent targets. Expanded police intelligence and surveillance operations, to include police informants and agent networks, may link criminal acts from robberies, kidnappings, terrorism, and extortion to insurgent activities.

Surveillance must be established and maintained over key individuals, groups, and activities of interest. Increased surveillance must be maintained over critical locations, especially government and civilian sources of weapons and ammunition. This includes maximum use of sensors, cameras and other electronic surveillance equipment to provide continuous coverage of suspected areas and routes used by insurgents. An additional means of gathering intelligence is to monitor local media (radio, newspaper) both for rumor control and counterpropaganda purposes as well as intelligence tip-offs. In addition, the public and private actions of influential local leaders provide additional insight. It is important to live forward with the local people and listen to what they are saying. Still other intelligence and surveillance measures include—

- Establish general covert surveillance measures at marketplace and stores.
- Use HUMINT collection teams (HCTs) or the HN police to recruit locals for surveillance and intelligence.
- Establish a system of block or village wardens with reporting procedures as well as incentives. Hold the wardens accountable for knowing what is going on in their block or village.

*Establish Control Measures*

Successful counterinsurgency operations typically use increased control measures to limit insurgent activities and their ability to hide within the population. Normal security functions must be performed efficiently and effectively, which is especially true if they are in any way part of the root causes of the insurgency. However, due to the insurgency, additional security measures and operational techniques must be implemented. Two such considerations are increasing the size of police/border patrols due to the threat and establishing QRFs to rapidly reinforce any patrol or site.

Combating an insurgency requires increased control measures to limit the insurgent's freedom of movement and their supporters. Control measures should be well planned and coordinated to ensure rapid and efficient operations, with a minimum of delay and inconvenience to the people. All control measures must be authorized by national laws and regulations, as well as be enforceable. Each control measure should be tailored to fit the situation and used to establish or reinforce the credibility of the Host Nation government.

The Host Nation government should explain and justify all control measures to the HN population. They should be the least restrictive to accomplish the purpose. Local civilians must understand that these measures are necessary to protect them from insurgent intimidation, coercion, and reprisals. Ideally, the local population and their leaders should accept the needed measures before implementation and that their support will minimize the inconvenience of the measures. These restrictions must be lifted as soon as the situation permits.

Once control measures are in place, the Host Nation government should implement a system of punishments for offenses related to them. These punishments should be announced and enforced equally. All inconveniences and discomforts these measures cause should be blamed squarely upon the insurgents. PSYOP products, widely disseminated to the population, can help ensure the measures have the intended effect and undermine popular support for the insurgency. Control measures fall under the populace control or resource control categories.

*Populace Controls*

Population control measures include—

- Curfews.
- Travel permits and passes.
- Movement restrictions.
- Restricted areas.
- Census or registration of residents.
- Block committee.
- National or regional identification system or ID cards.
- Licensing for jobs such as medical, security, construction, and drivers.
- Immigration restrictions

*Resource Controls*

Resource control measures include control of select resources to include foodstuffs, medical supplies, and key equipment through:

- Rationing or purchase permits
- Registration of firearms.
- Registration of automobiles and trucks.
- Export and import restrictions.

*Techniques*

Enforcement operations must be conducted both day and night. Checkpoints and roadblocks are set up to check and control the movement of personnel, vehicles, and material, and prevent actions that aid the insurgency. During counterinsurgency operations, checkpoints and roadblocks assist the government and counterinsurgent forces in maintaining the initiative against the insurgents by disrupting, interfering with, and deterring insurgent operations and disrupting the insurgents' decision-making cycle. It is important to conduct checkpoints and roadblocks with interpreters, HN police, or other HN security forces. Checkpoints and roadblocks used together can channel vehicles and personnel into a checkpoint or a search.

***Roadblock***

A barrier or obstacle (usually covered by fire) used to block or limit the movement of vehicles along a route.

***Checkpoint***

Checkpoints are manned locations used to control movement that may be established from 1 to 72 hours depending on the purpose of the operation.

***Inspections***

This includes random cordons and searches, both day and night, of homes, shops and buildings for arms, propaganda material and insurgents. The use of military working dog teams can be very effective in intercepting contraband. To decrease population resentment, it is important to ensure intelligence supports the inspections. An inspection can also be used as a subterfuge for a meeting with key people.

*Rewards for Cooperation and Compliance*

A separate, yet related, function is the rewarding of cooperation or compliance. The ultimate reward to the society as a whole is the removal or reduction of restrictions. However, this must be balanced with the increased potential of the insurgent to gain benefit from these reduced or removed restrictions.

- Cooperative towns, villages or neighborhoods may be collectively rewarded by essential service projects or economic stimulation projects.
- At the individual level, the government may provide monetary or material rewards for information on suspicious activities.

*Protection Measures*

A form of protection measures is internment and resettlement, which consist of those measures necessary to provide shelter, sustain, guard, protect, and account for people (enemy prisoners of war [EPWs] and civilian internees [CIs]), US military prisoners, and dislocated civilians [DC]. Resettlement of a population may vary from a geographic based resettlement to the resettlement of a specific population group within a defined area. Resettlement operations should be under Host Nation direction and control. For more information, see FM 3-19.40.

## Section VI—Phases

Major counterinsurgency tactical operations are long-term population security operations conducted in territory generally under Host Nation government control to establish, regain, or maintain control of those areas and to provide adequate security and control to the populace to allow restoration of essential services and improvements to the economy. They typically move through three phases. They combine offensive, defensive, and stability operations to achieve the stable and secure environment needed for effective governance, essential services, and economic development to flourish. At the operational level, the phases are—the initial response phase, the transformation phase, and the fostering sustainability phase. At the tactical level, these three phases may resemble the three parts of a *clear-hold-build* operation. Understanding this evolution and recognizing the relative maturity of the AO are important for the proper planning, preparation, execution, and assessment of COIN operations. It is also important to recall that the insurgent also operates in three stages—latent and incipient, guerrilla warfare, and war of movement. This knowledge allows commanders to ensure that their activities are appropriate to the current situation.

***Initial Response Phase***

The initial response phase generally reflects tasks executed to stabilize the operational environment in a crisis. During this phase, military forces perform stability tasks during or directly after a conflict where the security situation hinders the introduction of civilian personnel. Activities during the initial response phase aim to provide a safe, secure environment as well as to attend to the immediate essential service needs of the Host Nation population. At the tactical level, the initial response may appear to be similar to the clear phase of a *clear-hold-build* operation. Units may also incorporate strike operations and PRC operations during this phase. Against a determined insurgency, the initial response may last months or years.

### *Transformation Phase*

The transformation phase represents a broad range of post-conflict reconstruction, stabilization, and capacity-building. The transformation phase may be executed in either crisis or vulnerable states. Counterinsurgent forces are most active here, working aggressively along all lines of effort (LOEs). The desire in this stage is to develop and build enduring capability and capacity in the HN government and security forces. As civil security is assured, focus expands to include the development of legitimate governance, provision of essential services, and stimulation of economic development. Relationships with HN counterparts in the government and security forces and with the local populace are developed and strengthened. These relationships increase the flow of intelligence. This intelligence facilitates measured offensive operations in conjunction with the HN security forces. The Host Nation increases its legitimacy through providing security, expanding effective governance, providing essential services, and achieving incremental success in meeting public expectations.

At the tactical level, the transformation phase may look a lot like the *hold* phase of a *clear-hold-build* operation. Units may use Strike operations and PRC operations that complement their holding efforts. Once again, a tenacious insurgency may cause the transformation or *hold* phase to last months or even years.

### *Fostering Sustainability Phase*

Fostering sustainability phase encompasses long-term efforts that capitalize on capacity-building and reconstruction activities to establish conditions that enable sustainable development. This phase is characterized by the expansion of stability operations across contested regions, ideally using HN security forces. The main goal for this phase is to transition responsibility for COIN operations to HN leadership and security forces. In this phase, the multinational force works with the Host Nation in an increasingly supporting role, turning over responsibility wherever and whenever appropriate. QRF and fire support capabilities may still be needed in some areas, but more functions along all LOEs are performed by HN security forces with the assistance of multinational advisors. In this phase, the Host Nation has established or reestablished the systems needed to provide effective and stable governance that sustains the rule of law. The government secures its citizens continuously, sustains and builds legitimacy through effective governance, has effectively isolated the insurgency, and can manage and meet the expectations of the population.

At the tactical level, the *fostering sustainability phase* may resemble the *build phase* of a *clear-hold-build* operation. Due to the stable security environment, PRC operations may be relaxed. In ungoverned areas, there still may be a requirement for strike operations to disrupt the remnants of an insurgency. This phase progresses at the pace of the HN's ability to rebuild their infrastructure and institutions, assume control of their security, and provide legitimate governance to the people.

## SUMMARY

The foundations of COIN are the guiding principles for any unit conducting counterinsurgency operations. Once understood and applied they enable commanders to craft a coherent plan that achieves unity of effort amongst all organizations in the area of operations.

# 4. COMPREHENSIVE TACTICAL PLANNING IN COUNTERINSURGENCY

*"The fight against the guerrilla must be organized methodically and conducted with unremitting patience and resolution. Except for the rare exception, it will never achieve spectacular results, so dear to laurel seeking military leaders."*
Roger Trinquier. Modern Warfare—A French View of Counterinsurgency. 1964.

The aim of counterinsurgency operations is to set the conditions that eliminate insurgency by securing and garnering the support of the population, as well as increasing the legitimacy of the Host Nation government. To accomplish this, tactical COIN planning uses all capabilities, not just military capabilities. This chapter shows the tactical planning process of counterinsurgency operations through sections on basic tactical design, planning horizons, the military decision-making process (MDMP), troop-leading procedures (TLP), and targeting.

## Section I—Overview

The battlefield of the Twenty-First Century has proven to be complex due to increasing urbanization, globalization, and religious fundamentalism; multiple enemy, friendly, and neutral actors; and ambiguous guidance and direction. It is through planning that the commander is able to see the desired outcome, lay out effective ways to achieve it, and communicate his vision, intent, and decisions to his subordinates, focusing on the results he wants to achieve (FM 5-0). Progress in counterinsurgencies is slow and difficult to measure. It can continue for weeks, months, even years. With the complexity of counterinsurgency operations, commanders and staff have relied upon end states, the seven COIN lines of effort, measures of effectiveness and performance and tight planning horizons to impose order on the chaos and craft functional plans to guide counterinsurgent efforts to increase the legitimacy of the Host Nation government along multiple lines of effort.

### *End State*

At the tactical level, this is the set of *conditions* that, when achieved, accomplish the mission.

A *condition* is a specific existing circumstance, framed in military terms. that, when achieved, describes one aspect of the desired end state. Achievement of all of the conditions obtains the end state. For tactical commanders, the end state is typically a set of required conditions, usually for each LOE, that defines the achievement of the commander's tactical objective.

Historically, successful counterinsurgencies commonly focus on the political end state, unified in their approach, and flexible in addressing the core insurgency issues. They secure the people, enhance the legitimacy of the Host Nation government and, above all, show patience.

A military operation must be linked to a political end state, composed of various conditions that the populace supports.

**Unity of Action**

Close coordination and effective cooperation between the Host Nation (civil, police, and military), the US counterinsurgency force and all other coalition partners, allows the full strength of each to engage all levels of the insurgency. Unity of effort is key.

**Root Causes**

Root causes are the grievances of the people. Ultimately, for an insurgency to be successful, it must provide a solution to the key issues of the people. At the same time, for a counterinsurgency to be successful, it too must address these key issues.

**Protection of the People**

Most insurgencies use coercion and terror to gain support from the people and to inhibit the people's support of the government. The target of this coercion and terror is frequently the people connected to the government—police, local administrators and teachers. As such, if the counterinsurgent wishes to receive the support of the people they must not only protect their supporters and their families, but also their communities. Often, protecting and controlling the people enables the counterinsurgent to achieve the other conditions.

**Enhancement of the Legitimacy of the HN Government**

The US and other counterinsurgency partners must always focus on strengthening the Host Nation government's ability to defend itself and its populace from the insurgency.

**Patience**

There is no decisive battle in counterinsurgency operations. Often, it takes years to create an environment where a Host Nation government can and will defend itself and its people.

***Measures of Performance and Effectiveness***

Tactical units must measure progress toward mission accomplishment. Commanders continuously assess the operational environment and the progress of operations, and compare them to their initial vision and commander's intent. Units evaluate the operations progress or success through intelligence assets, patrol reports, engagement with local leaders, graffiti, or surveys. Commanders adjust operations based on their assessment to ensure objectives are met and the desired end state is achieved.

The assessment process measures the efforts that support operational and tactical objectives and progress toward the desired military end state of the counterinsurgency across all LOEs. The assessment process uses measures of performance and measures of effectiveness to gauge progress. They are defined as follows:

**Measure of Performance**

An MOP is criterion used to assess friendly actions that is tied to measuring task accomplishment (JP 1-02). MOPs confirm or deny that the task has been correctly performed. An example of a MOP is "How many people registered to vote at the school this week?"

**Measure of Effectiveness**

An MOE is criterion used to assess changes in system behavior, capability, or operational environment that is tied to measuring the attainment of an end state, achievement of an objective, or creation of an effect (JP 3-0). An example of a MOE is "Did the well project provide clean drinking water to the village?"

**Purposes**

Measures of effectiveness and of performance help commanders determine when all or part of the mission has been accomplished. The criteria used depend on the situation. Many times the MOP and MOE must be determined and evaluated by the HN government or security forces. They often require readjustment as the situation changes and objectives evolve. If an effect cannot be measured directly, then indicators of achieving the effect are measured. A measure of effectiveness or a measure of performance has four characteristics. They are—

*Measurable*

They require quantitative or qualitative standards that can be used to measure them.

*Discrete*

Each criterion measures a distinct aspect of the operation. Excessive numbers of MOEs and MOPs become unmanageable. At that point, the cost of collection efforts outweighs the value of assessing.

*Relevant*

Each MOE and MOP must be relevant to the result or outcome. The key is visualizing the desired result or outcome and identifying the most accurate and simplest indicator of it.

*Responsive*

MOEs and MOPs must detect changes quickly enough for commanders to respond immediately and effectively.

Commanders and staffs also develop a standard or baseline as a comparison and identify trends. From this information and analysis of why a trend is up or down, staffs can identify trouble spots and plan operations to reverse negative trends. They can also capitalize on positive trends by determining what is causing the positive increases and apply those tactics, techniques, and procedures more broadly.

Measures of effectiveness and measures of performance are included in the approved plan or order and reevaluated continuously throughout preparation and execution. Higher echelon staffs should ensure that the number of MOEs and MOPs do not overly burden lower echelons—especially battalion and below. Well-devised MOEs and MOPs, supported by effective management of available information, help commanders and staffs understand links between tasks, end state, and lines of effort.

### *Planning Horizons*

During operations, a headquarters sends a tactical unit an operations plan or operations order. Tactical units must then consider the scope of their mission and determine the planning horizons. A planning horizon is a point in time commanders use to focus the organization's planning efforts to shape future events. Planning horizons, which in major combat operations are measured in hours and days for tactical commanders may, in COIN operations, be measured in weeks, months, and years.

In contrast, tactical commanders, who have responsibility for terrain and a mission covering multiple objectives that must be achieved systematically, require extended planning horizons. Managing extended planning horizons have an impact on a tactical unit. The staff processes must be organized to facilitate multiple outlooks, as an extended outlook does not relieve the immediacy of current operations. However, this may be challenging for a company commander without a staff.

FM 5-0 defines three planning horizons—long-, mid-, and short-range plans. These horizons create a useful construct for operations executed by tactical units in a counterinsurgency. Additionally, the familiar quarterly training guidance and brief processes in FM 7-0 fit the long-term nature of counterinsurgency operations.

### Long-Range Planning

Long-range planning encompasses the range of time where the situation is too uncertain to plan for specific operations. Commanders must visualize what conditions they desire to exist, resulting from the cumulative effect of all their tactical objectives. Visualizing the time required to establish these conditions places approximate bounds on the extent of future planning. In general, when units are in a rotation and have a planned transfer of authority, the long-range plan should consider the conditions that the commander desires to exist at three months after the transfer.

The long-range plan provides a construct for a commander to describe his vision of the operational environment and their unit's role over time using lines of effort. In addition to describing the unit's tactical objectives and the conditions they create, the plan provides a description of major events. Providing a long lead time on major events can be critical to anticipating resources and identifying milestones, or identifying and completing a series of interrelated tasks well in advance of the event. Friendly forces as in a transfer of authority, the population as in elections, or the insurgency could drive the event as it changes in phases, strategies, or organization.

Management of long-range plans allows the unit to synchronize its efforts with adjacent units and the higher headquarters. When an objective requires the cumulative effect of hundreds of company, platoon and squad missions to achieve the desired end state, long-range plans using lines of effort become the tool to ensure unity of effort across the echelons. Assessment from current operations is analyzed to determine if the desired effect is being achieved, then the long-range plan adjusts the activities of current operations to ensure that the unit's efforts are directed towards the defined end state. Table 4-1 shows COIN long-range planning cycles.

**Table 4-1. COIN Long-Range Planning Cycle**

| Planning Action | Planning Guidance | Planning Horizon |
|---|---|---|
| Division Long-Range Plan and Long-Range Calendar | 3 months prior to start | Long-range plan horizon of 1 year Calendar at least 1 year |
| BCT Long-Range Plan and Long-Range Calendar | 2 months prior to start | Long-range plan horizon of 1 year Calendar at least 1 year |
| Battalion/Squadron Long-Range Plan and Long-Range Calendar | 1 months prior to start | Long-range plan horizon of 1 year Calendar at least 6 months |

**Table 4-2. COIN Mid-Range Planning Cycle**

| Planning Action | Planning Guidance | Planning Horizon |
|---|---|---|
| Division Quarterly FRAGO | 6 weeks prior to start of quarter | 3 months |
| BCT Quarterly FRAGO | 1 month prior to start of quarter | 3 months |
| Battalion/Squadron Quarterly FRAGO | 2 weeks prior to start of quarter | 3 months |
| Company plan and calendar | 1 week prior to start of quarter | 3 months |
| Quarterly assessment and backbrief | Prior to start of quarter | 3 months or more |

## Mid-Range Planning

Mid-range plans are derived from the long-range plan. The long-range plan narrows the scope to a frame of time where objectives or milestones that support objectives can be clearly defined and operations planned in detail along all lines of effort. The mid-range plans therefore refine and expand upon the appropriate portion of the long-range plan. Mid-range planning should not exceed the unit's capability to reasonably forecast events, assign resources and commit to a particular plan.

It is the mid-range plan that analyzes the mission in detail and produces the warning, fragmentary, and operations orders that drive daily actions. These plans represent a commitment of resources and initiate preparation by subordinate units. Table 4-2 shows various mid-range planning cycles.

## Short-Range Planning

Short-range planning represents the scope of detailed planning associated with a specific mission or one to four weeks of normal operations. With resources assigned through mid-range planning, short-range planning, the unit refines and expands the plan to include the tactical arrangement of forces, execution matrices, patrol schedules, ISR plans and convoys.

**Table 4-3. COIN Short-Range Planning Cycle.**

| Planning Action | Planning Guidance | Planning Horizon |
|---|---|---|
| Brigade/Battalion/Squadron Meetings, Calendars, and Framework FRAGOs | 3 to 4 days prior to execution | 1 to 4 weeks |
| Company Calendar and FRAGO | 1 to 2 days prior to execution | 1 week |

A framework order is a fragmentary order (FRAGO) that identifies and tasks units for missions for one to four weeks. It has also been called a "steady state order." An example is a battalion FRAGO that identifies the company that will patrol the route each day during the upcoming week. Table 4-3 shows possible short-range planning cycles for units.

## Section II—Tactical Design

During counterinsurgency operations, small units address immediate objectives through tactics such as a raid or cordon and search. These operations resemble the small unit actions of tactical units in major combat operations, and many of the planning tools apply directly. The counterinsurgent, however, must also accomplish objectives, which require time, assessment and continual redirection to accomplish tasks arranged in lines of effort. Thus, COIN is an iterative process. In addition to presenting commanders and staffs with objectives that must be accomplished both immediately and over time, counterinsurgencies typically present more tactical objectives to a unit than they can address simultaneously. The counterinsurgent must truly plan for full spectrum operations, transitioning from tasks that support one objective onto tasks that support a different objective often within the same day. A systematic approach must be developed that is tailored to the situation based on continuous assessment. While the complexities of major combat operations should not be underestimated, counterinsurgencies bring a different set of complexities that requires staffs to apply analytical tools to achieve mission success.

### *Considerations*

The long, mid and short-range planning horizons help units establish a routine to assess the success of their plan and revise the plan based on a changing operational environment. COIN is an iterative process. —No unit—company, battalion, or brigade—understands its AO well enough to craft a perfect plan across all seven LOEs. Like a rheostat, each must be balanced against the others. Success in the *Establish Civil Security* and *Establish Civil Control* lines of effort typically allows units to focus less on security and more on support to governance and restoration of essential services LOEs. The most effective C2 mechanisms often give one single, permanent, senior local government official overall responsibility for the counterinsurgency in their AO. These local officials in turn establish a local board, composed of representatives from the civil authority, the military, the police, the intelligence services, and the civil population, who manage all civil or military assets inside their AO. Additionally, tactical design in counterinsurgency must consider how to—

- Secure the populations and areas that remain loyal.
- Reclaim the populations and areas that support the insurgency.
- Eliminate the insurgency, politically, militarily and philosophically.
- Develop Host Nation military and police forces that—
    - Defend their own bases and other critical sites.
    - Protect and secure all elements of the populace equally.
    - Support the restoration of government presence and control.

- Aggressively oppose insurgency in order to neutralize the insurgent leadership, cadre, and combatants, through death, capture, co-opting individual, or forcing them to leave the area.
- Promote normalcy and stability.
- Protect basic services.
- Assist civic action projects.

- Help establish an HN legal framework and C2 mechanisms to aid implementation of the plan.
- Secure the critical infrastructure, to include governmental and societal critical sites.
- Win the information war, specifically—
  - Counter the insurgent's propaganda.
  - Conduct friendly information engagements.
- Continually analyze and assess the success of the plan.

### *The Seven Counterinsurgency Lines of Effort*

As discussed in Chapter 3, lines of effort (LOEs) are used to visualize, describe and direct operations when positional references to an adversary have little relevance, like in a counterinsurgency. Ideally, LOEs combine the complementary, long-term effects of stability tasks with the cyclic, short-term events typical of combat operations. Using lines of effort, tactical commanders develop tactical tasks and tactical missions, allocate resources, and assess the effectiveness of the operation. The commander may specify which line of effort represents the decisive operation and which are shaping operations. Commanders synchronize activities along multiple LOEs to achieve the conditions that compose the desired end state and do not view them as a sequential "road map" to success.

The seven counterinsurgency lines of effort—establish civil security, establish civil control, support Host Nation security forces, restore essential services, support to economic and infrastructure development, support to governance, and conduct information engagement—are intended as a guide for tactical units. LOEs can be combined, eliminated, or expanded depending on the insurgency and the overall situation. Employing the seven COIN LOEs adeptly requires a unity of effort with the HN and other nonmilitary agencies. Figure 4-1 shows a notional tactical unit's LOEs during a counterinsurgency. The subparagraphs that follow the figure summarize the seven counterinsurgency lines of effort and their first primary subtasks.

### Establish Civil Security

Civil security involves protecting areas, resources, and the populace from both external and internal threats (FM 3-07). Ideally, Army forces focus on the external threats while police and security elements address internal security against terrorists, criminals and small, hostile groups. However, during an insurgency, the Army may also address the internal security against criminals, terrorists, and guerrillas.

Most societal and government functions require a secure environment and obtaining civil security is often a prerequisite for other stability tasks to be effective. Although US and multinational forces can provide direct assistance to establish and maintain security, this situation is at best a provisional solution. Ultimately, the Host Nation must secure its own people. Typical civil security tasks include—

- Enforce cessation of hostilities, peace agreements, and other arrangements.
- Conduct disarmament, demobilization, and registration.
- Conduct border control, boundary security, and freedom of movement.
- Support identification.
- Protect key personnel and facilities.
- Clear explosive and CBRN hazards.

### Establish Civil Control

Civil control regulates selected behavior and activities of individuals and groups (FM 3-07). This control reduces risk to individuals or groups and promotes security. Civil control channels the population's activities to allow provision of security and essential services while coexisting with a military force conducting operations. Typical civil control tasks include—

- Establish public order and safety.
- Establish interim criminal justice system.
- Support law enforcement and police reform.
- Support judicial reform.
- Support property dispute resolution processes.
- Support corrections reform.
- Support public outreach and community rebuilding programs.

### Support to Host Nation Security Forces

The US military helps the Host Nation develop the forces required to establish and sustain stability within its borders. This assistance can include developing, equipping, training and employing HN security forces. It may expand to multinational operations where HN forces and US forces fight alongside one another. HN security forces include military forces, police, corrections and border guards. Typical civil control tasks include—

- Measure (assess) HN security forces.
- Organize HN security forces.
- Rebuild/build infrastructure for HN security forces.
- Train HN security forces.
- Equip HN security forces.
- Advise HN security forces.
- Mentor HN security forces.

### Support to Governance

Sometimes no HN government exists or the government is unable or unwilling to assume full responsibility for governance. In those cases, this LOE may involve establishing and maintaining a military government or a civil administration while creating an HN capability to govern. When well executed, these actions may eliminate the root causes of the insurgency. Governance activities are among the most important of all in establishing lasting stability for a region or nation. Stability operations establish conditions that enable interagency and Host Nation actions to succeed. Military efforts help to build progress toward achieving effective, legitimate governance by restoring public administration and re-sorting public services while fostering long-term efforts to establish a functional, effective system of political governance. Military actions help to shape the environment so interagency and HN actions can succeed. Typical support to government tasks include—

- Support transitional administrations.
- Support development of local governance.
- Support anticorruption initiatives.
- Support elections.

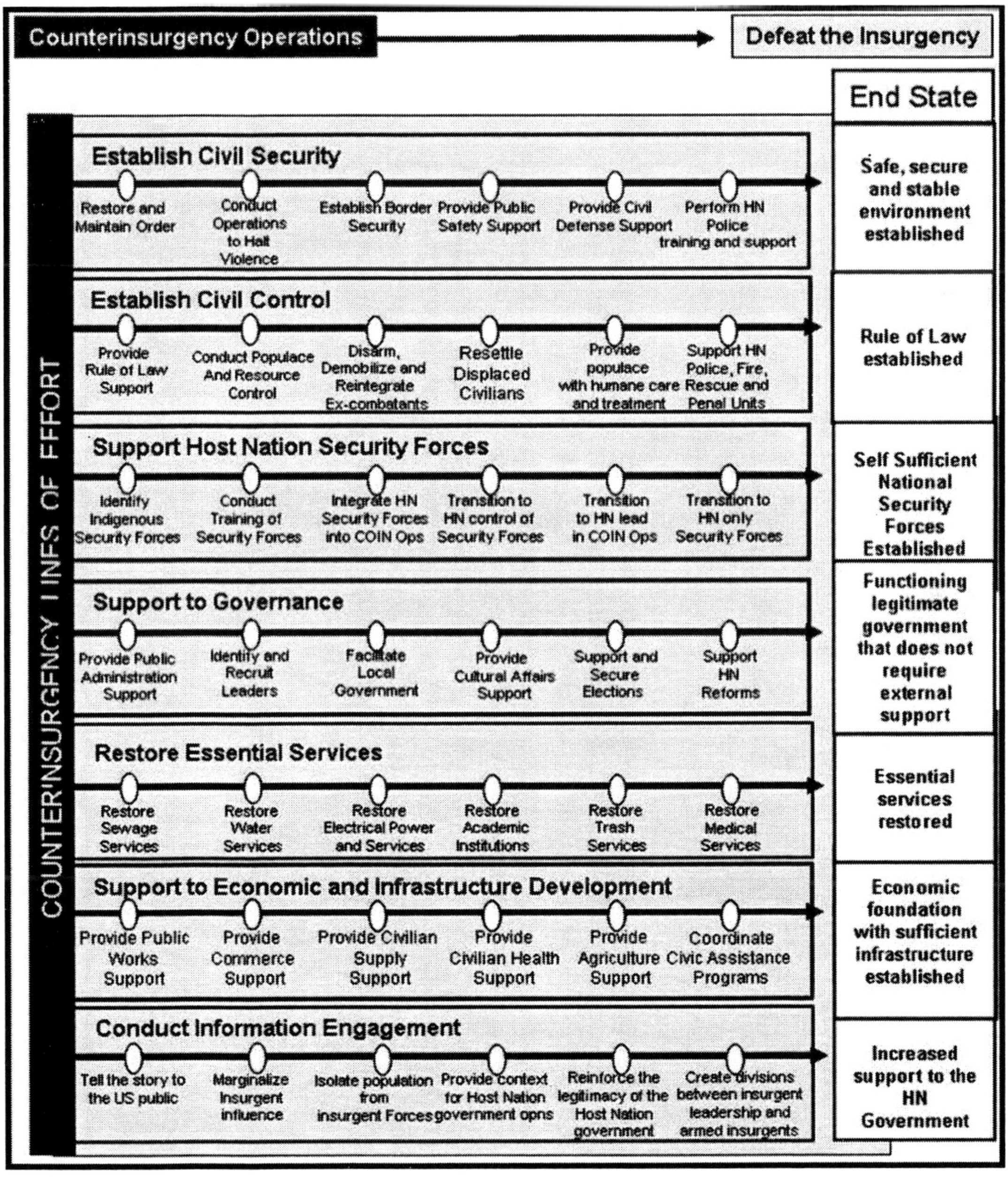


Figure 4-1. Lines of effort.

### Restore Essential Services

Army forces establish or restore the basic services that address the life support needs of the HN population and protect them until a civil authority or the Host Nation can provide them (FM 3-0). The counterinsurgent force works toward meeting the population's basic needs. Ideally, the military is simply providing the security for the humanitarian assistance by supporting other government, intergovernmental, and Host Nation agencies. When the Host Nation or other agency cannot perform its role, Army forces may provide the basics directly. Typical "restore essential services tasks include—

- Provide essential civil services.
- Tasks related to civilian dislocation.
- Support famine prevention and emergency food relief programs.
- Support public health programs.
- Support education programs.

### Support to Economic and Infrastructure Development

Military task executed to support the economic sector are critical to sustainable economic development (FM 3-07). The economic development LOE includes both short—and long-term aspects. The short-term aspect concerns immediate problems, such as large-scale unemployment and reestablishing an economy. The long-term aspect involves stimulating indigenous, robust, and broad economic activity. The stability a nation enjoys is often related to its people's economic situation and its adherence to the rule of law. However, a nation's economic health also depends on its government's ability to continuously secure its population. Support to economic and infrastructure development helps a Host Nation develop capability and capacity in these areas. It may involve direct and indirect military assistance to local, regional, and national entities. Typical support economic and infrastructure development tasks include—

- Support economic generation and enterprise creation.
- Support public sector investment programs.
- Support private sector development.
- Support agricultural development programs.
- Restore transportation infrastructure.
- Restore telecommunications infrastructure.
- Support general infrastructure reconstruction programs.
- Use money as a weapon system.

### Conduct Information Engagement

This is the integrated employment of public affairs to inform US and friendly audiences; psychological operations, combat camera, US Government strategic communication and defense support to public diplomacy, and other means necessary to influence foreign audiences; and, leader and Soldier engagements to support both efforts (FM 3-0). Information engagements are part of all military operations in an area. All leaders must be directly involved in conducting information engagements. These shape the information environment in three ways: by informing audiences using factual information (with public affairs), by influencing selected audiences in the area of operations (with psychological operations and face-to-face meetings), and most important, by persuading audiences through the actions of Soldiers.

*Considerations for Constructing an Information Engagement Plan*

Effective information engagement plans can rally popular support for counterinsurgent efforts, marginalize insurgent violent actions, and neutralize the insurgency's political decision-making capabilities. Everything that a tactical unit does or fails to do in COIN plays a part in its information engagement plan. Therefore, information engagements must be synchronized and nested through common multiechelon themes. Themes must be tailored to different target audience by the manner in which they are presented or distributed. In COIN, perception is often more important than truth.

***Successful Planning Characteristics***

To effectively craft an effective information engagement plan, tactical units must ensure that it is decentralized to the lowest level, executed in a timely manner, account for HN culture, be planned with HN counterparts, and easily be combined into all COIN operations. The commander should take an active interest in his information engagement plan. He does this through his information engagement intent, emphasis to his staff and soldiers, and assignment of information engagement responsibilities. The staff should be integrated into the information engagement plan and dedicated to information engagement success. Higher headquarters should prevent information fratricide by subordinate units through integrating mechanisms, which most often comes through daily or weekly synchronization meetings.

***Themes***

Tactical units most often implement their information engagement plans through the construct of themes that address target audiences. Themes may include simple messages that can easily be repeated (the "bumper sticker") and are often referred to as "talking points." Successful themes are those that –

- Keep objectives simple and achievable.
- Manage expectations and perceptions of the HN populace. Keep counterinsurgent goals attainable.
- Use previous HN contacts and relationships.
- Can be distributed using formal and informal methods by Soldiers at all levels. Formal methods include posting public notices, briefings to community leaders, and talking to local media. Informal methods include patrols talking to local citizenry, handing out leaflets, and using loudspeakers.
- Respond to events and insurgent propaganda rapidly, admit mistakes, and deals with the populace honestly.

***Target Audiences***

In COIN, themes must be tailored to resonate with the target audience. The tailoring process should take into account cultural understanding. In addition, audiences can be separated as far down as neighborhoods, ethnicity, religion, class, and other factors. The audience may include the US civilian population, international forums, US soldiers, HN population, and insurgents.

***HN Population***

Themes that target the HN populace should focus on rallying them against insurgent activities and highlight successes of the HN government and HN security forces. Both themes will reinforce legitimacy. These themes generally build cohesiveness among the population and counter insurgent divisive themes. Cohesive themes highlight—

- Commonality of ultimate goals.
- Commonality of labor, economic, and material problems.
- Ability to separate facts from perceptions and solve important problems.
- Ethnic similarities and common origins.
- Religion and social similarities.
- Traditional or historical evidence of unity.
- Failure of traditional solutions accompanied by awareness of a need for new solutions.
- Patriotism and nationalism.
- Ability to provide information to the government without fear of reprisal.

***Insurgent***

Insurgents manage perceptions through propaganda and disinformation. Rarely do they have to tell the truth. In technologically advanced insurgencies, filming an attack may be as important as the attack itself. Therefore, themes aimed at insurgents must divide insurgent groups into smaller groups and separate the insurgents from the population. Also, themes should counter insurgent propaganda. In many successful information engagement plans, reconciliation and amnesty for certain insurgents have been key components.

- Divisive theme categories.
  - Political, social, economic, and ideological differences among elements of the insurgents.
  - Leadership rivalries within the insurgent movement.
  - Portrait of insurgents as criminals, inept, and counter to the goals of the HN populace.
  - Danger of betrayal from among the insurgents.
  - Harsh living conditions of insurgents.
  - Selfish motivation of opportunists supporting the insurgents.
  - Separate foreign fighters from domestic groups using patriotism and nationalism, when appropriate.
  - Separate public from the perception that insurgents are noble.
  - Public outcry.
  - Reconciliation or amnesty.
- Countering insurgent propaganda
  - Proactive and designed to neutralize the psychological impact of future attacks.
  - Acknowledge the possibility, even likelihood, of future attacks.

- Criminal attacks against HN population.
- Foreign ties and leadership.
- Highlight Insurgent ineptitude.

*Public Affairs*

This is a commander's responsibility. Its purpose is to proactively inform and educate through public information, command information, and direct community engagement. Public affairs have a statutory responsibility to factually and accurately inform various publics without intent to propagandize or manipulate public opinion. Public affairs facilitates a commander's obligation to support informed US citizenry, US Government decision makers, and as tactical requirements may dictate non-US audiences. Public affairs and other information engagement tasks must be synchronized to ensure consistency, command credibility, and operations security as part of the planning process. (For more information, see JP 3-61, AR 360-1, FM 46-1 and FM 3-61.1.)

***Media***

In a COIN environment, the most difficult portion of information engagement—when necessary— involves dealing with the media. Media contacts normally should be handled by the appropriate public affairs officer (PAO). However, this is not always possible, and silence is not always the best solution. Refusal to speak with accredited members of the media may create strong negative impressions with strategic implications.

***Best Practices***

- Stress the human aspects of a story, including the impact of opposing operations on people, with which readers, viewers, or listeners can identify.
- Point out the needs of the unfortunate, and the fact that both Soldiers and HN counterparts are working to address those needs.
- If you do not know the answer, try to get it and then either inform the reporter, or refer the reporter to another source. This may establish your team as a helpful source and develop a relationship that can help ensure future balanced coverage.
- Resist the temptation to attack other groups or organizations, and avoid committing information fratricide.
- If questioned about another agency's activities, refer to that agency for comment. *Never* speak for other organizations.
- Keep trusted reporters and editors who cover the AO about significant activities there.
- Answer media inquiries promptly, accurately, and courteously.
- Learn reporters' deadlines and use them to the friendly force's advantage.
- Encourage media to see what HN and US Soldiers are doing.
- Avoid reacting emotionally to skepticism or hostility.
- Discuss issues calmly.
- Use facts to back up statements.
- Stay focused on the mission.
- Follow the policies of higher HQ and PAO for media interviews.

- Remain friendly, yet professional.
- Use reporters' names.
- Use clear, understandable language.
- Be prepared. Anticipate questions and think about various responses.
- Get to know the interviewers.
- Research the media's organizations and views.
- Find out how the reporter previously conducted interviews.
- Emphasize the interests of the local nationals or other beneficiaries of the mission.
- Avoid speculation.
- State only facts that can be verified.
- Quote statistics with care, as data is easily repurposed.
- Avoid repeating questions, especially those with incorrect or inflammatory language, as this could easily be misquoted or taken out of context.
- Refuse "off-the-record" discussions—there is no such thing!
- Avoid saying "No comment," which can make you sound evasive.
- Stay objective. Save personal opinions and beliefs for a more appropriate time and place.

*Psychological Operations*

These operations convey selected information and indicators to foreign audiences to influence their emotions, motives, objective reasoning, and ultimately the behavior of foreign governments, organizations, groups, and individuals. The purpose of psychological operations is to induce or reinforce foreign attitudes and behavior favorable to the originator's objectives (JP 1-02). Psychological operations can be directed at the civilian populace as a whole, specific groups, or individuals outside the United States and its territories. Psychological operations can influence and modify the behavior of foreign target audiences in support of US objectives. In a COIN operation, PSYOP seeks to increase popular support for the Host Nation government. Typically, PSYOP directs its messages at the populace. However, psychological operations may focus on convincing enemy fighters to surrender rather than risk destruction. Therefore, these capabilities may be integrated into counterinsurgency targeting. Psychological operations units may also be task-organized with maneuver forces.

***Population***

The PSYOP message to the population in COIN has three key facets:

- Obtain buy-in by COIN force for actions that affect the populace, such as—
    - Control measures.
    - Census.
- Win over passive or neutral people by showing HN legitimacy and commitment.
- Encourage locals to provide information about the insurgency to US or HN security forces.

***Insurgents***

The PSYOP message to the insurgents has three key facets:

- Divide insurgent leaders and guerrillas by emphasizing differences
  - In ideology within the insurgency.
  - In the degree of sacrifice required by different groups.
- Divide insurgents and mass base by emphasizing
  - Failures of the insurgency.
  - Successes of the government.
- Create a means for insurgents to abandon the movement and return to the society.

*Leader and Soldier Engagement*

The actions of leaders and Soldiers are the most powerful components of information engagement. Visible actions coordinated with carefully chosen, truthful words influence audiences more than either does alone. Local and regional audiences as well as adversaries compare the friendly force's message with its actions. Face-to-face interaction by leaders and Soldiers strongly influences the perceptions of the local populace. Meetings conducted by leaders with key communicators, civilian leaders, or others whose perceptions, decisions, and actions will affect mission accomplishment can be critical to mission success. These meetings provide the most convincing venue for conveying positive information, assuaging fears, and refuting rumors, lies, and misinformation.

*Techniques*

At the tactical level, there are three primary means a unit can use to disseminate its message: word of mouth, announcements, and town hall meetings.

***Word of Mouth***

This is the most basic form of sending and receiving information, and ultimately the form that every other method will become. Word of mouth is the quickest, most common, most inaccurate and most uncontrollable means of disseminating information; but it may be the best way to send a message. Rumors, spins, casual conversations and dinner-table discussions—whatever form they take, word of mouth travels like wild fire. It spreads out of control and the story grows and changes with each conversation. Everything a unit does is observed and discussed by the locals and spun by the enemy. Units should be prepared to counter false information. Patrols must interact with the populace. They must listen for rumors and correct the ones they hear, but do not waste time arguing about them. Units must spread a positive image and reinforce the good things the government is doing in the area to help the populace.

***Announcements***

Both written and verbal announcements are quick and controlled means of sending messages. Flyers, loud speakers and public speaking are useful ways of informing the populace of progress, incentive programs, civil projects and operations.

***Town Hall Meetings***

Town Hall meetings are an effective means of discussing points and counter points to coalition or government presence, operations and the unifying message. They tend to draw the people that are most interested in the issues and have the most legitimate grievances. This type of meeting also exposes the leaders of the community and the general opinion of the locals. These are planned meetings with an open forum. Units should exercise caution and not allow themselves to get trapped in arguments that take them off their message. Keep in mind that it is an open forum and the insurgents will ensure that people sympathetic to their cause are present and are fighting for their interests in the political arena.

## Section III—MDMP and TLP

The Army has two tactical planning processes: the military decision-making process (MDMP) and troop-leading procedures (TLP). The MDMP is more appropriate for headquarters with staffs; it provides a logical sequence of decisions and interactions between the commander and staff for developing estimates and effective plans orders. Leaders at company level and below use TLP to plan and prepare for an operation. The preferred method of planning is reverse planning. The commander visualizes and describes the end state and the staff plans from the operations end state, working backward in time. Counterinsurgency can create situations where the end state serves as a distant aiming point for a tactical objective. Over time and through continual assessment, the end state and conditions that define mission success become more clear. Forward planning involves starting with the present conditions and laying out potential decisions and actions forward in time, identifying the next feasible step, and the next after that. The counterinsurgent must effectively combine the two methods of planning to achieve both the immediate objective and those, which must be accomplished over time.

### *Military Decision-Making Process*

The military decision-making process is a planning model that establishes procedures for analyzing a mission; developing, analyzing, and comparing courses of action against measures of effectives and performance and each other; selecting the optimum course of action; and producing a plan or order. The MDMP helps organize the thought process of commanders and staff. It is the process used by tactical counterinsurgent forces to organize large amounts of information orchestrate the appropriate sequence of action to defeat the insurgency. The process of developing long-range plans uses the military decision-making process. The following are some key points in the development of the plan.

- Perform IPB (incorporate enemy and population throughout the process).
- Analyze the mission using the operational variables (PMESII-PT).
- Analyze the mission using the mission variables (METT-TC).
  - Use ASCOPE to understand `the civil considerations.
  - If not clear from higher, determine the problem. This may be an iterative process.
- Determine end state and conditions along nested LOEs.
- Determine objectives.
  - Specified (directed objectives and missions).

    - Implied (direct approach to insurgents, indirect approach to insurgents addressing prerequisites and root causes, supporting higher HQs objectives and end state).
- Organize objectives along LOEs and adjust LOEs and further define conditions.
- Identify a potentially decisive line of effort.
- Refine each objective and develop—
    - Decisive points.
    - MOEs.
    - MOPs.
    - Supporting objectives.
    - Targets.
- Frame the time and resources for near-term planning—
    - Events
    - Ability to predict or analyze
    - Combat power
    - Operating tempo
    - Phasing or timing.
- Pay attention to balanced development across all lines of effort.
- Prioritize supporting objectives and targets to resources.
- Develop order with tasks and ISR plan.
- Make current operations plans.
- Execute near-term missions.
- Assess current operations, new intelligence or new missions.
- Use this assessment to drive the process through another cycle.

Commanders modify the process as needed to fit the situation. While typical modification of the MDMP is driven by the need to save time, the counterinsurgent unit has two key reasons to modify the process:

- The end product of long-range planning is not an operations order that is prepared for execution in the traditional sense. The long-range plan provides a focus for short-range planning and ensures that the operations currently being conducted are having the desired effect on the end state.
- Each planning session within a counterinsurgency is interrelated to the planning that preceded it. Units will conduct a thorough MDMP initially and then build continuously on those products and their situational understanding as they progress through successive planning cycles. Units throughout their rotation continue to "learn and adapt." Through assessment, the commander will focus his staff on areas that require further detailed analysis and limit effort on areas where sufficient analysis has already been conducted.

This analysis is captured in the staff's running estimate. A running estimate is a staff section's continuous assessment of current and future operations to determine if the current operation is proceeding according to the commander's intent and if future operations are supportable. This, along with the commander's assessment and those of subordinate commanders, build a growing understanding of the unit's operations area and allows the unit to modify the MDMP, not out of necessity of time but through the ability to rapidly and accurately define problems and solutions.

Insurgencies are inherently complex and dynamic; they cannot be fully understood through a single cycle of the MDMP. It is the cumulative effect of analysis and planning that builds and refines knowledge of a unit's area of operations, captured in the running estimate over the course of many successive planning cycles that allows the successful defeat of the insurgency.

By developing a comprehensive and iterative plan the counterinsurgent force follows the conventional MDMP steps. However, fighting an insurgency, or more precisely conducting counterinsurgency operations, is not a conventional military operation. As such, the operational environment of counterinsurgency dictates some unique considerations to the conventional MDMP steps. These considerations apply primarily to four MDMP steps—receive the mission, mission analysis, COA development and COA analysis. Figure 4-2 highlights considerations in a COIN MDMP.

| *Input* | *Steps* | *Output* | *Coin* |
|---|---|---|---|
| Mission received from higher or developed internally * | **Step 1: Receipt of mission** | Commander's initial guidance *<br>WARNORD | Mission developed internally |
| | | **WARNORD** | |
| Higher HQ's order/plan<br>Higher HQ's IPB<br>Running estimates | **Step 2: Mission analysis** | Restated mission *<br>Initial commander's intent and planning guidance *<br>Initial CCIR *<br>Updated running estimates<br>Initial IPB products<br>Initial ISR plan<br>Preliminary movement | Use PMESII-PT to analyze the AO<br>Use the mission variables and civil considerations to develop better understanding of the AO<br>Account for HN security forces, multinational, paramilitary, and all US forces and agencies<br>Account for multiple enemy, multinational, and paramilitary forces |
| | | **WARNORD** | |
| Restated mission *<br>Initial commander's intent, planning guidance, and CCIR *<br>Updated running estimates<br>Initial PPB products | **Step 3: COA Development** | Updated running estimates and products<br>COA statements and sketches<br>Refined commander's intent and planning guidance * | Use components and manifestations of the insurgency in IPB<br>Develop COAs for HN security forces |
| Refined commander's intent and planning guidance<br>Enemy COAs<br>COA statements and sketches | **Step 4: COA Analysis (Wargame)** | War-game results<br>Decision support templates<br>Task organization<br>Mission to subordinate units<br>Recommended CCIR | War-game with four groups:<br>Enemy/population<br>Host Nation/US COIN forces<br>Modeling may be an appropriate wargaming tool |
| War-game results<br>Criteria for comparison | **Step 5: COA Comparison** | Decision matrix * | |
| Decision matrix | **Step 6: COA Approval** | Approved COA *<br>Refined commander's intent *<br>Refined CCIR *<br>High payoff target list * | |
| | | **WARNORD** | |
| Approved COA<br>Refined commander's intent and guidance<br>Refined CCIR | **Step 7: Orders Production** | OPLAN/OPORD | Ensure balance between LOEs<br>Ensure unity of effort<br>Using two languages increases the time required to produce and brief the order, and to conduct combined rehearsals |

* Commander's activity or decision

Figure 4-2. COIN MDMP.

*Step 1: Receipt of Mission*

As with any military operation, the MDMP begins upon receipt of an order, upon anticipating a change in mission or upon seizing an opportunity in the AO. As in conventional operations, the MDMP steps during this phase—alert the staff, gather the tools, update running estimates, perform an initial assessment, issue the initial guidance, and issue the initial warning order—are largely unchanged. Units should account for additional time when working with HN security forces and continue to refine their METT-TC analysis, especially the civil considerations using ASCOPE.

During receipt of the mission, two basic considerations apply. First, the strategic mission of defeating the insurgency applies to all levels at all times. This means that all operational or tactical missions, activities and tasks are nested within this mission and can have a direct, lasting impact on its success or failure. Second, it is here that the commander begins to visualize the operation. It is critical for the commander to create a comprehensive vision of the end state. From this desired end state, the staff develops the lines of effort that support creating the commander's vision. All lines of effort are focused on their own end states and those end states are nested with the commander's overall end state for the operation, which is nested with the overall mission of defeating the insurgency.

*Step 2: Mission Analysis*

The seventeen tasks within mission analysis differ little from those for conventional operations. For COIN, the staff considers the following factors:

***Analyze Higher Headquarters' Order***

In addition to analyzing the information the order, the staff must use the operational and mission variables to analyze the AO. Staff members pay special attention to civil considerations within the AO, using ASCOPE.

***Perform Initial IPB***

(See Appendix A for a detailed discussion of IPB in a COIN environment.)

***Determine Specified, Implied, and Essential Tasks***

The staff may find identifying the implied tasks in the ambiguous and complex COIN environment more difficult than the task would be for a conventional operation. They may have to conduct the MDMP many times as situations change.

***Review Available Assets***

The staff must remember to account for all HN security forces and for US, coalition, and Host Nation enablers.

***Determine Constraints***

The staff reviews the ROE.

***Identify Critical Facts and Assumptions***

The staff ensures that facts and assumptions include US, HN security forces, population groups as well as the enemy.

***Perform Risk Assessment***
(Same as for conventional operations; see FM 5-19.)

***Determine Initial CCIR (PIR and FFIR) and EEFI***
(No change from conventional.)

***Determine Initial ISR Plan***
The staff considers all HN security force assets.

***Update Operational Timeline***
The staff may decide to use the higher, operational, planning, enemy, and populace (HOPE-P) construct to develop the timeline.

***Restate Mission***
Although it is challenging to do for a COIN environment, the staff must use tactical mission task(s) as the verb(s) in the mission statement.

***Deliver Mission Analysis Briefing***
(No change.)

***Approve Restated Mission***
(No change.)

***Develop Initial Commander's Intent***
(No change.)

***Issue Commander's Planning Guidance***
(No change.)

***Issue Warning Order***
The staff must account for additional time needed to translate and brief a WARNORD in two languages.

***Review Facts and Assumptions***
(No change.)

*Step 3: COA Development*

The staff develops a COA to determine one or more ways to accomplish the mission. For all types of plans, the brigade and battalion must construct a COA that accounts for tasks across all seven COIN LOEs. In a time-constrained environment, leaders may be only able to craft one COA. The six steps of COA development remain the same, however the staff must—

***Analyze Relative Combat Power***
Combat power is the total means of destructive, constructive, and information capabilities that a military unit or formation can apply at a given time (FM 3-0). The staff compares

friendly combat power, including that of the Host Nation, all enablers, and insurgent combat power. The staff must also account for pro and anti government populations.

***Generate Options***

The staff ensures that they cover all options, including additional LOEs and HN actions.

***Array Initial Forces***

The staff arrays all enemy forces such as different insurgent groups as well as all friendly forces, to include Host Nation security forces. Small units may also have to develop plans to employ special teams such as military working dog teams, tactical site exploitation teams, UASs or attack, reconnaissance, and assault aviation.

***Develop the Concept of Operation***

The staff describes the operation from start to finish. Extremely important is construction of the battlefield framework to include, for example, objectives, phase lines, building numbers, and target reference points. It is important to establish this early to ensure subordinate units doing parallel planning use the correct graphic control measures.

***Assign Headquarters***

The staff assigns headquarters, including HN security, joint and coalition forces.

***Prepare COA Statement and Sketches***

The S3 develops statements and sketches that show how the unit will accomplish the mission and explains the concept of operation.

*Step 4: COA Analysis (Wargame)*

It is during COA analysis (wargaming) that the most significant and unique considerations apply. COA analysis (wargaming) includes rules and steps that help commanders and staffs visualize the flow of a battle. Wargaming focuses the staff's attention on each phase of the operation in a logical sequence. It is an iterative process of action, reaction, and counteraction. However, the standard wargaming methods of the belt, avenue-in-depth or box method may not capture the complexity of the COIN environment that commanders and planners must have to conduct successful counterinsurgency operations. The box method is the traditional method that comes the closest to facilitating COIN wargaming. The traditional wargaming process is shown in Figure 4-3.

To wargame an operation, the staff visualizes it from start to finish, evaluating the enemy's actions, and then modifying their plan. With only one COA, this step is often used to synchronize the plan across the warfighting functions. Analyzing a COA (modeling it) gives the staff a deeper understanding of the complex and adaptive OE of COIN, and how particular actions will affect it.

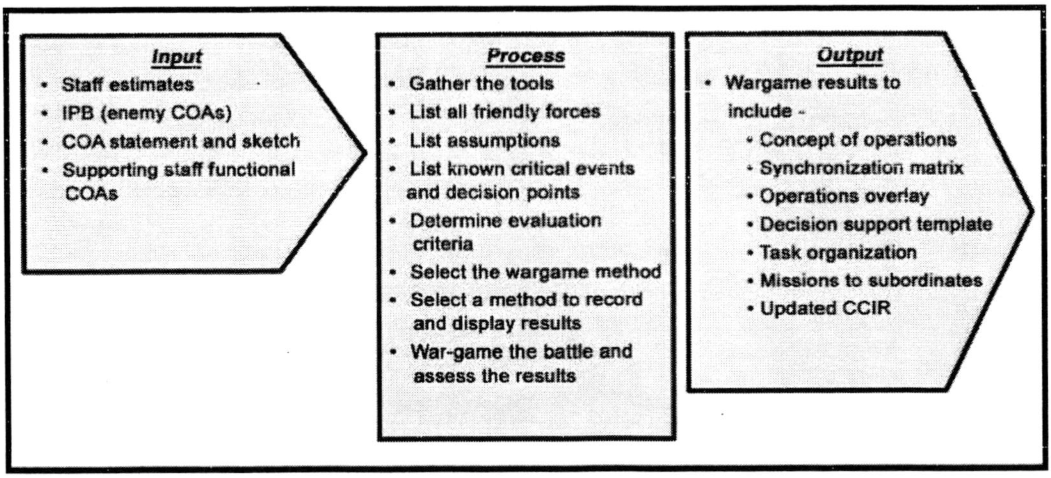


Figure 4-3. Course of action analysis (wargame).

***Eight Steps of COA Analysis (Modeling)***

Conduct the model using the same eight steps as laid out in FM 5-0 for the wargaming process— gather the tools, list friendly forces, list assumptions, list critical events, determine evaluation criteria, select the wargame method, select a method to record, and war-game the operation. All of these steps are still valid and essential. Within some steps, there are modifications that must occur to successfully model.

Different groups may conduct specific LOE modeling sessions prior to a larger group conducting a larger scale modeling session. For some LOE, the wargaming method may be more appropriate. Unlike wargaming, modeling is not a one-time event. Modeling sessions should be rerun as operations, events and the operational environment changes.

Steps one through four remain unchanged in this technique. Step five, determining evaluation criteria, can be complicated. This is where the planners develop the operation's measures of effectiveness and measures of performance. These measures are derived from the commander's end state and the conditions leading to the end state. These measures show when the unit is successful and achieves the commander's end state. Additionally, the measures should also tell commanders when they are gaining the ability to influence the system.

The number of insurgent attacks can be a misleading indicator. Measures of effectiveness are more accurate when there are multiple supporting measures that look at similar issues with the intent of seeing the whole. In a counterinsurgency, the three major areas to measure are the strength of the government, the support of the people, and the strength of the insurgency. By measuring these three areas through a variety of quantifiable measurements, such as the ability to complete an essential service, the number of tips reported to the police, and the volume and type of insurgent propaganda, the commander should have a good feel for success and workable measures of effectiveness.

For step seven, select a method to record and display the results; the recording method must capture the depth of the lessons that will be gained from the modeling and there must be a method to refine certain critical products. There are four areas where the planners should focus on capturing new information.

1. The planners must capture any greater depth of understanding for each of the variables.
2. They must record any new understanding of the variables and how they relate.
3. They must update the rules of behavior for each variable.
4. They must update the collection matrix.

An expanded synchronization matrix (Figure 4-4) captures the actions of US forces, HN forces and the insurgents using the warfighting functions (WFFs). It also shows the population groups across the political, economic, and social categories. Using a synch matrix to show these factors can reveal the complexity of the operational environment over time. Based on their deeper understanding of the system, new critical information requirements will arise that must be addressed in the collection matrix.

The last step, war-game the operation and assess the results, is the modeling session itself. The focus of the modeling method is to gain a better understanding of the complex system that makes up an insurgency and to accurately forecast how actions affect the system. Like wargaming, there are many different techniques to conduct the actual modeling session. For the purposes of this manual, the turn-based technique using four groups is used.

| ***Units*** | ***Week 1*** | ***Week 2*** | ***Week 3*** | ***Week 4*** |
|---|---|---|---|---|
| **US COIN Forces** | | | | |
| Movement and Maneuver | | | | |
| Fires | | | | |
| Intelligence | | | | |
| Sustainment | | | | |
| C2 | | | | |
| Protection | | | | |
| **HN COIN Forces** | | | | |
| Movement and Maneuver | | | | |
| Fires | | | | |
| Intelligence | | | | |
| Sustainment | | | | |
| C2 | | | | |
| Protection | | | | |
| **Population** | | | | |
| Political | | | | |
| Military | | | | |
| Economic | | | | |
| Social | | | | |
| **Insurgent** | | | | |
| Movement and Maneuver | | | | |
| Fires | | | | |
| Intelligence | | | | |
| Sustainment | | | | |
| C2 | | | | |
| Protection | | | | |

Figure 4-4. Example expanded synch(ronization) matrix.

***Modeling Technique***

During each turn, each group states their actions followed by the next group's actions. The four groups are the US COIN forces, the Host Nation (includes other friendly forces), all neutral elements (IPI, the population, NGOs, and so on), and the insurgents. Before beginning, the group must clearly explain the concept, rules, and limitations for the modeling session to everyone, to include specific guidance to key individuals. Two key parts of COIN modeling are modeling the individual LOE and modeling the overall area or sector affected. In a typical staff, the S-3 plays the role of US and HN forces, the S-9 plays the neutral elements, and the S-2 plays the role of the insurgents. The XO facilitates. The staff plays its respective parts as accurately as possible portray their group in the coming modeling session.

Like with wargames, the staff needs a visual reference to conduct the modeling process. This may be a sand table, a map, or some other visual board. Most of the items that compose this visual board come out of the mission analysis and must include more than just the hills and roads and other normal military aspects of terrain. For the model to be successful, it has to set the stage for the other planners to be able to see the world through their group's eyes. That means the visual board should include significant cultural sites and key social infrastructure sites. The more detailed the board, the more accurate and meaningful the modeling will be.

As operations unfold and more intelligence is gathered, the planners should update the rules of behavior and capabilities for each group. As the knowledge of the individual groups and the system as a whole is improved, the planners will be able to identify how to more effectively influence the system. This planning tool does not stop when the operation order is issued. Modeling is an effective way of understanding the counterinsurgency system and ongoing operations.

*Step 5: COA Comparison*

Staffs analyze and evaluate the advantages and disadvantages of each COA, as well as identifying the one with the highest probability of success against the most likely enemy COA. Additional COIN evaluation considerations may include—

- Effect on or role of the HN government.
- Effect on or role of the HN security forces.
- Effect on groups of the population.
- Impact on information engagement.

*Step 6: COA Approval*

Either after participating in the COA comparison process or upon receipt of a COA recommended decision, the commander makes a COA decision by approving a COA to execute.

*Step 7: Orders Production*

Units in COIN operations issue a warning order or WARNORD, and, after the commander selects a COA, issue a written operation order. However, additional time for orders production may be needed if the order must be translated into another language. To ensure nothing is lost in the translation, and ensure success in multinational operations, the staff must keep orders and rehearsals simple.

### *Troop-Leading Procedures*

Troop-leading procedures (TLPs) give small-unit leaders a dynamic framework for analyzing, planning, and preparing for an operation.

TLPs in a COIN environment differ little from those described in FM 5-0. The next sections will provide a brief description of the eight steps along with some additional considerations for COIN operations. Figure 4-5 shows addition COIN considerations during TLPs.

#### *Step 1: Receive Mission*

Receipt of a mission may occur with a written or verbal WARNORD, FRAGO, or OPORD. Due to the ambiguity in COIN operations, companies or platoons often start planning after identifying an insurgent vulnerability, receiving HUMINT, or after having a conversation with a local leader.

Upon receipt of a mission, leaders must rapidly assess the situation using METT-TC. Leaders must provide additional emphasis on analyzing civil considerations by using the ASCOPE method.

In a time constrained situation, an ongoing assessment of the AO using the operational variables (PMESII-PT) should provide relevant information to accomplish the mission. Based on current information, leaders estimate and allocate 1/3 of available time to plan and prepare for the mission, including time to accommodate the language barrier with HN security forces; and the remaining 2/3 of the available time for subordinates' planning and rehearsals. This remains relevant in a COIN environment.

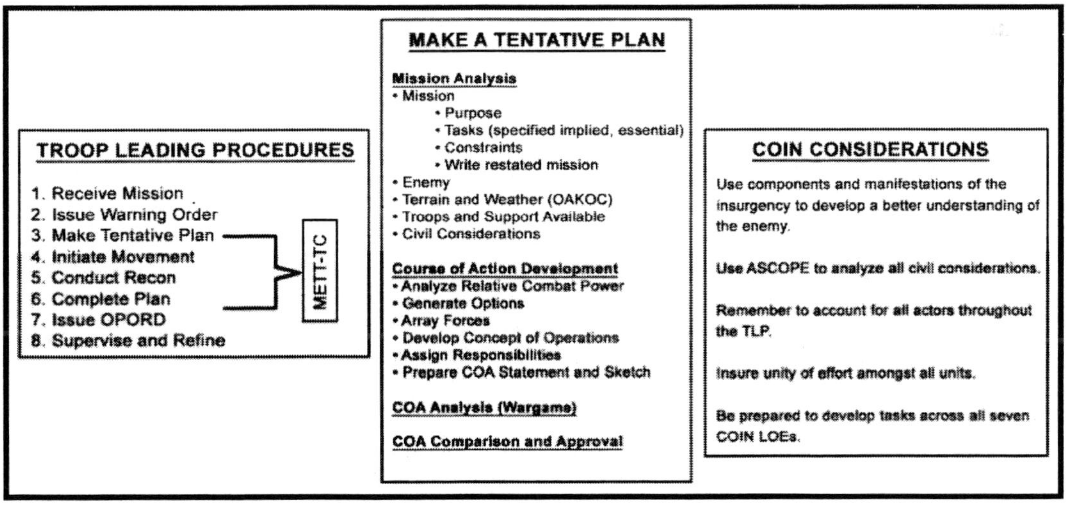


Figure 4-5. TLPs in COIN.

*Step 2: Issue Warning Order*

Once done with the initial assessment, leaders issue the best possible WARNORDs, often called WARNORD 1, with the information at hand. They also draw relevant information from their ongoing assessment of the AO using the operational variables (PMESII-PT). Typically, three WARNORDs are issued. Each builds upon the previous one. Normally, a WARNORD consists of—

- Mission.
- Time and place of the OPORD.
- Units participating in the operation.
- Specific tasks.
- Time line of the operation.

*Step 3: Make a Tentative Plan*

Once the WARNORD is issued, leaders build a tentative plan by combining mission analysis, COA development, COA analysis, COA comparison, and COA approval. Within companies and platoons, these steps are less structured and are often done mentally by the leader.

***Mission Analysis***

Leaders use the METT-TC format to conduct mission analysis as they expand on the hasty analysis they conducted in step one. After mission analysis is complete, an updated WARNORD, WARNORD 2 is issued. In a COIN environment, special consideration must be given to—

***Mission***

Units create their restated mission statement after analyzing higher headquarters' mission or analyzing their particular problem. It may take multiple times of conducting mission analysis to fully grasp the problem.

***Enemy***

Units use the tools from Chapter 2 (elements of an insurgency, dynamics of an insurgency, the six insurgent strategies, insurgent tactics and the strengths and vulnerabilities of insurgents) to help further identify the group or groups of insurgents they may encounter. Then, define composition, disposition, recent activities, most probable COA, and most dangerous COA.

***Terrain and Weather***

No significant changes from FM 5-0.

***Troops and Support Available***

Leaders must determine the amount and type of friendly forces available, especially Host Nation security forces, coalition forces and the nonmilitary enablers.

***Time***

Use the acronym HOPE-P (higher, operational, planning, enemy, and populace) to identify the various timelines.

***Civil Considerations***

Leaders must use the ASCOPE method to analyze areas controlled by various demographic groups to obtain a better understanding of the AO.

***COA Development***

Leaders use COA development to determine one or more ways to accomplish the mission. In a time-constrained environment, leaders may only be able to craft one COA. For long, mid, and short-range plans, companies will need to develop tasks along all lines of effort. The six steps of COA development in COIN are unchanged, but special consideration should be given to—

***Analyze Relative Combat Power***

Leaders compare friendly, including Host Nation, all enablers and insurgent combat power. Leaders may incorporate analysis of pro—and anti—government populations.

***Generate Options***

Options should always consider employment of HN security forces and the HN government.

***Array Forces***

Array all enemy forces, such as different insurgent groups, as well as all friendly forces to include Host Nation security forces. Small units may also have to develop plans to employ special teams such as military working dog teams, tactical site exploitation teams, UASs or attack, reconnaissance, and assault aviation.

***Develop Concept of Operations***

Leaders describe the operation from start to finish. Especially critical is the creation of graphic control measures such as phase lines and building numbers to assist with C2 and clarify roles and tasks. The COA must achieve unity of effort amongst all organizations.

***Assign Responsibilities***

Assign responsibilities for each task including those for HN security forces.

***Prepare a COA Statement and Sketch***

Leaders prepare a COA statement that describes all significant actions from start to finish, as well as a sketch.

***COA Analysis (Wargame)***

To wargame, leaders visualize the operation from start to finish, evaluate the enemy's actions and then modify their plan. With only one COA, often this step is used to synchronize the plan across the WFF.

***COA Comparison and COA Approval***

These two steps are normally done mentally by the leader during TLP.

*Step 4: Initiate Movement*

As in any military operation, leaders may choose to initiate movement necessary to enable mission preparation or position the unit for execution. In COIN, this may mean moving US forces to an HN security force patrol base to conduct combined planning and movement.

*Step 5: Conduct Reconnaissance*

Whenever time and circumstances allow, leaders personally observe the AO for the mission since there is no substitute for first hand information. Leaders may also consider using a map, Joint Land Attack Cruise Missile Defense Elevated Netted Sensor System (JLENs), UAS, or attack reconnaissance aviation to conduct the reconnaissance if time, OPSEC, or other considerations preclude a personal reconnaissance.

*Step 6: Complete Plan*

Leaders incorporate the results of the reconnaissance, update information, refine their products, make coordination with adjacent units, including Host Nation units, and issue the final WARNORD. Extra time may be needed to translate the order into the Host Nation's language.

*Step 7: Issue OPORD*

Leaders issue verbal or written orders to their subordinates, using the standard five-paragraph field order format. In addition, leaders use map, overhead imagery, or terrain model to enhance subordinates' understanding of the mission. Additional time may be required to brief the OPORD in two languages if HN forces are participating in the operation. Units who have experienced OPSEC issues with HN security forces may choose to issue the OPORD to the HN unit just prior to crossing the line of departure.

*Step 8: Supervise and Refine*

Throughout the TLPs, Army leaders monitor preparations, refine the plan, coordinate with adjacent units, and conduct rehearsals. COIN operations use the standard five types of rehearsals. They are the confirmation brief, the back brief, the combined arms rehearsal, the support rehearsal, and the battle drill or SOP rehearsal.

Additional time may be required to conduct rehearsals in two languages. Units who have experienced operational security (OPSEC) issues with HN security forces may choose to have HN forces conduct generic rehearsals, for example, clear a room or establish checkpoint versus the actual complete operation.

## Section IV—Targeting Insurgents

In major combat operations, targeting focuses on identifying capabilities or resources the maneuver commander must have effects for his operation to succeed, then attacking them with fire support, aviation, and close air support. During counterinsurgency operations, the role of targeting can include a broad range of both enemy and stability targets that use all seven COIN Lines of Effort to attack or influence them.

### *Role in COIN*

The expanded use of targeting in COIN gives the planner two key benefits: prioritization and synchronization. Because the counterinsurgent often faces multiple objectives, many without clear positional references, friendly courses of action may not be easily represented in terms of various forms of maneuver. Each may instead portray a series of targets that the unit must influence over a specified time.

The targeting process focuses operations and the use of limited assets and time. Commanders and staffs use the targeting process to achieve effects that support the objectives and missions during counterinsurgency operations. It is important to understand that targeting is done for all operations, not just attacks against insurgent. The targeting process can support PSYOP, civil-military operations, and even meetings between commanders and Host Nation leaders, based on the commander's desires.

The synchronization of RSTA/ISR assets with available combat power is the greatest contribution targeting brings to the counterinsurgent. The indirect nature of irregular warfare often creates limited opportunities to strike or influence targets. Integrating the intelligence process and the operations process through targeting can be used to ensure that maneuver units strike the right targets at the right time or are prepared to strike targets of opportunity.

**Personality Targets**

- Lethal
  - Insurgent leaders to be captured or killed
  - Guerillas
  - Underground members
- Nonlethal
  - People such as community leaders and those insurgents who should be engaged through outreach, negotiations, meetings, and other interaction
  - Insurgent leaders
  - Corrupt host-nation leaders who may have to be replaced

**Area Targets**

- Lethal
  - Insurgent bases or caches
  - Smuggling routes
- Lethal and Nonlethal Mix
  - Populated areas where insurgents commonly operate
  - Populated areas controlled by insurgents where the presence of U.S. or host-nation personnel providing security could undermine support to insurgents
- Nonlethal
  - Areas lacking essential services (SWEAT-MSO) that support the government

Figure 4-6. Lethal and nonlethal targeting.

Effective targeting requires the creation of a targeting board or working group, although in many cases, the targeting staff is similar to the planning staff. It is typically chaired by the XO or fire support coordinator, and includes representatives from across the staff including S-2, S-3, S-5, S-7, S-9, air liaison officer, and staff judge advocate. The goal is to prioritize targets and determine the means of engaging them that best supports the commander's intent and operation plan. The focus of the targeting cell, in a counterinsurgency environment, is to target people, both the insurgents and the population.

Effective targeting identifies the targeting options, both lethal and nonlethal, to achieve effects that support the commander's objectives. Lethal assets are normally employed against targets with operations to capture or kill. Nonlethal assets are normally employed against targets that are best engaged with PSYOP, negotiation, political programs, economic programs, and social programs. Figure 4-6 shows examples of potential targets.

***Counterinsurgency Targeting Process***

The targeting process comprises the four functions of decide, detect, deliver, and assess (D3A). Targeting is critically linked to the MDMP. The decide function derives critical information that develops from mission analysis through course of action approval. Both detect and assess functions are tied to the unit's RSTA/ISR plan which is driven by IPB, the MDMP, and tactical site exploitation. Figure 4-7 shows the links between the targeting cycle and MDMP.

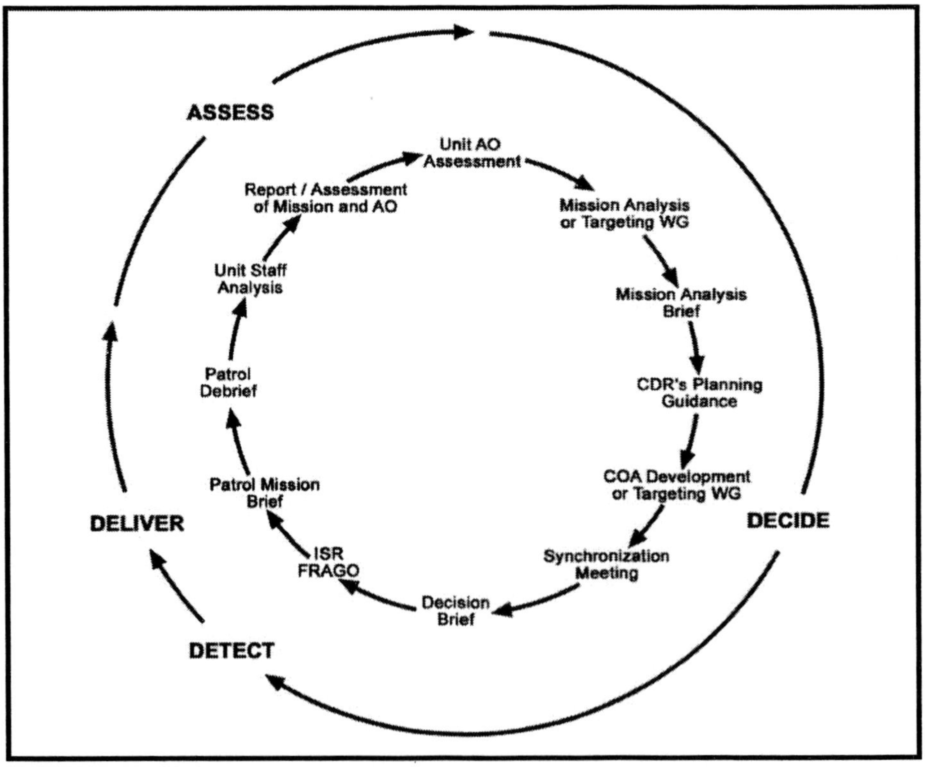


Figure 4-7. Targeting and MDMP.

The commander's guidance drives the targeting process. Commanders issue targeting guidance during *Decide*. Actions during *Detect* may give commanders the intelligence needed to refine their guidance. Target identification may be difficult once a counterinsurgency operation begins. The focus during *Deliver* should be on decisive points that the commanders can engage. Immediately after *delivering*, units *Assess* the effect.

The unit S2 must understand that the targeting process depends on the effective and timely use of the intelligence cycle and on the exploitation of objectives and detainees. Target development against insurgents results from complete and accurate situation development during peacetime, contingency planning, and analysis of the operational environment during conflict. The IPB supports the target development process and provides the commander with the intelligence needed to select valid target nodes to engage.

**Decide**

4-142. This function focuses and sets priorities for intelligence collection and both lethal and nonlethal plans. Intelligence and operations personnel, with the commander and other staff members, decide when a target is developed well enough to engage. Continuous staff integration and regular meetings of the intelligence cell and targeting board enable this function. Specifically, intelligence analysts need to identify individuals and groups to engage as potential counterinsurgency supporters, targets to isolate from the population, and targets to eliminate. During the decide activity, the targeting board produces a prioritized list of targets and a recommended course of action associated with each. Executing targeting decisions may require the operations section to issue fragmentary orders. Each of these orders is a task that should be nested within the higher headquarters' plan and the commander's intent. Targeting decisions may require changing the intelligence synchronization plan. The targeting working group participates in COA analysis and collaboratively develops the following decide function products:

*High-Payoff Target List*

An HPT list is a prioritized list of targets, by phase of the joint operation, whose loss to an enemy will contribute to the success of the mission (JP 3-60)

*Intelligence Synchronization Plan*

This is what the intelligence officer uses, with staff input, to synchronize the entire collection effort, to include all assets the commander controls, assets of lateral units, and higher echelon units, organizations, and intelligence reach to answer the commander's critical information requirements (FM 2-0).

*Target Selection Standards*

Target selection standards establish criteria, including accuracy, that must be met before an attack. For targets to be attacked using lethal means. requirements might include a picture, address, 8-digit grid, and enough evidence to prosecute. For targets to be attacked by nonlethal means, requirements may include background information on an individual, meetings he may attend, and known associates.

*Attack Guidance Matrix*

The attack guidance matrix lists which targets or target sets approved by the commander to act on, how and when to act on them, and the desired effects.

*Target Synchronization Matrix*

The target synchronization matrix combines data from the high-payoff target list, intelligence synchronization plan, and the attack guidance matrix. It lists high-payoff targets by category and the agencies responsible for detecting them, attacking them, and assessing the effects of the attacks.

*Targeting FRAGO*

The targeting FRAGO tasks units to execute the lethal and nonlethal plans.

## Detect

The detect function involves locating HPTs accurately enough to engage them. Targets are detected through the maximum use of all available assets. The S-2 must focus the intelligence acquisition efforts on the designated HPTs and PIR. Situation development information, through detection and tracking, will be accumulated as collection systems satisfy PIR and information requirements. Tracking is an essential element of the "detect" function of the targeting process. Tracking priorities are based on the commander's concept of the operation and targeting priorities. Tracking is executed through the collection plan, since many critical targets move frequently.

Detection at the tactical level is achieved through a variety of means such as a HUMINT source, an anonymous tip, UAS, a combat patrol, SIGINT, DOMEX, rotary wing aircraft, USAF aircraft or military working dog teams. The best means of detecting a target during an insurgency is HUMINT, though. As such, the detect activity requires a detailed understanding of social networks, insurgent networks, insurgent actions, and the community's attitude toward the counterinsurgent forces.

For a target that must be engaged by nonlethal means, the detect function may require patrols to conduct reconnaissance of a leader's home to determine if they are there, an assessment of a potential project, or attendance at a greeting to meet with a leader.

## RSTA and ISR Operations

Reconnaissance, surveillance and target acquisition (RSTA) and intelligence, surveillance, and reconnaissance are activities that synchronize and integrate the planning and operation of sensors, assets, and processing; exploitation; and dissemination systems in direct support of current and future operations. This is an integrated intelligence and operations function (JP 2-0 1). RSTA/ISR provides relevant information about all aspects of the operational environment. RSTA/ISR helps commanders assess the degree to which information superiority is achieved. RSTA/ISR also directly supports information engagements by defining targets in detail and by helping assess the effectiveness of friendly and adversary information engagements.

Reconnaissance and surveillance operations help to confirm or deny insurgent COAs and estimates of guerrilla capabilities and combat effectiveness. Reconnaissance and surveillance also confirm or deny assumptions about the operational environment and threat made during planning.

Doctrine requires the staff to carefully focus RSTA/ISR on the CCIR (PIR and FFIR) and be able to quickly retask units and assets as the situation changes. This ensures that the enemy situation drives RSTA/ISR operations. The S-2 and S-3 play a critical role in this challenging task that is sometimes referred to as "fighting ISR." Through RSTA/ISR, commanders and staffs continuously plan, task, and employ collection assets and forces. They collect, process, and disseminate timely and accurate information, combat information, and intelligence to satisfy the commander's critical information requirements and other intelligence requirements.

RSTA/ISR synchronization analyzes information requirements and intelligence gaps; evaluates available assets both internal and external to the organization; determines gaps in the use of those assets; recommends RSTA/ISR assets controlled by the organization to collect on the CCIR; and submits requests for information for adjacent or higher collection support (FM 3-0). Collection tasks linked to decision points play a critical part in the operation. The reconnaissance and surveillance plan must ensure that information tied to decision points reaches the commander and staff in time to support the decision. The reconnaissance and surveillance plan also ties directly into the targeting process. Collection assets are identified, prioritized, and planned to detect certain targets during the "decide" and "detect" phases of the D3A process. The collector's information, if it meets the established, criteria may trigger the deliver function. After delivery, the collectors, also identified in the "decide" phase, provide information to assess the attack. These may be the same collector or a different asset. If not, a different asset is tasked to do this. One tool to achieve this is a daily RSTA/ISR and operations synchronization meeting, which is discussed in the targeting section of this chapter. The RSTA/ISR synchronization plan is an output of this meeting, and coordinates and synchronizes the RSTA/ISR. (For more, see FM 2-0.)

### Deliver

The deliver function of targeting begins in earnest with execution. The targeting process provides speed and efficiency in the delivery of lethal or nonlethal fires on targets in accordance with the Attack Guidance Matrix or the targeting FRAGO. Within the deliver function, the system or combination of systems selected during the "decide" phase is employed.

For a target that requires lethal means, units may eliminate the target using a joint direct attack munition (JDAM) from a USAF aircraft, an Excalibur round, a Joint Tactical Attack Cruise Missile System (JTACMS), or a sniper. However, many times it is more important to capture the target, so commanders will choose to execute a raid or a cordon and search. For a target that requires nonlethal means, a commander may choose to use an information engagement to convince a local leader, or to conduct a project to garner the population's support through money or employment.

Target exploitation in the counterinsurgency environment is similar to that in law enforcement. An exploitation plan not only facilitates gathering evidence for future court cases, but also may lead to follow-on targets after successful exploitation. (See Chapter 6 for details on tactical site exploitation.)

| TIME | MON | TUE | WED | THU | FRI | SAT | SUN |
|---|---|---|---|---|---|---|---|
| 0600 | Commanders operations / intelligence update | | | | | | |
| 1200 | Daily ISR and operations synchronization meeting | | | | | | |
| 1800 | | | | COA Analysis for targeting | | Publish targeting order | |
| 2100 | Intelligence fusion / synchronization meeting | | | | | | |
| 2400 | Mission Analysis for targeting | | | COA Decision for targeting | | | |

NOTE: Update HPTL as needed.

Figure 4-8. Targeting battle rhythm.

**Assess**

At the tactical level, commanders use assessment to get a series of timely and accurate snapshots of their effect on the insurgent and the population. It provides commanders with an estimate of the insurgent's combat effectiveness, capabilities, and intentions, as well as an accurate understanding of the people. This helps commanders determine when, or if, their targeting efforts have been accomplished. The "assess" phase relies heavily upon MOEs and MOPs.

Producing the assessment is primarily an intelligence responsibility, but requires coordination with operations, civil affairs, public affairs, information operations, and PSYOP to be effective. As part of the targeting process, assessment helps to determine if another engagement of the target is necessary.

***Targeting Battle Rhythm***

During COIN operations, brigades and battalions typically use a one or two week targeting battle rhythm. The target cycle drives the tactical unit's daily and weekly operations. Figure 4-8 shows an example of a targeting cycle and battle rhythm.

Another important facet of the battle rhythm is the daily RSTA/ISR and operations synchronization meeting. Due to the air tasking order cycle, this traditionally synchronizes operations from 96 hours to 24 hours in the preparation to hand over to current operations. In counterinsurgencies, this meeting must synchronize all actions along the seven COIN LOEs. Especially critical is planning and requesting higher-level assets such as air force aircraft, jamming, UASs, ISR, and aviation ahead of the higher headquarters decision cycle. There are three outputs from this meeting—a 96-hour synchronization matrix, a 96-hour RSTA/ISR plan, and a daily FRAGO.

## SUMMARY

Planning in COIN uses tactical design, either MDMP or TLP, and targeting to ensure units achieve their end state over time. Throughout the COIN planning process, tactical units employ the seven COIN lines of effort to ensure that they achieve unity of effort, prioritization in accomplishing tasks, control of the population, and an increase in the Host Nation government's legitimacy.

## 5. Offensive Considerations in Counterinsurgency

> *"For it is in the nature of warfare…that the initiative must be maintained, that the regular army must lead while its adversaries follow, and that the enemy must be made to feel a moral inferiority throughout. There must be no doubt as to which side is in the ascendant, no question as to who controls the general course of the war…"*
>
> C. E. Callwell, Small Wars, 1896

Tactical units conduct offensive operations during COIN operations to destroy, disrupt, interdict, deny, or neutralize the elements of an insurgency in support of the lines of effort (LOEs). Tactical units may be tasked to conduct offensive operations as part of *clear-hold-build* operations, *strike* operations or *populace and resource control* operations. Units may be tasked to *isolate*, *disrupt*, or *fix* an insurgency's base, auxiliary, underground, leaders, or guerrillas. This chapter explores how offensive operations support the following seven LOEs in counterinsurgencies:

- Establish civil security.
- Establish civil control.
- Support Host Nation security forces.
- Support to governance.
- Restore essential services.
- Support to economic and infrastructure development.
- Conduct information engagements.

### Section I—Overview

In COIN, the characteristics of the offense apply to all offensive operations. Tactical units conduct all four types of offensive operations in the COIN environment—*Movement to Contact*, *Attack*, *Exploitation*, and *Pursuit*. Within the four types of offensive operations, units conducting COIN focus on specific tactics and techniques which include search and attacks, raids, cordon and searches, ambushes, sniper employment, site exploitation (SE), and COIN patrols. Additionally, US forces should conduct combined offensive operations with HN security forces at every opportunity in order to reinforce HN legitimacy, support HN security forces, and support the HN rule of law.

#### *Purpose in COIN*

Units conduct offensive operations to—

- Secure the populace continuously.
- Isolate the insurgency from populace.
- Prevent crime.

- Destroy, disrupt, interdict, deny or neutralize elements of the insurgency.
- Secure national and regional borders.
- Integrate with and support HN security forces.

### *Characteristics of the Offense*

The characteristics of the offense are surprise, audacity, tempo, and concentration. For COIN, an additional characteristic, flexibility, is added.

#### Surprise

Units achieve surprise by striking the enemy at a time, place, or manner in which he is unprepared. Total surprise is rarely attainable or essential in conventional operations but is often essential during COIN operations.

#### Concentration

This is the massing of combat power, particularly its effects, at the decisive point to achieve the unit's purpose. During COIN, insurgents avoid situations in which US/HN security forces could potentially mass combat power unless the potential collateral effects of use of that combat power will distance the population from the US/HN government. This outweighs combat losses. US/HN security forces always seek to mass nonlethal and lethal combat power, though not always visibly.

#### Tempo

This is the rate of military action relative to the insurgency. Tempo is not the same as speed. Successful COIN units control or alter tempo to maintain the initiative. Such action promotes surprise, enters the enemy's decision cycle, increases the protection of the attacking force, and decreases the insurgent's ability to defend or plan effectively.

#### Audacity

This is a simple plan of action, boldly executed. Audacity is critical to successful COIN offensive operations and is completely reliant on a thorough understanding of the operational environment. Creativity and mental agility are characteristics of an audacious counterinsurgent.

#### Flexibility

This is the ability of a military unit to adapt to unplanned or unexpected conditions of the operational environment to achieve its tactical purpose and support the LOE.

### *Types of Offensive Operations*

The types of offensive operations are movement to contact, attack, exploitation, and pursuit:

#### Movement to Contact

This develops the situation and establishes or regains contact with the enemy (insurgent) forces (FM 3-0). It also creates favorable conditions for subsequent tactical actions and stability operation. At a tactical level, every movement can be treated as a movement to contact, because of the lack of information concerning insurgent location, strength, capabilities, and intentions. Specific types of movements to contact include search and attack and cordon and search operations.

### Attack

This destroys or defeats enemy (insurgent) forces, seizes and secures terrain, or both (FM 3-0). Attacks require mobility, countermobility, and survivability supported by direct and indirect fires. Attacks may be hasty or deliberate, depending on the time available for planning and preparation. Commanders execute hasty attacks when the situation calls for immediate action with available forces and minimal preparation. They conduct deliberate attacks when they have more time to plan and prepare.

Tactical units normally conduct synchronized and special purpose attacks during COIN operations. Special purpose attacks are ambushes, spoiling attacks, counterattacks, raids, feints, and demonstrations. This chapter focuses on the tactics of raids, ambushes, and sniper employment.

### Exploitation

This rapidly follows a successful attack and disorganizes the enemy in depth (FM 3-0). Tactical commanders exploit successful offensive operations. In COIN, failure to exploit may allow the insurgent to egress, reposition, or disappear into the population. An example of a tactical unit conducting an exploitation in COIN would be sending a unit on a raid based on information and intelligence gathered on a cordon and search that occurred earlier in the day. Effective search procedures, tactical site exploitation, tactical questioning, and use of ISR assets are key to units being able to effectively conduct an exploitation.

### Pursuit

This is conducted to catch or cut off a hostile force attempting to escape with the aim of destroying it (FM 3-0). Pursuit operations begin when an insurgent forces attempts to conduct retrograde operations. Unlike conventional operations, where the enemy's transition to retrograde operations leaves him vulnerable to loss of internal cohesion and complete destruction, the insurgent's transition to retrograde operations may make it more difficult for tactical units to engage, capture,. or kill him. Successful pursuit of the insurgent relies on maintaining contact through surveillance assets, patrols, and HN security forces.

Tactical leaders must recognize the potential of the insurgent to conduct a baited ambush during retrograde operations. Critical to mitigating risk to friendly forces during a pursuit is maintaining one of the eight forms of contact (direct, indirect, nonhostile/civilian, obstacle, CBRN, aerial, visual, and electronic) and positioning of adjacent units such as aviation, HN security forces, surveillance assets, other ground forces, and quick reaction forces (QRF).

#### *Offensive Operations in Clear, Hold, Build Operations*

Offensive operations are the initial focus in clear, hold build, operations. These operations establish civil security and establish civil control. Both are pivotal in setting the conditions for balanced development across the seven COIN LOEs. Offensive operations will continue to be conducted in the *hold* and *build* phases. This keeps insurgents from reestablishing influence over an area and is based on a change in the insurgency's organizational and operational patterns.

### *Mission Variables in COIN*

In counterinsurgency operations, tactical units face a unique set of considerations based on the mission variables. Specific considerations for types of offensive operations will be discussed later. However, some considerations for all offensive operations include—

#### Mission

- Offensive operations should be based on the best intelligence available, while inflicting the minimal damage to the population, infrastructure, and local economy.
- To further gather intelligence, units must be prepared to conduct SE.
- Leaders should consider having a consequence management and a Perception Management plan, in case the offensive actions go poorly.

#### Enemy

- Leaders must pay careful attention to insurgent escape routes, as most insurgents will seek to flee from most. Other enemy considerations include—
  - The insurgent resistance in the direction of attack into the target area.
  - Insurgent resistance in the objective area.
  - Insurgent resistance at the target.
  - Insurgent resistance departing the objective.

#### Terrain and Weather

- Leaders often seek to conduct operations during limited visibility or early morning hours in order to surprise the targets.

#### Troops and Support Available

- Tactical units are often task-organized with additional teams or units (Chapter 3). Host Nation security forces are essential for every offensive operation.

#### Time Available

- Leaders allocate sufficient time to conduct the operation; in COIN, this should include time to conduct SE and tactical questioning.

#### Civil Considerations

- The level of inconvenience to the local populace should discourage insurgents and insurgent sympathizers from remaining in the locale and encourage the local population to provide information on the insurgents. The level of inconvenience should not be so great as to turn the local population towards active or passive support of the insurgency.

- Actions on the objective must include how to deal with nonhostile persons, bystanders, family members, and detainees.

## Section II—Civil Security and Control

Establishing civil security promotes a safe environment. Establishing civil control involves regulates selected behavior and activities of individuals and groups. It reduces risk to individuals and groups and provides security from both external and internal threats (FM 3-07). Together, actions along these LOEs in COIN often take the form of unilateral and combined offensive operations against insurgent leadership, guerrillas, underground, and auxiliary. These offensive actions help establish public order and safety. In COIN, offensive operations are more successful when supported by effective targeting.

### *Search and Attack*

This technique is used for conducting a movement to contact that shares many of the characteristics of an area security mission (FM 3-90). A search and attack is a specialized technique of conducting a movement to contact in an environment of noncontiguous AOs. In COIN, a search and attack uses multiple coordinated small-units (team, squad, or platoon) that conduct decentralized movement to find and attack the enemy. A commander normally employs this form of a movement to contact when the enemy is operating in small, dispersed elements. Often searches and attacks are used to support the establish civil security subtask of enforce cessation of hostilities, peace agreement, and other arrangements.

Search and attack operations are used in both urban and rural terrain. They are followed by operations supporting the other LOE such as restoring damaged infrastructure or conducting information engagements in a neighborhood. This tactic disrupts insurgent activities, while trying to solve some of the root causes of the insurgency. The search and attack is typically used during the *clear* phase of a *clear-hold-build* operation. It may also be used in a strike operation.

To develop a specific search and attack concept, the commander must understand the OE by using the operational variables, the mission variables, and mission analysis. The *troop-leading procedures* (TLP) and *military decision-making process* (MDMP) applied to the COIN OE serve as the planning foundation for commanders and leaders.

#### Considerations during Counterinsurgencies

Specific considerations using elements of the mission variables to a search and attack in COIN are—

##### *Mission*

Leaders determine whether the search and attack is enemy or terrain-oriented. If enemy-oriented, the search and attack should attack the enemy while inflicting minimal damage to the population, infrastructure, and local economy. If terrain-oriented, the search and attack must be prepared to locate base camps, caches, safe houses, or subterranean structures.

*Time Available*

The size of the area, especially the interior layout of urban buildings, impacts force size and search time.

## Phases

A search and attack has three basic phases: *organize*, *plan*, and *execute*.

*Organize*

The commander task-organizes his unit into reconnaissance, fixing, and finishing forces, each with a specific purpose and task. The size of the reconnaissance force is based on the available intelligence about the size of insurgent forces in the AO. The less known about the situation, the larger the reconnaissance force. The reconnaissance force typically consists of scout, infantry, aviation, and electronic warfare assets. The fixing force must have enough combat power to isolate insurgents once the reconnaissance force finds them. The finishing force must have enough combat power to defeat insurgents. The commander can direct each subordinate unit to retain a finishing force, or he can retain the finishing force at his echelon. The commander may rotate his subordinate elements through the reconnaissance, fixing, and finishing roles. However, rotating roles may require a change in task organization and additional time for training and rehearsal.

***Reconnaissance Force***

The reconnaissance force finds the enemy force using all means available. It can serve as an element of the fixing force or follow and assume the role of the attack force if sufficiently resourced. If the reconnaissance element makes contact without being detected by the insurgent, the commander has the initiative. In COIN, HN security forces are often the best suited to conduct the reconnaissance, if they have the training, equipment, and capability.

***Fixing Force***

Although sometimes included in the reconnaissance force in COIN, the fixing force develops the situation, and then executes one of two options based on the commander's guidance and the mission variables. The first option is to block identified routes that the insurgent can use to escape or use for reinforcements. The second option is to conduct an attack to fix the insurgent in his current positions until the finishing force arrives. The fixing force attacks if that action meets the commander's intent and it can generate sufficient combat power against the insurgents. Depending on the insurgent's mobility and the likelihood of the reconnaissance force being compromised, the commander may need to position his fixing force before his reconnaissance force enters the AO.

***Finishing Force***

The finishing force must possess and maintain sufficient combat power to defeat the insurgent templated. The finishing force may move behind the reconnaissance and fixing force or it may locate where it is best prepared to rapidly maneuver on the insurgent's location, by foot, vehicle, or air. The finishing force must be responsive enough to engage the insurgent before he can break contact with the reconnaissance force. The finishing force destroys or captures the insurgent by conducting hasty or deliberate attacks, or employing indirect fire, attack reconnaissance aviation, or close air support to destroy the insurgent. The commander may direct the finishing force to establish an area ambush and use his reconnaissance and fixing forces to drive the insurgent into the ambushes.

*Plan*

The commander establishes control measures that allow for maximum decentralized actions and small-unit initiative. Control measures facilitate the rapid consolidation and concentration of combat power before an attack. The minimum control measures for a search and attack are an AO, objectives, checkpoints, phase lines, limits of advance, and contact points. The use of target reference points (TRPs) facilitates responsive fire support once a reconnaissance force makes contact with the enemy. The commander uses objectives and checkpoints to guide the movement of subordinate elements. The commander uses other control measures as needed such as phase lines, restrictive fire lines, and marking systems.

***Zones***

The commander next determines how the area of operations will be broken down. Two options are—

***Multiple***

Assigning multiple small zones that keep subordinate elements concentrated and allow controlled, phased movement throughout the overall area. This facilitates overall control and allows subordinates to rapidly mass their combat power.

***Single***

Concentrate the main effort in one zone and use fire teams or squad patrols to reconnoiter the next zone. Once the main effort has completed a thorough reconnaissance of the initial zone, it then moves into the zone that the small units have reconnoitered, as they then move to their next zone. Small patrols provide the initial reconnaissance information, which commanders evaluate and then focus additional reconnaissance efforts.

***Orientation***

The commander determines how the search and attack will be conducted within the designated zones. The zones may be searched selectively or systematically. The commander must visualize, describe, and direct how subordinates will conduct the reconnaissance and how the attacking force will maneuver against the enemy. Two methods include—

***Decentralized Attack***

Each subordinate element is tasked to find, fix, finish, and exploit all enemy forces in their area within their capabilities. If more combat power is required, then the BCT will employ additional assets, the reserve, or adjacent units.

***Centralized Attack***

The commander retains control of the attack force while each subordinate element is tasked to find and fix the enemy in their AO. This method works well when insurgents use base camps.

*Execute*

The four typical steps in search and attack operations are enter the AO, search the AO, locate the enemy, and conduct the attack.

***Enter the AO***

Commanders determine how combined forces enter, conduct movement, and establish objective rally points (ORPs) and bases (patrol bases, combat outposts), or (joint security stations) within the AO by considering the eight forms of contact possible with the insurgent or the population. This technique allows commanders and subordinate leaders to identify their units' potential contact with the enemy and population throughout all phases of the search and attack. Leaders synchronize the actions of adjacent units and provide specific tasks to ensure subordinates understand actions on contact with both enemy and civilians within the ROE. Units may enter the area or zone by infiltrating as an entire unit and splitting or by infiltrating as smaller subordinate units via ground, air, or water (Figures 5-1 and 5-2).

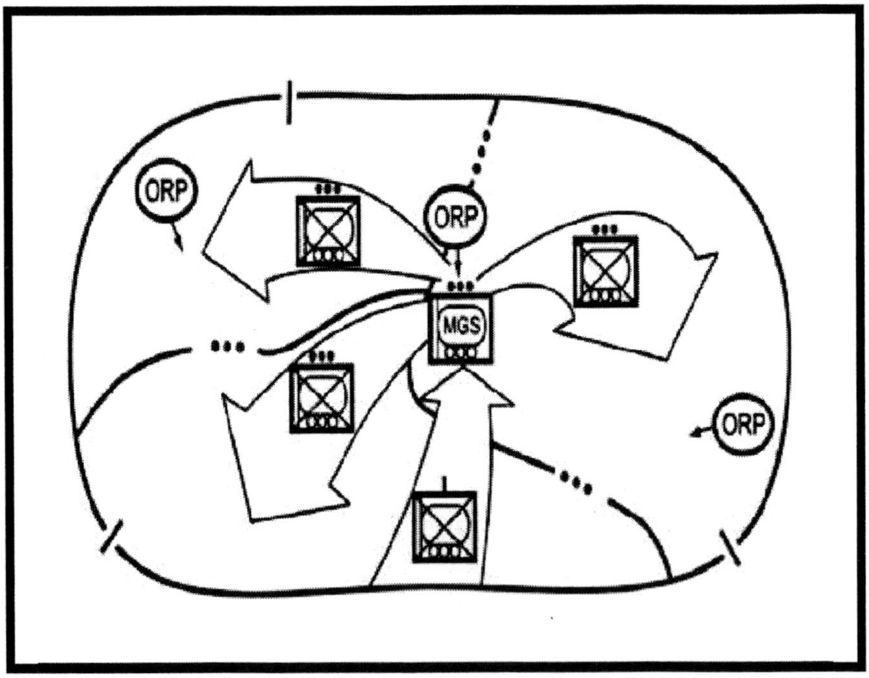


Figure 5-1. Infiltration by company.

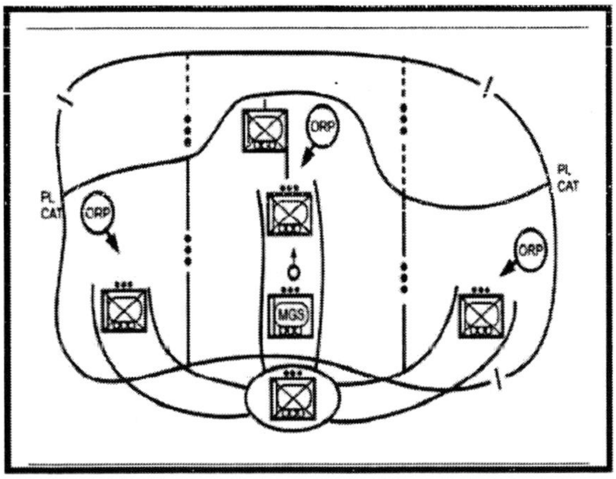


Figure 5-2. Infiltration by squad/platoon.

***Search the AO***

Reconnaissance elements search areas to locate the enemy without detection. This allows more time for leaders to plan and coordinate an attack. Generally, small units are used, since they move quickly and with more stealth among the population regardless of the AO. Once an element of the insurgency is discovered, the commander's concept, intent, and the situation on the ground dictates whether the reconnaissance element follows the insurgent or fixes the insurgent until the attack force is in position. Specific tasks may include route, area, and zone reconnaissance or other surveillance tasks.

***Locate Enemy***

Reconnaissance units must locate insurgent forces, tracks, or other indicators of direction or location. In rural and some border operations, well-trained trackers can identify and follow insurgent tracks that are hours or even days old. Units tracking the insurgent must be prepared to react to insurgent contact and avoid likely ambush situations. Leaders must ensure support for the reconnaissance force if it is compromised. In urban areas, tracking the insurgent is more difficult due to the nature of the terrain and the insurgent's use of the population. Leaders rely on HUMINT, a thorough knowledge of their AO, UAS, attack reconnaissance aviation, and sound communication and coordination with adjacent units to find the elusive insurgent.

***Conduct the Attack***

The attack in a search and attack has four elements:

***Concentrate Combat Power***

Once the insurgent is discovered, the plan must support the rapid concentration of combat power to fix and destroy the insurgent. Leaders at each echelon must plan to destroy the insurgent within their capabilities, or at least fix the insurgent.

***Fix the Enemy***

If the insurgent cannot be destroyed by the forces on hand, then the forces must fix the insurgent until finishing forces arrive. Fixing forces block egress routes with indirect fires, maneuver forces, obstacles or all three. They also suppress the insurgent's weapons systems, obscure his vision, and disrupt his command and control. Specific tasks may include establishing a blocking position, an ambush, or a support by fire position.

#### *React to Contact*

If a unit makes contact, it takes immediate action to fix or destroy the insurgent. The speed and violence of a hasty attack may compensate for the lack of a reconnaissance or combat power. However, this is rarely true against a prepared insurgent defense or during periods of limited visibility. Leaders should not assume the discovered insurgent force is alone; there may be mutually supporting positions or units.

#### *Finish the Enemy*

An initial attempt to finish the insurgent by a squad or platoon in contact may become a fixing effort for a platoon or company attack.

### *Cordon and Search*

A cordon and search operation is conducted to seal (cordon) off an area in order to search it for persons or things such as items, intelligence data, or answers to PIR. Effective cordon and search operations possess sufficient forces to both effectively cordon a target area and thoroughly search that target. Usually, this operation contributes to establishing public order and safety, a key *establish civil control* subtask. It is also one of the techniques used in the "clear" phase of a *clear-hold-build* operation.

*Cordon* is a tactical task given to a unit to prevent withdrawal from or reinforcement to a position. *Cordon* implies occupying or controlling terrain especially mounted and dismounted avenues of approach. *Search* implies the physical and visual inspection of an area. Both the object of the search and the physical area of the search influence the type and degree of the search (FM 3-90.5, FM 3-90.15, and FM 3.06.20), and for additional information on searches and site exploitation.

### Methods

The two basic methods of executing a cordon and search are—*cordon and knock* and *cordon and enter*. They differ in level of aggression. Based on the enemy SITEMP and identified operational risk, actual cordon and search operations vary between these two levels.

Key factors to consider in selecting the method to use include the enemy threat, the local populace support, the level of intelligence available, and the capabilities of the HN security forces. In both methods, the cordon is still established with as much speed or surprise as possible to isolate the objective. Both methods may require some integrated HN security forces or civil authorities to obtain the agreement by the occupants of the targeted search area. Figure 5-3 compares the characteristics of permissive and nonpermissive cordon and search operations.

#### *Cordon and Knock*

This is less intrusive than cordon and search. It is used when the populace is seen as friendly or neutral, when no resistance is expected, and when the goal is to disrupt and inconvenience the occupants as little as possible. One version of this is called the *tactical callout*. This is a procedure where occupants are asked to exit the before search forces enter. If occupants refuse to exit, or if the ground commander believes that the potential exists for an insurgent encounter, he may escalate to cordon and enter.

A second version of the cordon and knock is *cordon and ask*, which means occupants or the local Host Nation authorities are asked for permission to search a particular location. If permission is denied, no entry occurs. However, the cordon and knock and the cordon and ask require some degree of integration with HN security force or HN authorities to obtain the agreement by the occupants of the target to the subsequent search. At a minimum, a sufficient number of translators, preferably one with each element, is required.

*Cordon and Enter*

This approach is intrusive. The intent is to rapidly breach barriers to gain entry into the search area, typically using speed and surprise to allow the unit to quickly gain control. This action allows units to maintain the initiative over a potentially unknown insurgent force operating in the search area. Intrusive entry ranges from a Soldier simply opening a door without occupant permission, to mechanical ballistic, or explosive breaching. In addition, mounted units can use vehicles to breach. The cordon and enter approach does not explicitly require integrated HN security forces or HN authorities, because occupants' permission is not required. However, during a counterinsurgency, obtaining the leadership or direct support of the HN is always preferred. Commanders assume operational risk in COIN by foregoing these considerations. Some considerations when using the cordon and enter method follow. These considerations may be more or less important than capturing the target individual, site, or equipment. Gains in security by violent capture of a key insurgent leader may result in far more substantial losses along the other LOE:

- Risk to civilian occupants and bystanders.
- Collateral damage to infrastructure.
- Perception of the populace.
- Risk to Soldiers.
- Rehearsals.
- Level of training of breach element.
- Effects on subsequent tactical site exploitations.

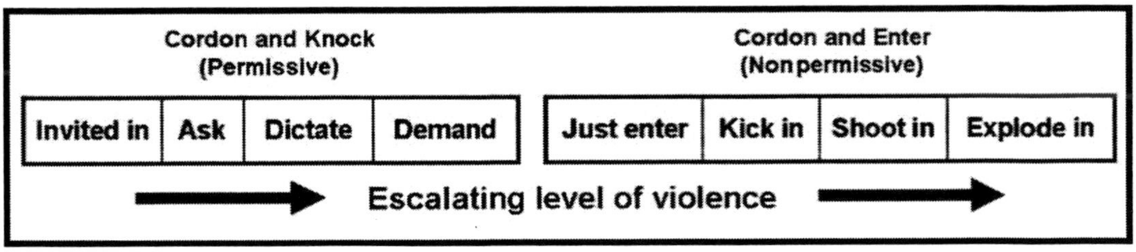


Figure 5-3. Comparison of cordon and search methods.

## Approaches

Leaders plan and execute cordon and search operations using either a systematic or selective approach. A systematic approach is the search of all buildings in the targeted area, while a selective approach is the search of specific locations within a targeted area. The approach used depends on numerous factors. However, the purpose of the operation is still to capture the designated personnel, site, or equipment.

## Considerations

If intelligence indicates enemy presence, and the local populace is either neutral or supportive of the insurgency, then the principles of speed and surprise are the keys to a successful cordon and search. Specific considerations using elements of the mission variables are—

### *Mission*

Leaders determine the focus and method of the cordon and search based on the anticipated threat and the level of violence in the area of operations.

### *Enemy*

Cordon elements cannot effectively block pedestrian egress or ingress. Therefore, commanders should consider how to best physically stop pedestrian traffic. Lethal fire is not a universal means of enforcing the nature of a cordon.

### *Troops and Support Available*

The size and composition of the cordon and search force is based on the size of the area to be cordoned, the size of the area to be searched and the suspected enemy SITEMP. Normally, a military commander, with the police in support, best controls a search involving a battalion or larger force. The police, with the military in support, best control a search involving smaller forces. Regardless of the controlling agency, HN police are the best choice for performing the actual search. However, they must be available in adequate numbers and be trained in search operations.

### ***Time Available***

As time available to plan and prepare for a cordon and search mission is generally limited, it is often necessary to conduct planning while reconnaissance and intelligence collection are ongoing. The size of the area, especially the interior layout of urban buildings, impacts force size and search time. Leaders should plan on allowing time for follow-on missions based on exploitable information.

### *Civil Considerations*

Cordon and search operations are a great opportunity for all Soldiers to conduct information engagements with the population. Each Soldier should know and understand the information engagement task and purpose.

**Phases**

The phases of a cordon and search are the planning phase, reconnaissance phase, movement to the objective phase, isolate the objective phase, search phase, and the withdrawal phase.

*Plan*

Establishing the cordon requires detailed planning, effective coordination, and meticulous integration and synchronization of available assets to achieve the desired effects. This requires the commander to consider both lethal and nonlethal effects. Each subordinate cordon position such as a traffic control point or blocking position must have a designated leader and a clearly understood task and purpose.

A cordon and search operation can usually support the conduct engagement LOE. Commanders must develop, integrate, and nest the information message in accordance with the purpose of the search. Often the best message in COIN is one's actions or that of the entire unit.

Search of an urban area varies from a few, easily isolated buildings to a large well-developed urban city. Leaders should divide the urban area to be searched into zones. Buildings should be numbered and assigned specific search parties for coordination and clarity

*Enablers*

Assets employed during the cordon and search may include tactical PSYOP teams (TPTs), tactical HUMINT teams (THTs), law enforcement professionals (LEPs), special advisors, attack, reconnaissance, and assault aviation, CAS, SIGINT enablers, MASINT enablers, military working dog teams, (MWDs) biometrics collection efforts, female searchers, and civil affairs teams (CATs, Chapter 3).

A TPT is an outstanding combat multiplier. Messages broadcast in the local language during cordon and search/knock operations facilitates situational awareness and understanding for the local inhabitants. These TPTs, using vehicle-mounted or man-pack loudspeaker systems, can help inform and control the population. In addition, the TPT conducts face-to-face communication along with disseminating handbills or leaflets explaining the purpose and scope of the cordon and search. This helps in gaining compliance by the local population.

THT is also an outstanding combat multiplier. THTs collect valuable information from individuals in the search area, provide a tactical questioning capability, and have additional language capabilities.

*Organization*

The typical cordon and search organization includes a command element, a cordon element, a search element, and a reserve element each with a clear task and purpose. Figure 5-4 displays a typical organization for search operations.

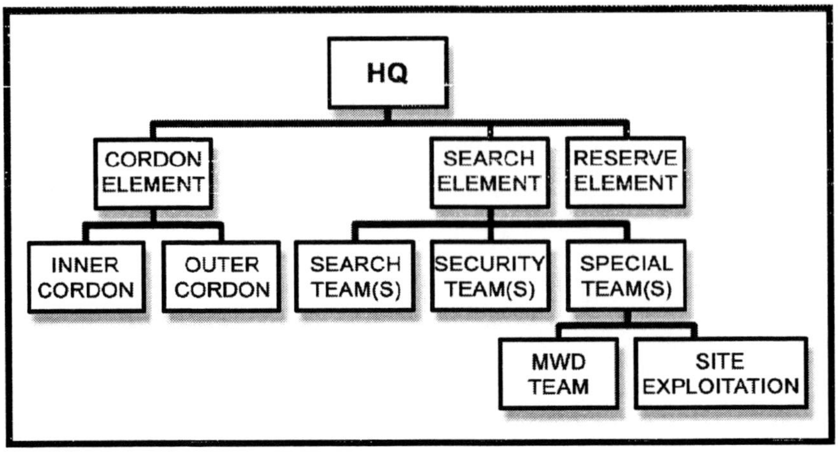


Figure 5-4. Typical organization for cordon and search operations.

***Command Element***

An overall commander controls the unit conducting the cordon and search. He identifies the subordinate element leaders.

***Cordon Element***

This force must have enough combat power to cordon off the area. An effective cordon that both prevents the egress of individuals from the search area and prevents outside support to the search area, is critical to the success of the search effort. Based on the mission variables (METT-TC), two cordons are often established: an outer cordon to isolate the objective from outside reinforcements or disruptions, and an inner cordon to prevent individuals from leaving or communicating with someone outside the search area. Both cordon elements must maintain 360-degree security. UAS, scouts, attack reconnaissance aviation, or sniper teams should be considered by tactical units for use in observing the objective area for enemy both before and during the operation.

***Search/Assault Element***

The search element conducts the actual search operation. A search may orient on people, on materiel, on buildings, or on terrain. Normally, it is organized into special teams. The most basic search team is a two-person team consisting of one person who conducts the actual search while another person provides immediate security to the searcher. Establish discipline and standardized search SOPs to ensure searches are thorough, PIR-focused, and of minimal risk to Soldiers.

All search elements must be prepared to handle male and female personnel, key equipment, hazardous materials (biohazards or other toxic elements), ordinance, and record key events. They must be trained to understand and on order execute information engagements, tactical site exploitation, detainee operations, and adjacent unit coordination. Search personnel must be trained to operate with HN security forces and within the established ROE. First aid and other medical training is critical. Soldiers must be proficient with signaling and marking devices as well as detection and recording equipment. Biometric and video/audio recording device proficiency is crucial in COIN search operations. Basic language training is essential to maintain effective searches and overall operational tempo.

Typical search teams are organized in two- to three-Soldier teams. Female Soldiers are a proven combat multiplier during search operations, because few cultures tolerate males searching females. Search teams clear each room or area in accordance with FM 3-21.8. Units should not confuse entry methods and their levels of aggression with the requirement to respect the Host Nation's people and homes. Typically, once a room is cleared, one team member provides security while the other(s) searches. All search element personnel are prepared to fight. Basic considerations for any search team include the following:

- Detailed instructions including prohibited items such as weapons, chemicals, medicines, and machine tools.
- Understanding of search restrictions and special considerations to include—
    - Searching of religious buildings.
    - Searching of females by female Soldiers.
    - Searching of historical, cultural, or governmental sites (unauthorized or hostile).
- Host nation security forces or local interpreters.
- Biometrics tools.
- Breaching kit.
- Vehicle access tools such as lock picks.
- Information engagement products and tools.
- Audio and video recording devices and data imaging devices.
- Markings and signaling techniques and any constraints.
- Respect for personal property.
- Tools to collect and record information for HUMINT.
- Necessity to maintain communication and report location.
- Standardization of maps, imagery, and labeling conventions.

***Reserve Element***

The reserve element or QRF must possess and maintain enough combat power to defeat the insurgent forces templated within the AO. The commander gives priorities for planning to the reserve that could include to be prepared to execute any of the subordinate unit missions. Priorities can also include additional missions such as CASEVAC or reinforcement. The reserve element leader focuses efforts on synchronized communications, rehearsals, battle tracking, and positioning before and during the operation.

*Reconnaissance*

Every target area should be reconnoitered prior to execution using many of the available resources. If the target is part of a unit's AO, then a patrol around the target may not be out of order. ISR assets, attack reconnaissance aviation, local nationals, and imagery are other methods for conducting reconnaissance. The reconnaissance plan must not provide the enemy with indicators of an impending cordon and search. Given the nature of COIN, the reconnaissance phase could last an extended period, as units identify the relative size and location of buildings, entry points, cordon position and avenues of approach. Further tools for objective analysis may be obtained from attack aviation photographs, maps, and local emergency services departments.

*Movement to the Objective*

The timing, routes, and execution of movement to the objective should consider the factors of METT-TC, and whether it should be simultaneous or phased. If contact is made in the movement, commanders should consider whether they wish to send forces forward to initiate the cordon.

*Isolation of the Objective*

Although analysis of the mission variables using METT-TC determines specifics, a unit typically establishes the outer cordon first, establishes the inner cordon second, and moves the search element to the objective last. Commanders should consider the value of using the opposite technique of forming the cordons following rapid movement to the objective to gain surprise. Timing is when executing either technique is important. The quicker these three events are accomplished, the less time personnel on the objectives have to egress, find concealment, or destroy materials or equipment.

***Position the Reserve Element***

The reserve element or QRF is a mobile force positioned in a nearby area, with multiple planned ground, water, or air routes to the objective area. Its mission is to aid the search and security elements if they require assistance or become unable to achieve their purpose.

***Establish the Cordon***

There are two techniques for emplacing the actual cordon positions: simultaneously and sequentially. Careful consideration must be given to both, because each has advantages and disadvantages. Units establishing a cordon position themselves to be able to block movement to and from the objective area. This may be by observed fire, but usually it will be by physically controlling routes. Cordon positions should be occupied rapidly just prior to the search element reaching the objective. Establishing the cordon during a period of limited visibility increases movement security but makes control difficult. Cordon positions, once occupied, will be detected by locals as they conduct their daily business.

Both the outer and inner cordon leaders must maintain situational understanding of not only their AOs, but also each other's cordon and the progress of operations of the search element. In doing so, they can anticipate insurgent activity, control direct and indirect fires, and achieve their task and purpose.

The various positions of the outer and inner cordons may include, vehicle mounted platoons or sections, dismounted platoons or squads, interpreters, detainee security teams, crowd control teams, tactical PSYOP teams, observation posts, traffic control points or blocking positions, Host Nation security forces (military or police), and aviation assets.

The outer cordon usually focuses on traffic control points and blocking positions, while the inner cordon focuses on overwatching the objective and preventing exfiltration or reposition of persons within the search area. Figure 5-5 shows the typical establishment of a cordon and Figure 5-6 shows the details of an inner cordon in an urban setting. Note the technique of assigning each building a number to increase clarity and coordination between units.

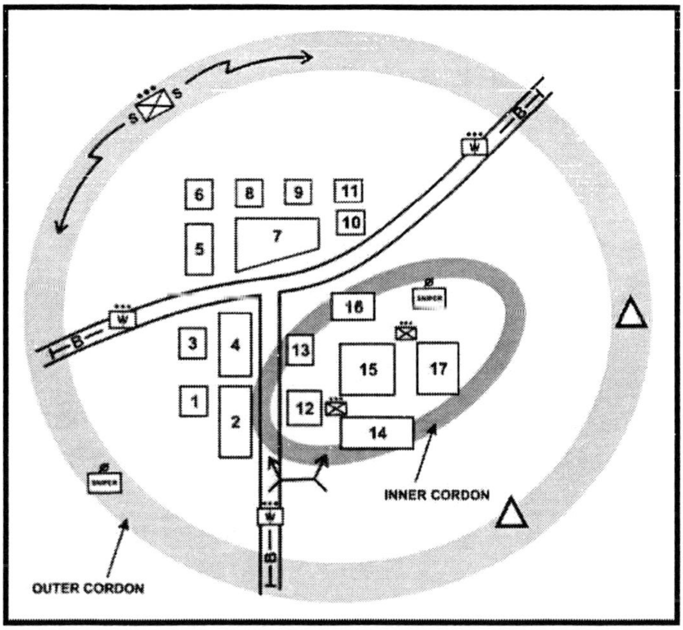


Figure 5-5. Typical establishment of an urban cordon.

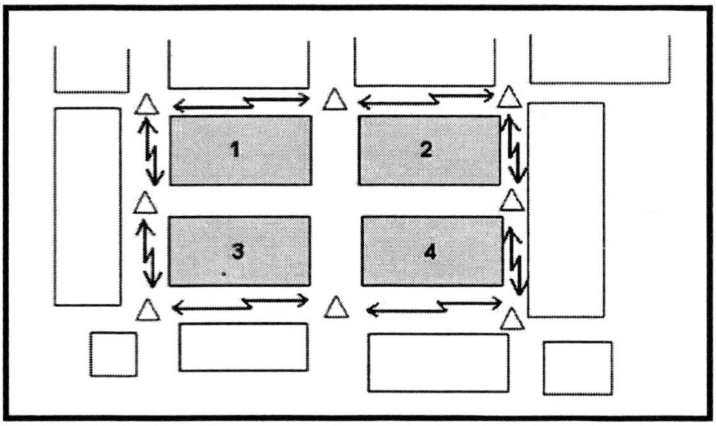


Figure 5-6. Urban inner cordon.

*Search*

A search may be oriented toward people, materiel, buildings, or terrain. It usually involves both HN police and military personnel. It must be a systematic action to ensure that personnel, documents, electronic data, and other material are identified, evaluated, collected, and protected to develop intelligence and facilitate follow-on actions.

The tempo at which a search operation is conducted should be slow enough to allow for an effective search, while not so slow that it allows the insurgent force time to react to the search. Search teams must consider a return to an area after an initial search. This can surprise and remove insurgents who may not have been detected or may have returned. All searches should create pressure on insurgents and sympathizers to not stay in the area, but not inconvenience the local residents to the degree that they will collaborate with the insurgents.

Special laws regulate the search powers of military forces. Misuse of search authority can adversely affect the outcome of operations and future legal proceedings; therefore, all searches must be lawful and properly recorded to be of value. These laws must be disseminated to the population to ensure understanding and compliance. Additional information on searches can be found in FM 3-06.20. Search teams must have instructions for three basic categories:

***Personnel***
This includes both male and female and both persons of interest and other persons.

***Physical Items***
This includes weapons, equipment, documents, computers, and cameras.

***Information Mediums***
This includes data inside computers, cameras, and cell phones.

*Withdrawal*
During this phase, the unit may be the most vulnerable. To mitigate risk, a commander may choose to—

- A relief in place.
- Stay-behind elements to cover the withdrawal.
- Different routes and timing.
- Simultaneous or phased withdrawals.

***Search Operations***

A search is the deliberate examination of a person, place, area or object using Soldiers, animal or technological sensors to discover something or someone. Examples include searches of enemy or detained personnel, military objective areas, personnel or vehicles at a checkpoint, and lines of communication.

A search is conducted under a wide variety of situations and for a wide variety of purposes. Typically, a person is searched in order to find something that is concealed. A place, area, or object, such as a car or desk, is searched for something that may or may not be concealed. For more on vehicle searches, see Chapter 7. Communication objects, such as letters, books, computers, cell phones, and other media and signaling tools, are searched to discover information.

During a counterinsurgency, the rules of engagement and various agreements between the Host Nation and US counterinsurgent forces often describe search situations, and may limit search methods.

During a search, it is important to keep the local population informed, as much as tactically possible, that search contributes to their safety and security. This communication should begin during the actual search, if possible, but is often accomplished after the search by follow-up patrols. Follow-up patrols can not only aid in mitigating some of the negative aspects of the search but also see if missed individuals have returned to the searched area. Follow up patrols which include civil affairs teams or tactical PSYOP teams provide a great capability to conduct consequence management, assisting in the achievement of information engagement, often through reinforcing themes, and collecting information for development into intelligence.

## Techniques

A search can orient on people, materiel, buildings, or terrain (FM 3-21.8) Key basic considerations for conducting a search in various includes searching individuals, tactical site exploitation, aerial searches, searching subterranean areas, searching individuals, detention of individuals, tactical questioning and detainee processing.

### Individuals

Any individual can be an insurgent, auxiliary, or member of the mass base. However, searchers must avoid mistaking all suspects for the enemy. Because there may be little or no Host Nation personnel identification procedures, identifying the correct person as an insurgent may be very difficult. It is during the initial handling of individuals about to be searched that the greatest caution is required. During the search, one member of a search team always covers the other member who makes the actual search. When females have to be searched, every precaution is made to prevent violating local customs and mores. If female searchers cannot be provided, consider using the medic to search female suspects.

#### *Teams*

Soldiers conduct individual searches in search teams that consist of the following:

#### *Searcher*

Actually conducts the search. This is the highest-risk position.

#### *Security*

Maintains eye contact with the person being searched.

#### *Observer*

Supervises search and warns of suspicious behavior or actions.

#### *Methods*

The most common search methods used to search an individual are *frisk* and *wall* searches. A third, less common method, used in very select situations, is the strip search.

#### *Frisk Search*

Quick and adequate to detect weapons, evidence, or contraband. A frisk search is more dangerous because the searcher has less control of the individual being searched.

#### *Wall Search*

Affords more safety for the searcher by leaning the suspect against any upright surface, such as a wall, vehicle, tree, or fence. The search team places the subject in the kneeling or prone position if more control is needed to search an uncooperative individual.

***Strip Search***

Considered only when the individual is suspected of carrying documents or other contraband on his or her person. This extreme search method should be conducted in an enclosed area and by qualified personnel when available.

***Search with Sensors***

Metal detectors or thermals can identify hidden items.

***Population Control***

Three basic methods are used to control the population during a search of an urban area: assembly of inhabitants in a central location, restriction of inhabitants to their homes, and control of the heads of the households.

***Assemble Inhabitants in a Central Location***

This method moves inhabitants from their homes to a central area. It provides the most control, simplifies a thorough search, denies insurgents an opportunity to conceal evidence, and allows for tactical questioning. However, this method has the disadvantage of taking the inhabitants away from their dwellings and possibly encouraging looting, which, in turn, engenders ill feelings. A specific element must be identified to control the centralized inhabitants. A TPT, using a loudspeaker, can facilitate assembly by giving specific instructions to the inhabitants of the search area.

***Restrict Inhabitants to Their Home***

This technique prohibits movement of civilians, allows them to stay in their dwellings, and discourages looting. The use of a TPT to broadcast "stay-indoors" messages facilitates clearing the streets of civilians and aids in restricting their movement. The security element must enforce this restriction to ensure compliance. The disadvantages of this method are it makes control and tactical questioning difficult, and gives inhabitants time to conceal contraband in their homes.

***Control Heads of Households***

The head of each household is told to remain in front of the house while everyone else in the house is brought to one room. The security element controls the group at the central location, controls the head of each household, and provides external security for the search team. When dealing with the head of a household, it is important to explain the purpose of the search using an interpreter. During the search, the head of the household accompanies the search team through the house. This person can be used to open doors and containers to facilitate the search. It is important for the head of the household to see that the search team steals nothing.

**Houses or Buildings**

The object of a house search is to look for contraband and to screen residents to determine if any are guerrillas, auxiliaries, members of the underground or the mass base. A search party assigned to search an occupied building should consist of at least one local police officer, a protective escort for local security, and a female searcher. If inhabitants remain in the dwellings, the protective escort must isolate and secure the inhabitants during the search.

Forced entry may be necessary if a house is vacant or if an occupant refuses to allow searchers to enter. If the force searches a house containing property while its occupants are away, it should secure the house to prevent looting. Before US forces depart, the commander should arrange for the community to protect such houses until the occupants return.

Try to leave the house in the same or better condition than when the search began. In addition to information collection, the search team may use digital cameras or video recorders to establish the condition of the house before and after the search. All sensitive material or equipment found in the house should be documented before it is removed, to include date, time, location, the person from whom it was confiscated, and the reason for the confiscation. The use of a digital camera can assist in this procedure. For a detailed search, the walls and floors must be searched to discover hidden caches.

### *Site Exploitation*

Tactical leaders plan, resource, direct, and supervise tactical site exploitation efforts during all COIN offensive operations. Site exploitation (SE) is the systematic action executed with the appropriate equipment, to ensure that personnel, documents, electronic data, and other material at a site are identified, evaluated, collected, and protected in order to gather intelligence and facilitate follow-on actions. It is a means by which tactical units exploit and analyze the insurgent after collecting biometric, physical, digital, and spoken data. In COIN, insurgents who are captured many times are prosecuted within the HN rule of law and SE is an excellent means of providing courts with evidence, especially when properly recorded. SE contribute to the decide, detect, assess activities of D3A targeting cycle. Once collected, analyzed, and assessed, it may lead to future operations. More information on SE can be found in CALL product 07-26. Figure 5-7 shows an example SE site sketch.

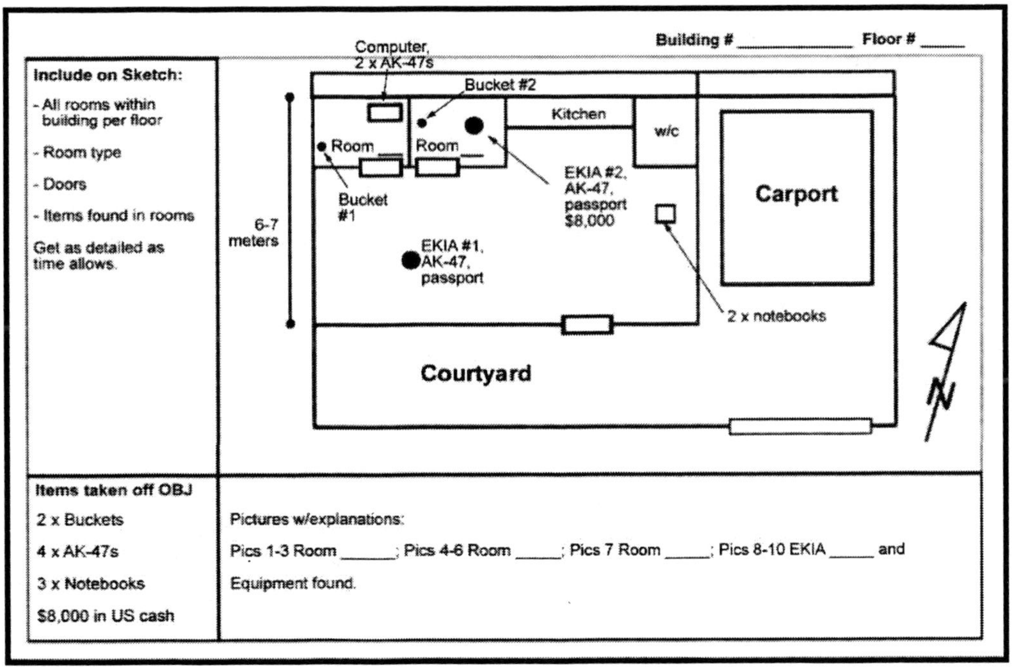


Figure 5-7. Example site exploitation sketch.

*Conduct*

Leaders ensure that Soldiers methodically and effectively identify, preserve, and collect evidence while maintaining its integrity. They strive to prevent damage or corruption from foreign materials, undocumented chains of custody, or loss. They consider how to mitigate risk by allowing minimal personnel to operate in the area, by minimizing time on target, by concealing movement of evidence, by avoiding patterns, and by maintaining an obvious respect for civilians and their belongings.

An important SE resource, biometric assets can measure humans by face, fingerprint, hand geometry, handwriting, iris, retina, vein, voice, and DNA. Other SE resources include search, detention, and marking tools; collection containers; marking materials; photographic, video, and voice recording devices; linguists; and artificial sources of light. In addition, leaders—

- Initially evaluate the situation.
- Decide whether the unit must perform an expedient or formal site exploitation.
- Properly bag and tag evidence.
- Record sworn statements, from Soldiers and locals, for entry into the legal system. (Using evidence kits is the best and easiest way.)
- Photograph captured contraband evidence with the suspect for judicial proceedings.
- Ensure that photos of people, materiel, and other items of potential intelligence interest and evidence collection are documented on and adjacent to the site.
- Use document and material exploitation (DOMEX), which includes hasty analysis of pocket contents, electronic mirror-imaging media, and evacuating data for further, more detailed analysis.
- Process detainees into detainee holding area (DHA), to include collection of biometric and computer database information, by special BCT-level teams.
- Ensure that pre- and post-bomb blast forensics and signatures collected from IED factories, routinely collected by weapons intelligence teams (WITs), are entered into the system for comparison and analysis with detainee records.

### Techniques

Basic SE techniques include search methods, searches of individuals, detention of individuals, tactical questioning, and debriefing.

*Methods*

Search methods must include providing security for search team, ensuring integrity of site, conducting a methodical search, and coherently documenting effort for later review. The type of search team used depends on many factors such as available forces, HN capabilities, and purpose of the search. Most of these are covered in METT-TC. The following SE on-target checklist is not all inclusive:

- Search all rooms or caves, to include roof, yard, any subterranean areas, and associated vehicles for—
  - ID cards.
  - Weapons.

  - Computers.
  - Documents.
  - Digital media.
  - Propaganda.
  - Cellular and satellite phones.
  - Large amounts of money.
- Search other likely hiding places, which may include—
  - Appliances (refrigerator—ice cube trays, under and inside back housing).
  - Furniture hide spots (taped under furniture, hollow legs, inside cushions.
  - Floors (hollow flooring, removable wood boards, removable tiles, under rugs).
  - Gardens, false wall locations, chimney hide locations.
- Search all vehicles for weapons and photograph the weapons.
- Positively identify (PID) the target.
- Photograph all individuals.
- Complete packets on all individuals.
- Verify individual by checking ID cards.
- Perform a gunpowder test or use X-spray on all individuals.
- Photograph each detainee with evidence.
- Collect and document all evidence, and take it to the transport vehicle.
- Load all detainees for transit to FOB, combat outpost, patrol base, or DCP.

*Detention*

*Detainee* refers to any person captured or otherwise detained by an armed force (JP 1-02). The reason for capture or detention of a detainee, and his ultimate disposition and categorization, depends on the situation. Before detaining someone, a leader must first decide what to do with the person afterward. Does the leader want to search, question, process, or release the person?

AR 190-8, FM 3-19.40, and, international law (including the *Law of War* and the Geneva Conventions) cover policies, procedures, and responsibilities for administering, treating, protecting, securing, and transferring custody of detainees. They also cover other planning factors as well as regulatory and legal requirements concerning detainees.

***Rules***

The six rules for processing detainees follow:

- Search the detainee thoroughly and disarm him.
- Silence the detainee.
- Segregate the detainee from other detainees by sex and rank.
- Safeguard the detainee from harm while preventing him from escaping.
- Speed the detainee to the designated detainee collection point.
- Tag the detainee with key information. Use approved format if possible. The tag includes the date of capture, location of capture (grid coordinate), capturing unit, and special circumstances of capture (how the person was captured).

***Protected Status***

Once the suspected insurgents are under friendly control, they assume the protected status of detainees. This term includes any person captured or otherwise detained by armed force. Under the law of war, leaders and Soldiers are personally responsible for detainees under their control. Mistreatment of detainees is a criminal offense under the Geneva Convention and the 1996 War Crimes Act. One of the most conspicuous violations of the Geneva Convention is the unauthorized photography of detainees.

*Note:* Avoid photographing detainees for nonmilitary or unofficial purposes. Doing so is unauthorized.

*Tactical Questioning*

Units recognize value of timely information and intelligence during COIN operations and thus may tactically question an insurgent on the objective. Units designate or construct a detainee holding area, ensuring detainees are unable to communicate in any manner. Leaders then conduct tactical questioning of priority detainees away from the group, as they carefully gather facts and details required to establish consistencies or inconsistencies. Leaders then sort detainees into those to be taken off target, those without further value, those no longer of interest, and those who require immediate battlefield interrogation by qualified personnel. Leaders wanting the ability to interrogate personnel during a potential operations should request and incorporate qualified interrogators into their mission task organization to facilitate timelier information or intelligence.

***Necessity for Soldier Presence***

The terminology may change, but the need for Soldiers at the point of capture or point of detainment to ask questions remains. Trained interrogators are seldom on hand, but Soldiers are always present. Experiences from recent operations show that US forces using immediate tactical questioning techniques with their Soldiers on target find important information that leads to actionable intelligence. Tactical questioning must not violate the Law of War or any legal agreements (SOFAs and coalition agreements). Typical questions should focus on discovering other nearby insurgents, their intentions, their equipment, how they are financed, or their means of support.

***Backtracking of Route***

A successful technique to determine where the individuals were before they came to the point of capture is to verbally backtrack their route. The detainee is questioned as to when he/she arrived, how he/she traveled to the point of capture (foot or vehicle), and from what direction. With a general direction and a means of travel, the route can be developed. Using a map, the Soldier asks leading questions to determine the route. The detainee identifies significant terrain features seen at specific locations, such as rivers, bridges, key buildings, or hills. Talking jogs the detainee's memory. Gradually, the detainee reveals their route (where they originated). Considerations for successful tactical questioning follow:

- Know your linguist; use more than one to double-check the integrity and accuracy of information and reliability of linguist.
- Have a basic knowledge of language(s) of detainee(s).

- Learn to identify physical behavior and posture, resistance or defensive postures; use multiple observers.
- Study your target before he becomes a detainee on the objective.
- Study behavior, values, and interests of others in the OE *before* conducting tactical questioning.
- Consider placing uniformed linguist in the area of detainees and allow the detainees to communicate so that the uniformed linguist can collect information (deception collection).
- Conduct questioning of one individual in a separate room from other detainees so that if the detainee answers he maintains plausible deniability with members of his community.
- Commanders can prepare four to five questions related to information requirements. These can be briefed as part of the patrol order and used by leaders during operations.
- Prepare a tactical questioning plan for information that you believe the detainee may have.

***Aerial Searches***

Unmanned aerial systems (UAS), close air support (CAS), and rotary wing aircraft (attack and reconnaissance aviation) can be used as observation platforms. CAS aircraft and attack reconnaissance aviation can also provide commanders an aerial search capability with different search perspectives and constraints. Rotary wing aircraft are an effective means of conducting mounted search patrols, specific population control measures, and security operations.

Helicopter-mounted patrols may reconnoiter an assigned area or route in search of insurgent elements. They may conduct snap checkpoints on roads to interdict insurgent mounted and dismounted movement. When the element locates a known or suspected element, it can instruct attack aviation teams to engage the insurgent element or it may also choose to land and attack the enemy with a dismounted assault. This technique can be useful in open rural areas unless an air defense threat is present. Use of aerial patrols should be used in operations when sufficient intelligence is available to justify their use or friendly ground-based operations have become predictable to the insurgents. Such patrols are most effective when used in conjunction with ground operations.

In aerial or air/ground search operations, helicopters insert troops in an area suspected of containing insurgents. With the helicopters overwatching from the air, Soldiers search the area. Soldiers remount and the process is repeated in other areas. Members of aerial patrols should be trained in tracking procedures to follow insurgents to their base or safe houses using terrain, deception, and stand-off capability of aviation optics in conjunction with ground and other technological assets. Leaders must plan for the evacuation of prisoners, casualties, and materials, both by air and ground.

*Subterranean Area Searches*

In both urban and rural areas, insurgents must remain undetected to survive. Therefore, insurgents use all means of concealment available. Subterranean areas, in both rural and urban areas, reduce the chance of detection and facilitate insurgent movement. Subterranean areas include natural caves, basements, manmade underground bunkers, tunnels, holes, and sewer systems. Underground sewers and tunnels may also be used in the attack of targets and for egress after an attack. See FM 3-34.170, Engineer Reconnaissance, for a discussion of tunnel (and subterranean) detection, reconnaissance, maneuver, and destruction.

### *Signs of Use of Subterranean Area*

Certain signs may reveal the often signal or identify that insurgent forces within a certain area are using subterranean structures. These indicators include—

- Movement of insurgents in a specific direction when spotted by aircraft.
- Sniper fire occurring from areas where there are no obvious avenues of withdrawal.
- HUMINT reports of subterranean areas
- Failure of cordons to prevent withdrawal or infiltration of insurgent forces
- Turned or managed soil far away from places of habitation or daily labor.
- Operations where insurgents inflict casualties and withdraw without detection or engaging COIN forces.
- The smell of burning wood or food cooking in an uninhabited area.
- Mounds of dirt, and dirt of different colors, which might indicate digging.
- Trails to water sources in uninhabited areas that may indicate personnel requiring water.

### *Methodical and Coordinated Approach*

Searching an area where suspected subterranean facilities are located requires a methodical and coordinated approach. The size of the surface area and the suspected size of the subterranean determine the size and the strength of the unit assigned. The unit is task-organized for subterranean search operations, and is divided into five elements: C2, security, search, guard, and reserve. The C2 element often remains with the reserve element.

### *Narrowing of the Search*

To detect or locate subterranean, leaders first reduce the geographical area of interest to smaller areas of probable locations. Acquiring existing blueprints, maps, imagery, video, aerial photographs, and hydrology analysis tools; actively observing for indicators of probable subterranean access locations; and questioning the local population as to the existence or specific knowledge of any subterranean. Overhead imagery may produce results if the appearance of the surface and vegetation are changed or if deductions about substructure can be made from analysis of existing or historical terrain.

### *Security*

Perimeter and flank security is imperative. A slow, methodical search is conducted in the area of operations, with each search team systematically searching every square meter. The security element moves toward the limits of advance of the search area. Deliberate search techniques emphasize where to look for the insurgent locations that provide him with observation, cover, concealment, and an egress route.

*Signs of Tunnels*

Several visual signs help in detecting the actual tunnels. Visual inspections often disclose the general area of a tunnel, but not its precise location. The keys to finding a tunnel system are a thorough terrain analysis (OAKOC) and an equally thorough physical ground search.

***Rural Visual Indicators***

Visual indicators in a rural operational environment include—

- Air holes.
- Worn places on trees the insurgent uses as handholds.
- A small trail, much like a game trail, through the brush into a clump of small trees
- Cut trees and Limbs tied near a treetop to conceal the use of a tunnel from aircraft.
- Slight depression in or around a group of small trees.
- A lone individual, especially a female, in the area.
- Fresh cooked food with no one attending the site.
- Fresh human feces in an area.

***Urban Visual Indicators***

Visual indicators in an urban operational environment include—

- Sewer, storm drain, or utility grates or manhole covers
- Disturbed soil in mature gardens
- Presence of flooring materials in homes, businesses, and other structures not under construction.

### *Raid*

This is an operation to temporarily seize an area in order to secure information, confuse an adversary, capture personnel or equipment, or to destroy a capability. It ends with a planned withdrawal upon completion of the assigned mission (JP 3-0). A raid is conducted to destroy a position or installation, destroy or capture insurgents or equipment, free friendly prisoners, or seize possible intelligence; and is followed by a rapid withdrawal. By capturing insurgents, information can be developed into intelligence and confiscating contraband can contribute to improved public order and safety. It is often used as part of a strike operation. For additional information on raids, see FM 3-21.8 and FM 3-21.10.

A raid in a COIN environment can differ from a raid in conventional operations since the requirement for minimizing collateral damage may be a significant factor. In addition, the time on the objective prior to withdrawal may be greater, due to the requirement to conduct a detailed SE. As in all raids, the success of the raid is based on accurate, timely, and detailed intelligence and planning.

### Considerations

Raids in COIN could have lasting effects on the population and the insurgents. Specific considerations using elements of the mission variables are—

*Mission*

In a COIN, raids target insurgents, terrain, intelligence, or equipment. Missions are often executed in conjunction with a form of cordon to prevent enemy escape into the population. Units should plan and rehearse according to target intelligence, the commander's intent, and the purpose of the raid.

*Enemy*

The objective of the raid may be a valuable asset the insurgency is prepared to defend. Often, the insurgent will have additional forces in the area positioned to alert, react, facilitate egress, or conduct combat. Effective reconnaissance can increase awareness of these factors.

*Time Available*

Leaders should plan on allowing time for follow-on missions based on exploitable information. Also, units should not stay on the objective too long.

**Techniques**

The minimum task organization for a raid is a headquarters section, an assault element, and a security element. Depending on METT-TC variables, a support element may be formed separately or integrated into the assault element. Additional specialty teams should consist of detainees or EPWs, SE, or CASEVAC teams. Squads are typically too small to execute raids, especially in urban AOs. Figure 5-8 shows the concept of operations for a typical raid.

A typical raid during a counterinsurgency is executed in five phases—insertion, seal off the objective, assault the objective, secure the objective, and withdrawal. The following are so

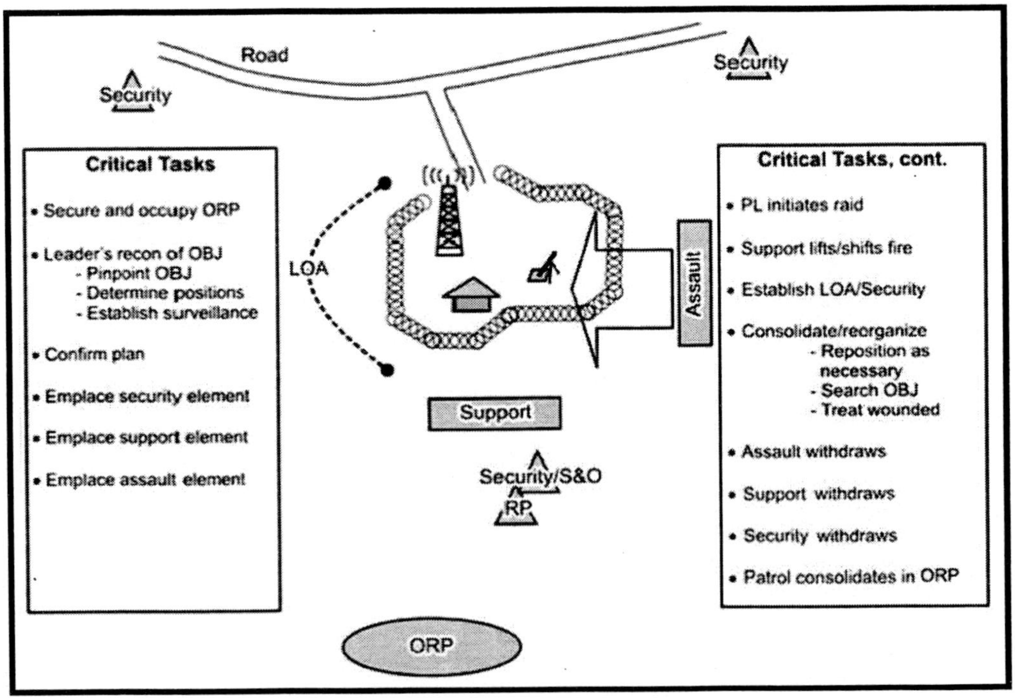


Figure 5-8. Example raid concept.

*Insert*

Insertion in COIN is less difficult than infiltration, especially in urban terrain, due to constant civilian interaction. Units should use whichever method is most likely to achieve surprise. Deception, such as disguising intent with other activities such as establishing a checkpoint, may be useful. Other considerations include—

- Insertion by airborne or air assault can enhance surprise.
- Launch the raid at an unexpected time or place by taking advantage of darkness and limited visibility and moving over terrain that the enemy may consider impassable.
- Infiltration of a sniper team before the raid.
- Avoid detection in rural areas through proper movement techniques and skillful camouflage and concealment to include taking advantage of natural cover of the terrain.
- In urban areas, avoiding detection is difficult, therefore planning to delay detection or awareness of the objective is better. This can be accomplished by rapid movement or deception.

*Seal Off the Objective*

Units should ensure insurgents can neither leave nor reinforce the objective. In many cases, this may be the decisive operation. Some considerations include—

- All forces must understand and adhere to rules of engagement (ROE) and escalation of force (EOF) procedures.
- Establish an outer cordon to block avenues of approach into the objective areas.
- Support force provides initial overwatching fire for the assault force (can be aviation).
- Use of aerial UAS or aviation support to maintain observation as needed.

*Assault the Objective*

Any insurgent element at or near the objective is overcome by surprise and violence of action. Some considerations include—

- Time the assault as close as possible to the execution of the cordon.
- Perform quick, violent, precise, and audacious actions that focus full combat power at the decisive point.
- Breach rapidly—if possible, the breach should be the first overt action of the raid.
- If fires are used, the support element either provides a heavy volume of fire or precision fires dictated by civil considerations. Fires must be closely controlled to ensure precision using FCMs, marking, and signaling. On order or as planned, fires are lifted and shifted to support the assault element by suppressing enemy fire from the objective.

*Secure the Objective*

Units secure the objective by detaining insurgents, controlling personnel on or near the objective, clearing the objective of other threats, conducting SE, and setting conditions that prevent insurgent fires from outside the objective.

*Withdraw*

As planned, the unit withdraws from the objective area. A support force may provide suppressive fires for withdrawal or provide escort away from the objective. Commanders consider use of stay behind measures or devices to monitor backfill of insurgents or population support for the insurgency. A raid in a COIN environment may modify phase five and withdraws at the discretion of the commander.

### *Ambush*

An ambush is a form of attack by fire or other destructive means from concealed positions on a moving or temporarily halted enemy (FM 3-90). Ambush patrols are combat patrols with missions to establish and execute ambushes to harass or destroy insurgents or capture personnel and equipment. (For further information on ambush, see FM 3-21.8 and FM 3-21.10.) By eliminating insurgents, an ambush contributes to improving public order and safety, a key civil security subtask. It may be used in *clear-hold-build* operations or strike operations.

#### Types

The two types of ambushes follow:

- A *point ambush* involves elements deployed to support the attack of a single killing zone.
- An *area ambush* involves elements deployed as multiple, related, point ambushes.

#### Categories

Based on the amount of preparation time, ambushes can be hasty or deliberate.

- A *hasty ambush* is an immediate action drill of a friendly force with little or no information on the insurgent force. The discovery of a nearby insurgent element, usually moving, provides a limited opportunity for a friendly force to hastily occupy a position from which to ambush the guerrilla.
- A *deliberate ambush* is a planned operation against a specific insurgent force. Sufficient detailed information of the enemy force, such as the size, nature, organization, armament, equipment, route, direction or movement timeline is available to permit the detailed planning of an ambush.

#### Characterizations

An ambush is characterized as either near or far. These are based upon the proximity of the insurgent to friendly forces.

- A *near ambush* is an ambush with the assault element within hand grenade distance of the kill zone (less than 50 meters). Close terrain such as urban, jungle, and heavy woodlands may require this positioning. It may also be appropriate in open or mountainous terrain in a "rise from the ground" ambush.
- A *far ambush* is an ambush with the assault element beyond reasonable assaulting distance of the kill zone (beyond 50 meters).This location may be appropriate in open terrain offering good fields of fire.

## Control Measures

The ambush commander's control of all elements at the ambush site is critical. This includes the initial occupation, time in position, execution, and withdrawal. Commanders should develop control measures for the—

- Occupation.
- Execution.
- Fire control measures.
- Fratricide and collateral damage prevention, especially if assault element or a nonlinear ambush.
- Initiation of assault and actions on the objective.
- Tactical site exploitation.

## Considerations

Well-planned and well-executed ambushes are a useful offensive technique to employ against insurgents. It is an effective technique to interdict and disrupt movement of insurgent forces within an area. Specific considerations using elements of the mission variables are—

### *Mission*

5-150. Tactical units should attempt to ambush insurgents in manners they least expect and in ways that minimize compromise by and risk to the HN populace. It is difficult for US forces to emplace an ambush in populated areas due to size of units, appearance, and insurgent presence amongst the population.

### *Enemy*

Commanders ensure ambush plans are flexible as to allow adjustment and initiative at the ambush site.

### *Troops and Support Available*

A small ambush party is generally more practical but likely less secure. The size of the party depends on the size of the unit targeted, the estimated insurgent strength in the area and an analysis of operational risk. In COIN, some popular units used to execute ambushes are small capture teams (SCTs), small observation teams (SOTs) and small kill teams (SKTs). Most SCTs, SOTs, and SKTs in an urban AO conduct area ambushes.

## Techniques

An ambush in COIN has five basic phases: planning, organizing, moving, occupying, and executing.

### Plan

Key steps in planning a deliberate ambush include—

- Determine the target and purpose of the ambush (kill or capture).
- Determine the advantages and disadvantage for executing the ambush during limited visibility, amongst the population, or vicinity of sensitive sites.
- Determine if the ambush will be a point ambush or an area ambush.
- Select the appropriate ambush formation.
- Determine if the ambush will be a near ambush or a far ambush.
- Determine communication requirements.
- Determine weapons requirements and limitations.
- Determine compromise contingency plans.
- Determine what, if any, special equipment is required. This includes money for damage compensation or the need to video or photograph the area for documentation.
- Withdrawal.

*Organization*

An ambush patrol is organized in the same manner as other combat patrols to include a headquarters, an assault element, a support element, and a security element. If an ambush site is to be occupied for an extended period, double ambush forces may be organized to allow continuous coverage. One ambush force occupies the site while the other conducts routine maintenance, rests, and eats at the objective rallying point or alternate and supplementary concealed locations. They alternate on command, usually after no more than eight hours.

*Movement*

Deliberate ambushes should include an objective rally point (ORP). Units should plan movement to the ORP, from the ORP to the ambush site and back and withdrawal from the ambush site is back to the ORP or to another final destination. In addition, leaders should plan movement that allows the unit to enter the ambush site from the rear and avoid moving into the kill zone or across the suspected route of the enemy force to be ambushed. In COIN, the presence of people near the ambush positions requires a carefully planned movement. Units should use maps, imagery, video, HUMINT, and aerial photographs to analyze the terrain and HN population. If possible, units conduct ground reconnaissance and avoid selecting obvious ambush sites. Surprise is even more difficult to achieve in these areas. An ambush site should provide—

- Clear fields of fire.
- Concealed positions.
- Canalization of the insurgents into the killing zone.
- Little or no cover and concealment in the kill zone.

- Covered routes of withdrawal (to enable the ambush force to break contact and avoid pursuit).
- No egress route for the insurgent force.
- A defensible position if compromised.

*Occupation*

In COIN, especially urban environments, it can be extremely difficult to occupy ambush sites or positions uncompromised due to locals. As a rule, the ambush force occupies the ambush site at the latest possible time permitted by the tactical situation and the amount of site preparation required. This not only reduces the risk of discovery, but also reduces the time Soldiers must remain still and in position.

***Techniques***

Some techniques include using—

- Stay-behind elements.
- Alternate infiltration methods such as HN security force vehicles.
- Subterranean (urban) or subterranean structures (rural).
- Feints.
- Deception.

***Firing Positions***

The unit typically moves into the ambush site from the rear. Security elements are positioned first to prevent surprise while the ambush is being established. Position automatic and precision fire weapons so each can fire along the entire kill zone. If this is impossible, then ensure that automatic weapons have overlapping sectors of fire. The point is to cover the entire kill zone and to achieve a large volume of near simultaneous concentrated fires into the kill zone, fires that can inflict maximum damage on the insurgent. The unit leader then selects a position where he can see when to initiate the ambush. Claymores, explosives, and M203 grenade launchers are examples of what may be used to cover any dead space left by the automatic and precision fire weapons. All weapons are assigned sectors of fire to provide mutual support. Multiple positions also provide interlocking or overlapping support. The unit leader sets a time by which positions are to be prepared.

***Kill Zone***

If Soldiers must enter the kill zone to place booby traps, special-purpose munitions, or expedient devices, they must remove any tracks or signs that might alert the insurgents and compromise the ambush. Under a strict ROE, units may choose to record the ambush using a video cameras.

*Execution*

A clear target engagement criteria is all that is needed to execute an ambush. Audible and visible signals such as whistles and pyrotechnics must be changed often to avoid establishing patterns, or alerting the insurgents to friendly actions or positions.

- A signal by the security force to alert the patrol leader to the insurgent's approach may be given by hand-and-arm signals, radio, as a quiet voice message, transmission of a prearranged number of taps, or by signaling with the push-to-talk switch or field telephone when there is no danger that wire between positions will compromise the ambush.

- A signal to initiate the ambush given by the patrol leader or a designated individual may be a shot or the detonation of mines or other types of explosives. The ambush should be initiated with a mass casualty-producing weapon (claymore, machine gun, or similar system).
- A signal for lifting or shifting fires may be given by voice command, whistles, or pyrotechnics. All fire stops immediately so the assault can be made before the insurgent can react.
- A signal for withdrawal may also be by voice command, whistles, or pyrotechnics.
- Surprise must be achieved or the attack is not an ambush. Surprise allows the ambush force to seize and retain control of the situation. Units achieve surprise by careful planning, preparation, and execution. Concealment and fire discipline are also critically important.

### Formations

Whether independent or part of an area ambush, a point ambush is positioned along the expected avenue of approach of the insurgent force. The selection of the type of ambush formation is important, because it determines the volume of concentrated fire required to isolate, trap, and destroy the insurgents. The formation to be used is determined by carefully considering possible formations and the advantages and disadvantages of each in relation to terrain; conditions of visibility, forces, weapons, and equipment ease or difficulty of control; force to be attacked; and overall combat situation. Types of ambush formations include linear, L-shaped, Z-shaped, T-shaped, V-shaped, triangle, and box.

#### *Linear*

In a linear ambush, the attack element is deployed generally parallel to the insurgent force's route of movement (road, trail, and stream). This positions the attack element parallel to the long axis of the kill zone and subjects the insurgent force to heavy flanking fire (Figure 5-9). An advantage of the linear formation is its relative ease of control under all conditions of visibility. The size of the force that can be trapped in the kill zone is limited by the area the attack element can effectively cover with highly concentrated fire. The force is trapped in the kill zone by natural obstacles, mines, booby traps, or expedient devices, and direct and indirect fires.

A disadvantage of the linear formation is the chance that lateral dispersion of the force may be too great for effective coverage. The linear formation is appropriate in close terrain that restricts insurgent maneuver, and in open terrain where one flank is restricted by natural obstacles, mines, booby traps, special-purpose munitions, or expedient devices. Similar obstacles and casualty producing systems can be placed between the attack element and the kill zone to provide protection from insurgent counter ambush measures. When a destruction ambush is deployed in this manner, access lanes are left so the force in the kill zone can be assaulted. The line formation can be effectively used in a "rise from the ground" ambush in terrain seemingly unsuitable for ambush.

*L-Shaped*

The L-shaped formation (Figure 5-10) is a variation of the linear formation. The long side of the attack element is parallel to the kill zone and delivers flanking fire. The short side of the attack element is at the end of, and at right angles to, the kill zone and delivers enfilading fire that interlocks with fire from the other leg.

The L-shaped formation is flexible. It can be established on a curving stretch of a trail, near a stream, or at a sharp bend in a trail or stream. When appropriate, fire from the short leg can be shifted to parallel the long leg if the insurgent force attempts to assault or escape in the opposite direction. In addition, the short leg prevents escape in that direction or reinforcement from that direction. Positive means of controlling fires, such as aiming stakes, are needed to prevent the fire from one leg hitting Soldiers positioned on the other leg.

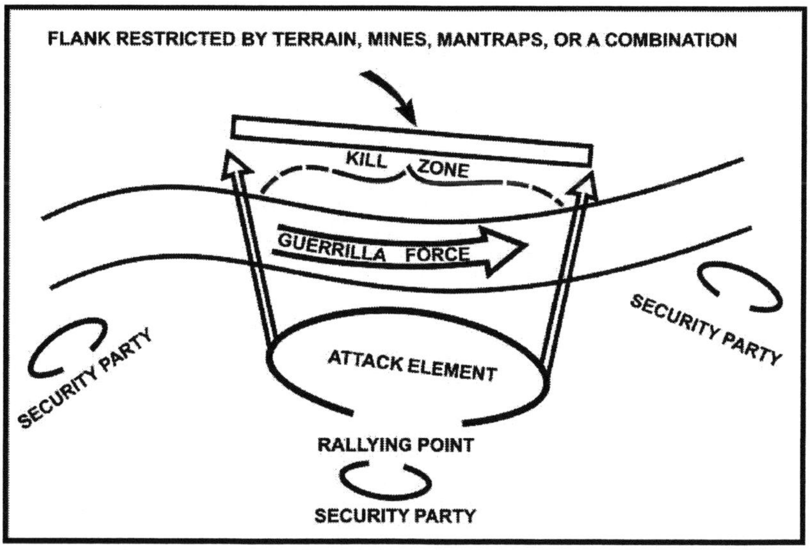


Figure 5-9. Linear formation ambush.

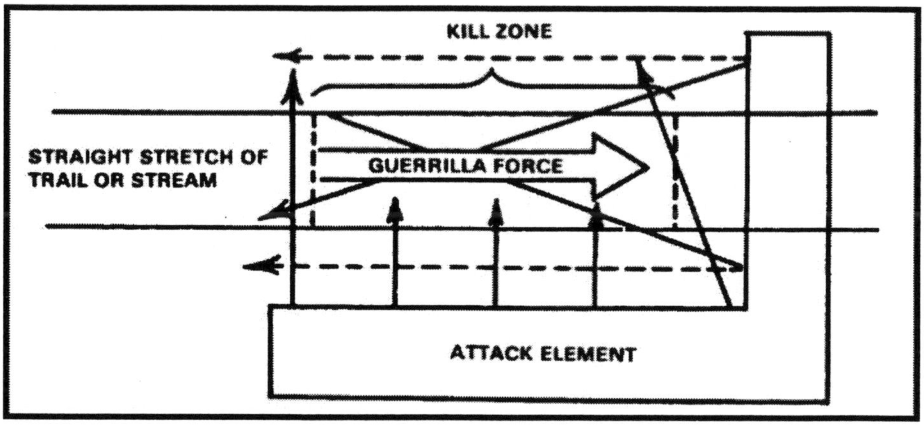


Figure 5-10. L-shaped formation ambush.

*Other*

Other traditional ambush formations that are highly METT-TC dependent and usually better for rural operations, include the following.

***Z-Shaped Formation***

The Z-shaped formation is a variation of the L-formation. The attack force is deployed as in the L-formation, but with an additional side so that the formation resembles the letter Z. The additional side may serve to engage a force attempting to relieve or reinforce the guerrillas, restrict a flank, prevent an envelopment of the ambush force.

***T-Shaped Formation***

In the T-shaped formation, the attack element is deployed across, and at right angles to, the route of movement of the hostile force so that the attack element and the target form the letter T. This formation can be used day or night to establish a purely harassing ambush. It can be used at night to interdict movement through open, hard-to-seal areas.

***V-Shaped Formation***

The V-shaped attack element is deployed along both sides of the insurgent route of movement so it forms a V. Care is taken to ensure that neither group fires into the other. The V-formation is suited for open terrain, but can also be used in the jungle.

***Triangle Formation***

The triangle is a variation of the V-formation that can be employed in three ways. The most common technique is the closed triangle. The attack element is deployed in three groups, positioned so that they form a triangle. An automatic weapon is placed at each point of the triangle and positioned so it can be shifted quickly to interlock with either of the others. Elements are positioned so their fields of fire overlap. Mortars may be positioned inside the triangle. When deployed in this manner, the triangle ambush becomes a small unit strongpoint that is used to interdict night movement through open areas when insurgent strategy is likely to be from any direction. Advantages include ease of control, all-round security, and guerrillas approaching from any direction can be fired on by at least two automatic weapons. Disadvantages include the requirement for an ambush force of platoon size or larger to reduce the danger of being overrun by a large guerrilla force; one or more legs of the triangle may come under guerrilla enfilade fire; and lack of dispersion, particularly at the points, increases danger from guerrilla mortar fire.

***Box Formation***

The box formation is similar in purpose to the triangle ambush. The unit is deployed in four elements positioned so each element becomes a corner of a square or rectangle. Advantages and disadvantages are much the same as the triangle formation.

## Variations

Common ambush variations during counterinsurgency operations include—

**Baited Trap**

A variation of the area ambush is the “baited trap” ambush. A central kill zone is established along the insurgent’s avenue of approach. Point ambushes are established along the routes over which units supporting or reinforcing the insurgent force will have to approach. The insurgents in the central kill zone serve as “bait” to lure relieving or reinforcing insurgent units into the kill zones of the outlying ambushes. Items such as infrastructure, sensitive equipment, caches, and security measures can be used as bait.

*Spider Hole*

This type of point ambush is designed for open areas that lack the cover and concealment and other features normally desirable in a good ambush site. This technique is effective in less populated rural areas or urban areas with subterranean. Concealed in a “spider hole," a type of covered and concealed foxhole, the attack element is deployed in the formation best suited to the overall situation. Soil is carefully removed and positions expertly camouflaged. This ambush takes advantage of the tendency of patrols and other units to relax in areas that do not appear to favor ambush. The chief disadvantage of this technique is that the ambush element’s vulnerability if compromised.

***Sniper Operations***

Sniper operations are effective for both insurgent and counterinsurgent in the COIN environment. Snipers, if employed correctly, are a COIN force multiplier and provide the commander an immediate means to enter the insurgent decision cycle.

Small kill teams (SKTs), small observation teams (SOTs), and small capture teams (SCTs) differ by task, manning, and equipment. SKT focuses on destroying of insurgent elements. SCTs focus on the capture of those elements. SOTs are for reconnaissance. In COIN, more may be gained by capturing than by killing the insurgent. However, SKTs can give commanders a deterrent to insurgent activity.

Sniper teams sometimes as part of SKTs, SOTs, or SCTs, are employed in populated urban areas or large rural areas with adequate fields of fire. For mission success, all three teams employ specific point and area weapons systems. All three also rely on redundant communications, marking materials, detailed infiltration plans, exfiltration, mutual support, security, compromise contingencies, and engagement criteria

**Considerations**

Sniper employment in a COIN must be carefully considered. The infiltration or exfiltration of snipers, SKTs, and SCTs must be meticulously planned with the understanding of the habits, behaviors, and density of the local population. Besides the basic mission of precision fires and surgical elimination of individuals, they are valuable for ISR collection, route and area security augmentation, countersniper operations, and counter mortar or rocket operations. Compromise and contingency planning is critical to sniper employment. Commanders base this detailed planning on a thorough analysis of the mission variables with specific attention to terrain and civil considerations. Specific considerations using elements of the mission variables are—

*Mission*

- Snipers must have clear engagement criteria and a thorough understanding of the ROE. Engagement criteria must be refined so the individual sniper can make the correct engagement decision while the target is engageable.
- Once an engagement area and a sniper position are identified, then leaders should determine a mutually supporting position for a security team to occupy.
- Current and planned locations of sniper teams should be tracked by the headquarters responsible for the area in which the snipers are operating. Often this means designating their locations as a no fire area that can be tracked by all forces operating in the area to prevent fratricide.

*Enemy*

- Sniper positions need to be defendable in case of compromise by the enemy.

*Terrain and Weather*

- Snipers, small kill teams, small observation teams, and small capture teams can aid commanders in denying terrain and freedom of movement to insurgent elements while providing security to sensitive or critical sites in both rural and urban areas.

*Troops and Support Available*

- In addition to a robust communications capability, commanders resource these elements with linguists, medical personnel, surveillance equipment, and sustainment resources. The QRF is typically involved in the detailed planning of these operations and can be deployed forward to minimize reaction time.
- SKTs, SOTs or SCTs are task-organized with sniper-qualified personnel or squad designated marksman, weapons squads, and specialty skills personnel. They may be comprised of sniper-qualified individuals or squad designated marksman.

**Techniques**

All sniper operations in COIN should consider sniper positions, sniper security sniper insertion and sniper extraction.

*Positions*

- Tactical leaders must continually analyze the OE for potential sniper team overwatch positions. Leaders reconnoiter to determine the suitability of potential positions. Leaders reconnoiter to determine the suitability of potential positions. When employing sniper teams, leaders ensure each position provides mutual support against a threat.
- The insurgent often will engage from areas of passive support, and where he has a sense of security. Detailed reconnaissance and accurate knowledge of the OE allows

leaders to recognize advantages and disadvantages of different positions and types of areas. There are two basic areas for sniper positions in an urban environment—residential areas and industrial areas. Both areas include the use of elevated positions or ground level positions. Roofs are not always the best place to operate since roofs are often not the highest location in the area and are often wide open with little cover. Successful position selection techniques in COIN are—

***Rent Uninhabited Dwellings or Other Structures***

- Teams are careful not to establish a pattern or expose the property owner to the insurgency. Owners may compromise the snipers.

***Positions in Locally Occupied Homes***

- Teams, however, must consolidate the family in one part of the house with a security element. When exfiltrating the residence, compensation should be provided to the owner.

***Stay-Behind Opportunities***

- Teams may have the opportunity to achieve surprise if insurgents attempt to return to areas after another counterinsurgent unit has conducted an operation.

*Security*

- Sniper teams should move with a security element (squad or platoon) whenever possible. This allows the teams to reach their area of operation faster and safer than if alone. The security element also protects snipers during operations. When moving with a security element, snipers follow these guidelines:
  - The leader of the security element leads the sniper team.
  - Snipers must appear to be an integral part of the security element. To do so, each sniper carries his weapon system in line with and close to his body to hide the weapon's outline and barrel length. Snipers also conceal from view all sniper-unique equipment (optics and ghillie suits).
  - Sniper uniforms must be the same as that of security element members.
  - Snipers and element members maintain proper intervals and positions in the element formation.

*Infiltration*

- The key to sniper infiltration is undetected occupation of hide positions. An insertion may be as simple as a drop off in close proximity and moving dismounted to the position. It can be as complex as conducting a cordon and search/knock and leaving a stay-behind team. Planning considerations for insertion include the mission variables and a detailed understanding of the terrain and people. Planning considerations include—streetlights, barking dogs, nightlife establishments, insertion vehicle noise (HMMWV engines, vehicle ramps hitting the ground), and local nationals who may sleep on their roof or outdoors during hot summer months.

*Exfiltration*

- Exfiltration is just as important; snipers do not want to compromise a position you may want to use again in the future. The amount of time spent in a Sniper Position is dependent on the situation. Sniper teams should have a preplanned emergency exfiltration route to a safe zone, known by all supporting elements. The snipers can withdraw dismounted, during hours of limited visibility, to a designated rally point, where the QRF can retrieve them
- Some units use a quick reaction force (QRF) as a means of reinforcement or emergency extraction, as well as maintaining or regaining contact with insurgents engaged by the snipers. When properly employed, the US sniper teams can contribute significantly to the fight by overwatching key areas, serving as an economy of force, eliminating insurgents, and causing uncertainty within the insurgents. For further information on conventional sniper employment techniques, see FM 3-21.11.

***COIN Patrols***

A patrol is a detachment sent out by a larger unit to conduct a combat or reconnaissance operation. A patrol may be a fire team, squad, platoon, or company. Conventional patrolling doctrine applies to counterinsurgency operations, but some modifications must be made to account for the insurgent's activities and the operational environment. Aggressive patrolling in an area greatly reduces the insurgents' freedom of movement, disrupts operations, and weakens their influence on the local population. Furthermore, Patrolling becomes more significant in counterinsurgency operations because of the difficulty in locating and identifying insurgent forces. This section discusses the important role patrols play in defeating the insurgent force. There are two types of patrolling in COIN operations: Reconnaissance patrols and combat patrols. Patrols are an integral part of *clear-hold-build* operations, *strike* operations, and PRC operations.

## Considerations

Specific considerations using elements of the mission variables for COIN patrols are—

*Mission*

Leaders must brief all patrol members as to the task and purpose of the operation. Every COIN patrol has a task and purpose that is nested within one of the LOEs.

Leaders must plan patrol routes carefully and coordinate in detail with higher, lower, and adjacent units, to include Host Nation security forces, aviation elements, fires, ISR elements, and reserve forces. All patrols conduct rehearsals and a patrol brief at a minimum.

*Enemy*

Small-unit patrols are more effective than larger unit patrols against insurgent activities. This is because they can cover more territory than a large unit, and are more difficult to track and predict. This keeps the insurgent off balance.

*Terrain and Weather*

Patrol leaders must learn and know the routes, terrain, and weather implications on the HN population, the counterinsurgent, and the insurgent. For example, urban patrol techniques will differ from rural patrol techniques.

*Troops and Support Available*

Communication between patrol vehicles, riflemen, and higher headquarters is essential. Within the patrol, radios, data transfer devices, voice commands, and visual signals may be used. Vehicular-mounted radios and data systems are usually the best means for communication within the patrol and to higher headquarters. Aircraft may be used to relay radio messages for long distance patrol communication. Blue Force Tracker text messages or single channel TACSAT provide excellent means of maintaining long haul communications.

Tactical units may be involved in patrolling in one of three ways: patrol as a complete unit; provide subordinate unit for patrols (as directed by higher), or send out patrols to support their own operation. Normally, the planned action at the objective determines the type of the patrol, usually categorized as either combat or reconnaissance. Patrols may be mounted or dismounted or a combination.

*Civil Considerations*

Patrols are often the easiest way for tactical units to engage the HN population across multiple LOEs.

***Mounted Patrols versus Dismounted Patrols***

Patrols may be mounted or dismounted. Mounted patrols allow greater coverage of distances than dismounted patrols but sacrifice interaction with the populace and the opportunity to conduct more effective information engagements. Mounted patrols can operate in insurgent controlled areas too dangerous for dismounted patrols while carrying more or heavier equipment, weapons, and ammunition.

Dismounted patrols can be physically demanding and patrol members must be in good shape. Additionally, contact with insurgents in close combat is physically demanding. The patrol leader ensures patrol members carry only mission essential equipment.

A mounted patrol is prepared in the same manner as a dismounted patrol. Leaders ensure vehicles are mechanically fully mission capable and properly supplied with fuel, oil, ammunition, and water. Drivers and other personnel receive the same patrol brief as dismounted patrol members.

A mixture of mounted and dismounted patrolling can provide greater flexibility. The dismounted element can interact with the population. The mounted element provides increased firepower, communications, and CASEVAC.

***Urban Patrols***

The basics of patrolling remain the same in both urban and rural environments; however, the differences between the urban and the rural environment require specific patrol considerations. Urban areas have a high population density and patrols must be prepared for population related incidents. The population may interact with the patrol in many ways, such as asking for medical attention or demonstrating against the presence of the patrol itself. Basic urban patrolling consideration include—

- While contact with insurgents may happen, contact with elements of the population is certain.
- If available, armored vehicles should be ready to rapidly reinforce urban patrols to provide additional firepower.
- At least one Soldier in each squad or team should be dedicated to scanning the rooftops and upper level windows.
- Actions at a halt must include 360-degree security. Soldiers should seek cover and face out. Cover in an urban environment may be a light pole, a building corner or even a parked car.
- Urban patrols should thoroughly scan the far side of all open areas, since insurgents will use them to achieve stand-off.
- If contact is likely, then the patrol should move by bounds. Moving by bounds, with one element overwatching another element, is used in urban terrain just as in rural terrain.
- React to contact in an urban environment often includes the basic elements of a cordon and search. Once contact is made, return fire is initiated and simultaneously the area should be cordoned to prevent the insurgent's escape.

***Combined Patrols***

The combined patrol is a patrol conducted with Host Nation security forces and US units. They are an important piece in increasing the legitimacy of the HN government and improving the skills of the HN security forces.

*Mutual Support*

Commanders consider mutual support when task-organizing forces and assigning areas of operations. Mutual support is support units render each other against an enemy, because of their assigned tasks, their position relative to each other and to the enemy, and their inherent capabilities (JP 1-02). Mutual support has two aspects: supporting range and supporting distance.

***Supporting Range***

Supporting range is the distance one unit may be geographically separated from a second unit yet remain within the maximum range of the second unit's weapons systems (FM 3-0). It depends on available weapons systems and is normally the maximum range of the supporting unit's indirect fire weapons. For small units such as squads, sections, or platoons, it is the distance between two units that their direct fires can cover effectively. If one unit cannot effectively or safely fire in support of the other unit, they may be out of supporting range even though their weapons have the requisite range.

***Supporting Distance***

Supporting distance is the distance between two units that can be traveled in time for one to come to the aid of the other. It is a function of terrain and mobility, distance, enemy capabilities, friendly capabilities, and reaction time. During counterinsurgency operations, commanders should always consider supporting distance. Units maintain mutual support when one unit can draw on another unit's capabilities for support.

## Types of Patrols

There are two types of patrols:

*Reconnaissance Patrols*

This type of patrol collects information and confirms or disproves the accuracy of information previously received. They are used to locate insurgent units and base camps, to reconnoiter specific locations, locate leaders, and gather intelligence. Reconnaissance patrols provide the commander with timely, accurate information of insurgents, the population, and the terrain. This information is vital in making tactical decisions. Leaders must ensure that no pattern is established that would allow an insurgent force to ambush reconnaissance units. Reconnaissance patrols are further classified into two types.

***Route Reconnaissance Patrols***

These are a form of reconnaissance that focuses along a specific line of communication, such as a road, railway, or cross-country mobility corridor. It provides new or updated information on route conditions, such as obstacles and bridge classifications, insurgent and civilian activity, and traffic patterns along the route. A route reconnaissance includes not only the route itself, but also all terrain along the route from which the insurgent could influence the friendly force's movement.

***Zone Reconnaissance Patrols***

These are conducted to obtain information on enemy, terrain, people, and routes within a specified zone. The commander may require information of an extended area, or may desire information of several locations within an area. A zone reconnaissance patrol secures this information by reconnoitering the area, maintaining surveillance over the area, or by making the coordinated area reconnaissance of designated locations within the area.

***Area Reconnaissance Patrols***

This is conducted to obtain information on a specific location or small specific area, usually a known or suspected position or activity. An area reconnaissance patrol secures this information by reconnoitering the location or by maintaining surveillance over the location.

In addition to reaching the objective without discovery, a reconnaissance patrol also tries to conduct its reconnaissance or surveillance without being discovered. Stealth, patience, and maximum use of concealment are mandatory. A reconnaissance patrol must be prepared to fight to protect itself.

Continual technological improvements have a significant impact on reconnaissance abilities. Sensors and video cameras can be emplaced to be remotely monitored. Computer and electronic technology must be leveraged to extract information from a wide array of technological systems.

***Day and Night Patrols***

These use about the same techniques as other patrols. The main differences are—

- *Day Reconnaissance* requires greater use of concealment. The patrol is more likely to be seen than at night and usually will not be able to move as close to the objective.
- *Night Reconnaissance* requires stealth. Sounds carry farther at night, and reduced visibility usually requires a closer approach to the objective.

*Combat Patrols*

In counterinsurgency operations, the term security patrol has often been used; however, this is still a combat patrol. A combat patrol provides security and harasses, destroys, or captures enemy troops, equipment, and installations. A combat patrol also collects and reports information, whether related to its mission or not. Combat patrols in a counterinsurgency include raids, ambushes, security, saturation, and satellite patrols. Regardless of the name, all counterinsurgency combat patrols have the general mission of seeking out and attacking targets of opportunity.

***Raid and Ambush Patrols***

Raids and ambushes were discussed earlier in Chapter 5.

***Security Patrols***

The difference between a security patrol and a raid or ambush patrol is that combat activity is not the primary mission of the security patrol. This is true even if combat is expected during the patrol. Security patrols normally seek to control critical roads and trails, maintain contact between villages and units, provide security for friendly forces, provide security in rural areas, and interdict insurgent routes of supply and communication.

***Saturation Patrols***

This is when units use numerous combat patrols to saturate an area of suspected insurgent activity by moving over planned and coordinated routes, which are changed frequently to avoid establishing patterns. Saturation patrols are extremely effective against insurgents. Use of saturation patrols results in the following:

- Denial of an area to an insurgent force as it seeks to avoid contact with the saturation patrols.
- Ability to harass insurgent forces.
- Opportunity to discover insurgent forces.
- Chance to gain an intimate knowledge of the area of operations.
- Chance to reassure the local population that the government provides protection and security.

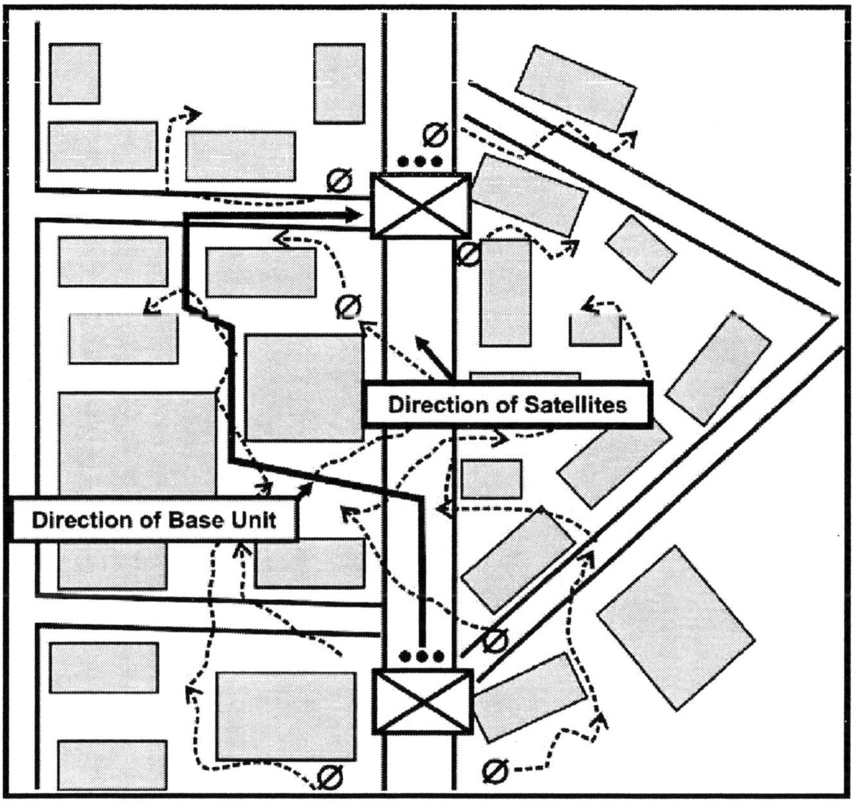


Figure 5-11. Satellite patrol movement.

### *Satellite Patrols*

This patrol technique adds depth to a patrol, deters ambushes, and provides patrols with a maneuver element on enemy contact. Figure 5-11 shows a satellite patrol moving through a built-up area.

The satellite patrol uses a base unit to control smaller units, or satellites, that leave and return to the base unit. The advantage of this technique is the unpredictability, to the enemy, of the route, size, locations, and the patrol's overall axis of advance. Satellite patrols are given either an area or an axis of movement. As with all other patrols, they should have a specific task and purpose. Units have specific requirements including—

### *Organization*

At a minimum, the patrol has one base and one satellite unit.

### *Size*

The size of the base unit and satellites is METT-TC dependent. Normally, a satellite unit consists of either a squad or a fire team. All units must be able to defend themselves until reinforcements arrive.

*Command and Control*

The base unit is under the direct control of the senior leader and must have radio communications with each of the satellites units. This facilitates control and actions if contact is made. The base unit sets the pace and maintains the general direction of the patrol. Controlling multiple small satellite patrols is difficult and requires an experienced leader and excellent communications.

*Movement*

All units must know the overall route and if possible, left and right boundaries. Both the base unit and the satellite units move in ways to confuse the enemy as to the patrol's actual axis of advance. Standard movement techniques are still used. Satellites move away from the base unit for limited periods of time to inspect potential ambush sites, dead spaces, parallel roads, or other assigned missions. The time that the satellite is separated from the base unit should be prescribed by the patrol leader prior to departure.

*Training*

Units may experience initial difficulty with this technique because of the dispersed, unpredictable, and seemingly random movement of the satellite patrols. To properly execute the technique, units must train and practice.

*Actions on Contact*

The unit in contact reacts normally. All other units move towards the unit in contact. The satellite patrol leader coordinates, as needed, their routes, actions, and linkup.

*Patrol Debrief*

One of the best ways to turn information into intelligence is to conduct a patrol debrief following every patrol. When the patrol is over, the unit has not completed its mission. The leader must ensure that all the information collected during the patrol is turned over to the appropriate staff section for evaluation. Additionally, all patrol members must be debriefed to collect any information not already identified. The unit must also conduct an after-action review of the entire mission from start to finish. The unit must record what operations were executed correctly and what could have been done better. This information must be passed on so that others can learn from the operation. Finally, unit members must account for and clean their equipment in order to be ready for the next operation. The debrief, at a minimum, should include—

- Specifics on the five W's (who, what, when, where, why) and how.
- Photos or sketches.
- Answers to priority intelligence requirements (PIRs) and other information requirements.
- Actionable intelligence.
- Recap of route.
- Reports of enemy contact.
- Engagements conducted:
    - Who engaged the guerrilla force?

  - What concerns have the HN population brought to the patrols attention?
  - What promises did the US make as a result of these concerns?
- Tips or actionable information for intelligence.
- Noteworthy observations (propaganda, graffiti, and so on).
- Changes in the HN population, which leaders can use to improve understanding of community dynamics. These changes may be detected in—
  - People's attitudes toward coalition forces.
  - Local infrastructure.
  - Civil leadership.
  - Local organization.
  - Civil institutions.

## Section III—Other Lines of Effort

Offensive operations enable and complement the other lines of effort—support HN security forces, restore essential services, provide support to economic and infrastructure development, provide support to governance, and conduct information tasks. Without the increased security resulting from offensive operations, units may not be able to accomplish any significant tasks in the other LOE.

### *Support Host Nation Security Forces*

The preferred method is to conduct all offensive operations combined with HN security forces to improve their tactical and technical competence and develop their professionalism. This technique also takes advantage of both forces' strengths and capabilities, especially the HN security forces' language skills and cultural knowledge. An example of this is a combined cordon and search of a suspected insurgent village where US forces establish the outer cordon and the HN security forces conduct the actual search of the village.

### *Support to Governance*

In COIN, tactical units must consider how offensive operations can be used to assist their efforts to strengthen the local government. For instance, a unit might conduct a combined cordon and search based on a tip provided by the local mayor, assist in securing polling sites during local or national elections, or set the conditions for completion of a government project. Units must remember that all offensive operations support one or another group's political agenda. Eliminating one insurgent group may lead to another group filling the power vacuum.

### *Restore Essential Services*

Tactical units must consider how to incorporate essential service projects as a complementary portion of their offensive operations. For instance, delivering water to a neighborhood without clean drinking water can be useful in several ways. For example, it may afford the opportunity to gather intelligence for an upcoming raid, reconnoiter for a cordon and search, serve as a deception for the emplacement of a sniper, or allow the covert positioning of a force for a later attack. Units must be careful about too closely linking projects and offensive operations to ensure that they do not inadvertently decrease the population's support.

***Support to Economic and Infrastructure Development***

Tactical units conducting COIN must consider how offensive operations can assist in stimulating the economy of their local area. For instance, a raid on an insurgent group that has been extorting money from the local factory to finance their operations may dramatically increase the capital available to conduct repairs to the factory. This in turn creates jobs in the village. As in essential services projects, the economic project may serve as a means to gather intelligence for an upcoming raid, reconnoiter for a cordon and search, serve as a deception for a sniper team, or be used to position forces for an attack. Units may decide not to conduct an offensive operation due to the economic impact of the operation.

***Conduct Information Engagement***

All tactical unit offensive operations must consider and incorporate information engagements. Options range from simple explanation of the purpose of a raid to the head of the household, to handing out leaflets as part of the outer cordon, to using tactical PSYOP teams to keep the population from interfering with operations, to attending a city council meeting to explain a recent raid. In each endeavor, the counterinsurgent force must use information tasks to increase the HN government's legitimacy, isolate the insurgent, and garner the support of the population.

## SUMMARY

Offensive operations are a critical part of counterinsurgency and help Host Nations establish civil security and civil control. Key operations include search and attacks, cordon and searches, searches, raids, ambushes, and COIN patrols, especially in the "clear" phase of a *clear-hold-build* operation. Offensive operations can support other LOEs such as support Host Nation security forces and restore essential services. Commanders conduct offensive operations in COIN with constant considerations of the population in support of the seven COIN LOEs.

# 6. DEFENSIVE CONSIDERATIONS IN COUNTERINSURGENCY

*"A defensive attitude is almost always to be deprecated and only under certain special circumstances is it to be recommended. The operations of regular troops in such warfare (small wars) must never be allowed to stagnate; the troops must be up and doing, striking their adversaries when these attempt resistance, hunting them down when they shun combat."*

COL C. E. Callwell, *Small Wars*, 1896

Insurgents will not passively let counterinsurgent forces disrupt their plans and isolate them from the populace. They will attack counterinsurgent forces when they feel that they have a good chance of success. Additionally, the insurgents will choose targets with an eye towards reducing the legitimacy of the established government. The conduct of defensive operations in COIN is very much like that in a conventional defense. However, in COIN, the counterinsurgent must be concerned with securing the HN population, HN government, and infrastructure. This chapter will examine defensive operations, defensive operations as they apply specifically to establishing civil security and civil control, defensive operations as they apply to the other lines of effort, and security operations.

## Section I—Overview

Tactical formations conduct three types of defensive operations in every environment—area defense, mobile defense, and retrograde. At the tactical level, counterinsurgency units focus mostly on executing an area defense, although, there may be instances where mobile defense or retrograde are appropriate.

### *Area Defense*

This type of defensive operation concentrates on denying enemy forces access to designated terrain for a specific time rather than destroying the enemy outright (FM 3-0). In counterinsurgencies, both the insurgency and the counterinsurgent are vying for the support of the population as if it were terrain. The focus of the area defense is on retaining terrain where the bulk of the defending force positions itself in mutually supporting, prepared positions. In counterinsurgencies, commanders choose an area defense to secure the population by living among the population employing the seven COIN lines of effort. Examples of an area defense in counterinsurgencies are a combat outpost next to a city market, a permanent squad position at an electrical plant, and a checkpoint near a police station.

### *Mobile Defense*

This type of defensive operation concentrates on the destruction or defeat of the enemy through a decisive attack by a striking force (FM 3-0). The mobile defense focuses on defeating or destroying the enemy by allowing him to advance to a point where he is exposed to a decisive counterattack by the striking force. Mobile defense operations are less common in counterinsurgencies, because insurgents rarely mass. However, strike operations can be useful in destroying guerrilla bases. Another example of a mobile defense at the tactical level is the use of quick reaction forces to exploit tactical intelligence.

### *Retrograde*

This type of defensive operation involves organized movement away from the enemy (FM 3-0). The enemy may force these operations, or a commander may execute them voluntarily. In counterinsurgencies, retrograde operations are generally conducted intentionally after a unit is transitioning responsibility of an area of operations to a Host Nation security force or a relief in place with another US unit.

### *Major Counterinsurgency Operations*

In *clear-hold-build* operations, defensive operations are foremost in the "hold" phase, although operations occur during all three phases. These operations help to establish and maintain civil security and civil control, enabling units to set conditions across the seven COIN LOEs. Defensive operations in the "hold" phase deal primarily with securing the HN population and isolating insurgents from their support.

In strike operations and populace and resource control operations, defensive operations are conducted in concert with stability and offensive operations. In strike operations, units conduct defensive operations to establish bases where they can conduct offensive operations. In PRC operations, units conduct defensive operations to protect the populace and its materiel resources from insurgents, to deny insurgents access to the populace and materiel resources and to identify and eliminate the insurgents, their organization, their activities, and influence while doing so.

## Section II—Civil Security and Control

Tactical units conduct most defensive operations under either the establish civil security or establish civil control lines of effort. In COIN operations, especially *clear-hold-build* operations, the most common method that tactical units employ to secure the population is the establishment of bases. Units use site selection techniques to decide where to put bases and focus security. Certain defensive actions like counter—ambush, counter—IED, and counter—sniper clearly contribute to increased civil security or increased civil control as they reduce violence against HN and US forces, as well as the HN population. Throughout, units must use appropriate protection measures.

### *Site Selection*

Brigades, battalions, and companies routinely possess enormous AOs during counterinsurgencies, some as large as US states. Within these large AOs, there can be hundreds or thousands of essential service structure, governmental infrastructure, economic assets, population centers, and important leaders that may be targets for the insurgency. There are many techniques to conduct security operations at the tactical level. Tactical units need a framework to decide how best to secure the HN population, and HN infrastructure.

It is not possible to protect every asset and every point. Eventually, the insurgent will attack. A primary goal is to protect those vital assets and points to such a degree that, if they are attacked, the damage to them is minimal, while the insurgent's losses are heavy. The CARVER-P Vulnerability Prioritization Matrix is a method for units to help the HN government in deciding what HN assets to protect.

### Carver-P Risk/Vulnerability Prioritization Matrix

This matrix considers targets within the HN population, government, and infrastructure. It allows leaders to identify target vulnerability, determine corresponding risk, and then prioritize security assets. The matrix is based on seven criteria represented by the acronym CARVER-P (criticality, accessibility, recuperability, vulnerability, effects, recognizability, and psychological impact). This matrix derives from the Army risk-assessment process. Figure 6-1 shows an example format of the Carver-P matrix for the vulnerability criteria. Figure 6-2 shows an example of the same matrix completed for a power plant.

Complete the *CARVER-P Vulnerability Prioritization Matrix* using these five steps:

### *Step 1—Identification*

Identify and continually reevaluate the key structures, capabilities, organizations, and individuals in the AO that the insurgents may target. Record each target's name and location on a separate Carver-P Risk/Vulnerability Prioritization Matrix. Figure 6-2 showed an example matrix completed for a power plant.

**VULNERABILITY PRIORITIZATION MATRIX for** *[Name]* at *[Location]*

| *CARVER-P* | *Why and How* | *Val* | *Controls or Mitigation* | *PRI* |
|---|---|---|---|---|
| **C**riticality | | | | |
| **A**ccessibility | | | | |
| **R**ecoverability | | | | |
| **V**ulnerability | | | | |
| **E**ffect | | | | |
| **R**ecognizability | | | | |
| **P**sychological impact | | | | |
| **Total Value (Lower is better—max value is 21)** | | | | |

Figure 6-1. Example format for Vulnerability Prioritization Matrix.

**VULNERABILITY PRIORITIZATION MATRIX for** *Power Generation Plant* at *Location XYZ*

| *CARVER-P* | *Why and How* | *Val* | *Controls or Mitigation* | *PRI* |
|---|---|---|---|---|
| **C**riticality | Provides power for all of city | 3 | 1. Full-time military security: established<br>2. Full-time police security: established<br>3. Boundary fencing and barrier plan: building | 1 |
| **A**ccessibility | Covers large area; hard to defend; few or no barriers | 2 | | |
| **R**ecoverability | Some repair parts on hand | 2 | | |
| **V**ulnerability | Vulnerable to small-arms fire | 3 | | |
| **E**ffect | Overwhelming propaganda effect for insurgents if destroyed | 3 | | |
| **R**ecognizability | Easy to recognize | 3 | | |
| **P**sychological impact | Would show population that local government is not in control | 3 | | |
| **Total Value** *(Lower is better—max value is 21)* | | 19 | | |

Figure 6-2. Example completed CARVER-P Vulnerability Prioritization Matrix.

*Step 2—Evaluation*

Evaluate this potential target using the Criteria Evaluation Tool shown in Figure 6-3. This tool provides a simple means to determine risk either using the four generic risk statements of each criteria, or by developing or modifying similar statements.

*Step 3—Analysis*

In the Criteria Evaluation Tool (Figure 6-3), for each criteria, choose an appropriate risk statement, which corresponds to a risk level. Then, explain why and how you assessed the risk level, assign a numerical value for each of the criteria and, if needed, identify the control and mitigation for each assessment. Transfer this information to the matrix and sum the values. This sum represents the assessed desirability of the potential target from the insurgent's perspective. The higher the number the more likely the insurgent will attack the target.

*Step 4—Facilities Category*

Using this sum, identify the facilities category (Figure 6-4). Both the identification and the prioritization of vital assets and key points should be the responsibility of the Host Nation. However, US commanders should participate in the planning process for assets within their area of operations. To assist in this process, it is useful to have a simple method of categorizing the facilities to be considered. Figure 6-4 shows a suggested method for categorizing facilities.

**CRITICALITY CRITERIA.** Criticality refers to *asset value*. This is the primary consideration in targeting. A target is critical when its destruction or damage has a significant impact on military, political, or economic aspects of a community.

- High effect on output, production, service, or mission ☐ **High Risk**
- Moderate effect on output, production, service, or mission ☐ **Moderate Risk**
- Negligible effect on output, production, service, or mission ☐ **Low Risk**
- No significant effect on output, production, service, or mission ☐ **No Risk**

| Why and how | Value | Controls and Mitigation |
|---|---|---|
| | | |

**ACCESSIBILITY CRITERIA.** An asset is *accessible* when an insurgent can reach the target with sufficient personnel and equipment to accomplish its mission.

- Highly accessible: standoff weapons can be employed ☐ **High Risk**
- Moderate accessibility: heavy barriers or protection can be employed ☐ **Moderate Risk**
- Little accessibility: some barriers or protection employed ☐ **Low Risk**
- Inaccessible or accessible only with extreme difficulty ☐ **No Risk**

| Why and how | Value | Controls and Mitigation |
|---|---|---|
| | | |

Figure 6-3. (Continued).

**RECUPERABILITY CRITERIA.** An asset's *recuperability* is measured in time. That is, how long will it take to replace, repair, or bypass the destruction of or damage to the target?

| | |
|---|---|
| High effect on replacement, repair, or substitution | ☐ **High Risk** |
| Moderate effect on replacement, repair, or substitution | ☐ **Moderate Risk** |
| Some effect on replacement, repair, or substitution | ☐ **Low Risk** |
| No effect on replacement, repair, or substitution | ☐ **No Risk** |

| Why and how | Value | Controls and Mitigation |
|---|---|---|
| | | |

**VULNERABILITY CRITERIA.** An asset is *vulnerable* if the insurgent has the means and expertise to attack the target.

| | |
|---|---|
| Vulnerable to small-arms fire, light antiarmor fire, or charges of 5 to 10 pounds | ☐ **High Risk** |
| Vulnerable to medium antiarmor fire, bulk charges of 10 to 30 pounds, or very careful placement of smaller charges | ☐ **Moderate Risk** |
| Vulnerable to heavy antiarmor fire, bulk charges of 30 to 50 pounds, or requires special weapons | ☐ **Low Risk** |
| Invulnerable to all but the most extreme targeting measures | ☐ **No Risk** |

| Why and how | Value | Controls and Mitigation |
|---|---|---|
| | | |

**EFFECTS CRITERIA.** The effect of an asset attack is a measure of possible military, political, economic, psychological, and sociological impacts at the target and beyond.

| | |
|---|---|
| Overwhelmingly positive effects for insurgent; no significant negative effects | ☐ **High Risk** |
| Moderately positive effects for insurgent; few significant negative effects | ☐ **Moderate Risk** |
| No significant effects; neutral | ☐ **Low Risk** |
| Overwhelmingly negative effects for insurgent; no significant positive effects | ☐ **No Risk** |

| Why and how | Value | Controls and Mitigation |
|---|---|---|
| | | |

**RECOGNIZABILITY CRITERIA.** An asset's *recognizability* is the degree to which an insurgent or intelligence-collection and reconnaissance assets can recognize it under varying conditions.

| | |
|---|---|
| The target is clearly recognizable under all conditions and from a distance. Requires little or no training for recognition | ☐ **High Risk** |
| The target is easily recognizable at small arms range. Requires a small amount of training for recognition | ☐ **Moderate Risk** |
| The target is hard to recognize at night or in bad weather, or might be confused with other targets or target components. Requires some training for recognition | ☐ **Low Risk** |
| The target cannot be recognized under any conditions, except by experts | ☐ **No Risk** |

| Why and how | Value | Controls and Mitigation |
|---|---|---|
| | | |

**PSYCHOLOGICAL IMPACT CRITERIA.** An asset's *psychological impact* is the degree to which it affects the local population and that population's perception of the government.

| | |
|---|---|
| Overwhelmingly negative effects on population's perception of government | ☐ **High Risk** |
| Moderately negative effects on population's perception of government | ☐ **Moderate Risk** |
| Little negative effects on population's perception of government | ☐ **Low Risk** |
| Overwhelmingly negative effects on population's perception of government | ☐ **No Risk** |

| Why and how | Value | Controls and Mitigation |
|---|---|---|
| | | |

Figure 6-3. Criteria evaluation tool.

| | |
|---|---|
| **Category A** (15-21 Total Value) | Assets in Category A include those whose interruption or denial would disrupt the functions of governments or the economy, and impair the morale of the population and its confidence in the government's abilities to preserve law order. |
| **Category B** (8-14 Total Value) | Assets in Category B include those whose interruption or denial would disrupt the functions of governments or the economy. They can also damage the morale of the population and its confidence in the government's abilities to preserve law order. |
| **Category C** (0-7 Total Value) | Assets in Category C include those whose interruption or denial would seriously disrupt the function of governments or the economy, and affect the morale of the population. |

Figure 6-4. Facility categories.

*Step 5—Prioritization*

Compare this target to others to determine priority.

### *Protection*

*Protection* is the preservation of the effectiveness and survivability of mission-related military and nonmilitary personnel, equipment, facilities, information, and infrastructure, deployed or located within or outside the boundaries of a given operational area (JP 3-0). The elements of protection are the same in concept between conventional operations and COIN. However, one important aspect is that, counterinsurgents are responsible for protecting local citizens. By ensuring that the local population is secure, counterinsurgents increase their effectiveness in all other tasks and gain allies who will provide information and cooperation. This increases the security of tactical units.

### Technique Considerations during Counterinsurgencies

Individual and small groups of Soldiers, leaders, and civilians are targets for the insurgents. This targeting not only includes lethal operations, such as attacks against an individual, but also nonlethal operations such as kidnapping, subversion or character attacks. Established individual protection measures remain the first line of defense. Basic site-protection operations, a building block to secure the population, include the following:

*Static Posts or Bases*

Each static post has a full-time detachment at the location of the vital asset. The size of the detachment is determined by the size of the vital asset, the threat, and the distance to the nearest available reserve. Consideration must be given to conserving manpower by employing surveillance devices and intruder alarms.

*Observation Post*

Although an OP has too little combat power to secure the vital asset, it can immediately call for support.

*Vehicle Patrols*

Vehicle patrols may be used to give periodic coverage to many low category vital assets. Timings for patrols must be varied to prevent the likelihood of ambush and to retain the element of surprise. Patrols should be strong enough to deal with anticipated threats and they must be supported by a local reserve.

*Foot Patrols*

Foot patrols also give periodic coverage to low category vital assets. Foot patrols will be particularly valuable at vital assets where movement is congested, observation is difficult, and concealment is easily afforded to enemy forces, such as those in busy city blocks.

*Airmobile Patrols*

Airmobile patrols may be used as a supplement to vehicle and foot patrols. They will be valuable for checking vital assets over extended distances or where access is difficult.

**Random Anti-Terrorism Measures**

One means to disrupt insurgent attack plans is to implement random antiterrorism measures (RAMs). These consistently change the look of a security posture to defeat surveillance attempts and introduce uncertainty to a site's overall security plan, thereby making it difficult for the enemy to accurately predict friendly actions. Examples of RAMs are—

- Moving Jersey barriers, vehicular barriers, Class IV objects, or objects in other classes to route traffic around base.
- Starting random security patrols in the surrounding blocks.
- Installing floodlights that operate at random times.
- Changing guard shift at random times.
- Changing access time for entry points.
- Changing access procedures at random.
- Changing the way personnel are searched on a random basis.
- Observing surrounding areas with UAS or JLENS at random times.

***Armor Protection***

Vehicle and personnel armor protection saves lives and makes it more difficult for an insurgent to conduct a lethal attack. However, the added weight will cause vehicles, especially the engine, to wear out much faster than normal. Individual body armor also contributes substantial weight to the Soldier and degrades his maneuverability and endurance.

*Hardening*

- Hardening vehicles or static sites—Increases protection.
- Makes an insurgent attack more difficult.
- Uses natural or man-made materials to protect personnel, equipment, or facilities.
    - Concrete or expedient barriers.
    - Sandbags, walls, shields, berms, or some other type of physical protection.
- Protects resources from blast, direct and indirect fire, heat, radiation, or electronic warfare.

- Concrete and expedient barriers.
- Is intended to defeat or negate the effects of an attack.

*Combat Identification*

CID is the process of attaining an accurate characterization of detected objects in the OE sufficientto support an engagement decision (JP 3-0). These objects, all of which are potential targets, must be quickly discriminated as friendly, enemy, or neutral. This is necessary to ensure that enemy forces can be destroyed, friendly entities can be identified and prevented from becoming fratricide victims, and neutral entities can be identified to prevent collateral damage. Common combat identification techniques include the use of—

- Glint tape.
- Infrared lights and strobes.
- Heat sources.
- VS-17 panels.
- Flares.

***Counterinsurgency Bases***

US counterinsurgency forces operate within a Host Nation and must have a base from which to operate. Typically, bases try to secure the population and isolate the insurgency from its support. A *base is* a locality where operations are projected or supported (JP 1-02). All bases must be securable and defendable. Bases vary in accordance with the size of the unit occupying the base and the mission of the units using the base. All types of bases require clear command relationships. In counterinsurgencies, forward operating bases (FOBs), combat outposts, and patrol bases are three types of bases.

**Forward Operating Bases**

Normally, each AO has at least one FOB. The size of the area, its physical characteristics, and the number and size of the units operating within the area often require additional operating bases. The FOBs established by a brigade or battalion are often semipermanent and provide deployed units with command, control, and communications facilities; sustainment; personnel systems support; staging areas; and intelligence activities. They provide units with relatively secure locations from which to plan and prepare for operations. During counterinsurgency operations, they also aid in limiting insurgent mobility nearby, and providing security to the local population.

Some differences exist between brigade and battalion FOBs. Brigade FOBs are larger than battalion FOBs, and they provide a rear location for elements of battalions such as forward support companies. A battalion FOB is normally staffed with the minimum personnel needed to operate and provide security. It should also maintain two methods for sustainment: road and either air or water.

**Combat Outposts**

This is a reinforced observation post that can conduct limited combat operations (FM 3-90). In counterinsurgency operations, combat outposts are often company and platoon-sized bases inside of insurgent-influenced territory. They represent a cornerstone of counterinsurgency operations, in that they are a means to secure the population. Located in strategically important areas, a combat outpost provides security in its immediate area and direct contact with the local populace. These benefits are unavailable from remote bases. Although the strategy carries with it potential downsides in terms of increased protection concerns and limiting flexibility, the bases provide a huge increase in overall security in the area.

Properly placed combat outposts often increase overall security. Emplacing a company or platoon combat outpost in sector is a deliberate operation requiring detailed planning and additional logistical support. The unit must first decide the task and purpose of the outpost by analyzing their sector.

*Purpose*

Outposts may be employed—

- To secure key lines of communication or infrastructure.
- To secure and co-opt the local populace.
- To gather intelligence.
- To assist the government in restoring essential services.
- To force insurgents to operate elsewhere.

*Priorities of Work*

For the initial establishment of combat outposts, priorities of work need to be considered. Some considerations include—

- Ensuring the position is free of noncombatants. Removing them from the area of operations before occupying the position.
- Selecting key weapons and crew-served weapon positions to cover likely mounted and dismounted avenues of approach.
- Clearing fields of fire. Prepare loopholes, aiming stakes, sector stakes, and target reference point markings. Construct positions with overhead cover and camouflage.
- Identifying and securing supra- and subsurface avenues of approach such as rooftops, sewers, basements, and stairwells.
- Constructing barriers and emplacing obstacles to deny the enemy any access to streets, underground passages, and buildings, and to slow his movement in general.
- Integrating barriers or obstacles with key weapons.
- Improving and marking movement routes between positions, as well as to alternate and supplementary positions.
- Stockpiling ammunition, food, firefighting equipment, and drinking water.

## Patrol Bases

When a patrol halts for an extended period, it takes active and passive measures to provide maximum security by occupying a patrol base. A patrol base can be permanent or temporary.

*Situations*

Common situations that require establishing a patrol base include—

- A requirement to cease all movement to avoid detection.
- A requirement to hide the unit during a lengthy, detailed reconnaissance of the objective area.
- A need to prepare food, maintain weapons and equipment, and rest after extended movement.
- A need to formulate a final plan and issue orders for actions at the objective.
- A requirement for reorganization after a patrol has infiltrated the enemy area in small groups.
- A need for a base where several consecutive or concurrent operations, such as ambush, raid, reconnaissance, or surveillance patrols, can be conducted.

*Purposes*

In counterinsurgency operations, collocating patrol bases in population centers enables combined forces—

- To deny the insurgent access to the local population.
- To influence and assist the local government.
- To provide security.
- To help Host Nation security forces provide their own unaided security.

*Methods of Establishment*

Patrol bases, in the current fight, can be established using either of two methods. The same priorities of work described for combat outposts apply also to patrol bases:

- Move in with the indigenous population. The advantages of the first method are that Soldiers will have more direct contact with the local government, the locals will identify combined forces with the emerging Host Nation government, and the construction will be less intensive. The disadvantages are that Soldiers may live in unsanitary conditions, the mass base or auxiliary may inform insurgents about outgoing patrols with relative ease, attacks on the base will have collateral damage considerations, and houses are often not suited for defense.
- Build a new patrol base. Although more isolated from the population, new patrol bases are usually on chosen ground and, therefore, easier to defend. Additionally, they are far more resource and personnel intensive during construction. It is generally advisable to set aside detailed planning time before sending a combined force to occupy the terrain.

***Planning Considerations for a Base Defense***

Regardless of the ongoing operation, the type of base, or the location of a base, the characteristics of the defense do not change. The best technique for base defense is the perimeter defense.

**Terrain**

Proper evaluation and use of the terrain in the area is essential to hold down the number of additional forces required for base defense. Key terrain factors to consider include the following:

- Use of the terrain's natural defensive characteristics.
- Use of artificial obstacles to enhance the terrain's natural defensive characteristics.
- Control of all roads and waterways leading into the base.
- Control of military lines of communications and civilian commerce routes.
- Control of land areas surrounding the base to a range beyond that of enemy mortars and rockets.

**Host Nation Security Forces**

The base commander should consider the integration of Host Nation security forces in the overall base defense effort. Particular emphasis is on integration of host country forces in patrol and populace control activities. Both host and third country forces provide local security for their own units. However, to ensure maximum benefit, all such local plans should be coordinated and integrated with the base master defense plan.

**Communication**

Control is the key to a successful base defense. To achieve the necessary control, a communication capability must be established between the base defense operations center and sector commanders, and between the sector commander and his bunkers, towers, and reserve. Bunkers or Towers within each section can communicate laterally within the sector, and flank bunkers of one sector can communicate with flank bunkers of adjacent sectors.

**Sustainment**

Depending on the mission and status of the battalion, the type of transport available, the weather, and the terrain, resupply may be by air or ground. The availability of landing zones and drop zones protected from the enemy's observation and fire is the main consideration if selecting organizing aerial resupply.

**Protection**

All units in the base area are responsible for preserving its fighting potential. Protective measures reduce the probability (and the effects) of damage caused by hostile action. Responsibility for the conduct of protective measures is assigned to fire fighting units, chemical units, medical units, and other units. In addition, all units assigned to the base are tasked to conduct activities such as dispersion, camouflage, blackout, field discipline, and use of shelters.

**Security**

Early warning of pending actions ensures the base commander time to react to any insurgent threat. Outposts, patrols, ground surveillance and countermortar radar, military working dogs teams, and air reconnaissance and surveillance provide early warning.

Information provided by civilians and actions of indigenous personnel near the base are excellent indicators of pending enemy actions. All-round security is essential.

### Defense in Depth

Alternate and supplementary positions, observation posts, and mutually supporting strong points in front of the base forward defense area extend the depth of the defense. The commander plans fires throughout the defensive area up to the maximum range of available weapons. Portable obstacles may be placed around critical targets during reduced visibility to disrupt the enemy's plan and add depth to the defense.

### Patrols

Base defense operations to counter small groups of enemy forces include aggressive, frequent patrolling by squad—and platoon-size forces to detect, capture, or destroy small groups of insurgents. Dogs, if available, may be used to add security and additional detection ability to patrol operations. Populated areas near the base are searched, and surprise checkpoints are established along known or suspected routes of insurgent movement.

### Maximum Use of Offensive Action

Since the objective of the base defense is to maintain a secure base, the defender maximizes the use of offensive actions to engage enemy forces outside the base. On initial occupation of the base site, friendly forces take offensive actions to destroy enemy forces in the immediate area. The area commander employs patrols, raids, ambushes, air attacks, and supporting fires to harass and destroy any remaining enemy force. Once the enemy has been cleared from the area, the base can be defended by a smaller force. The base commander maintains constant liaison with major tactical unit commanders in the area to stay abreast of efforts to remove the threat.

### Mutual Support

Defending forces are positioned to ensure mutual employment of defensive resources that include fires, observation, and maneuver elements. Mutual support between defensive elements requires careful planning, positioning, and coordination because of the circular aspects of the base area. Surveillance, obstacles, prearranged fires, and maneuvers are used to control gaps.

### All-Around Defense

In defensive planning, the base commander must be prepared to defend against enemy attack from any direction. Plans are sufficiently flexible, and reserves are positioned to permit reaction to any threat. Base defense forces are assigned primary and alternate positions and sectors of responsibility. All personnel are assigned duty stations or shelters.

### Responsiveness

Attacks against a base may range from long-range sniper, mortar, or rocket fire to attacks by suicide bombers or major forces. The insurgent has the advantage of deciding when, where, and with what force he will attack. The defender positions his forces and plans fires and movement so he can respond to the widest possible range of enemy actions. The defender prepares plans, to include counterattack plans, and rehearses, evaluates, and revises them as necessary.

## Quick Reaction Force

A QRF is a designated organization for any immediate response requirement that occurs in a designated area of operation (FM 3-90.6). A QRF increases the overall flexibility of a base defense and is available for contingencies. Usually a battalion will maintain a platoon sized QRF.

### *Combat Outpost Construction Considerations*

Building a combat outpost is a complex task that must be well thought-out, with a clear vision from the beginning for expansion and development. It is always best to have trained engineers, either military or civilian construct the base. Heavy consideration must be given to using local companies or personnel in constructing the base. This will benefit the local economy; however, it will also increase the security risk. Figure 6-5 shows a typical combat outpost construction in Iraq. The following considerations are critical:

## Protection

Protection involves enemy and security considerations.

*Enemy*

- Coverage of dead spaces.
- Creation of a safety zone to prevent rocket attacks.
- Emplacement of IEDs along routes.
- Observation.

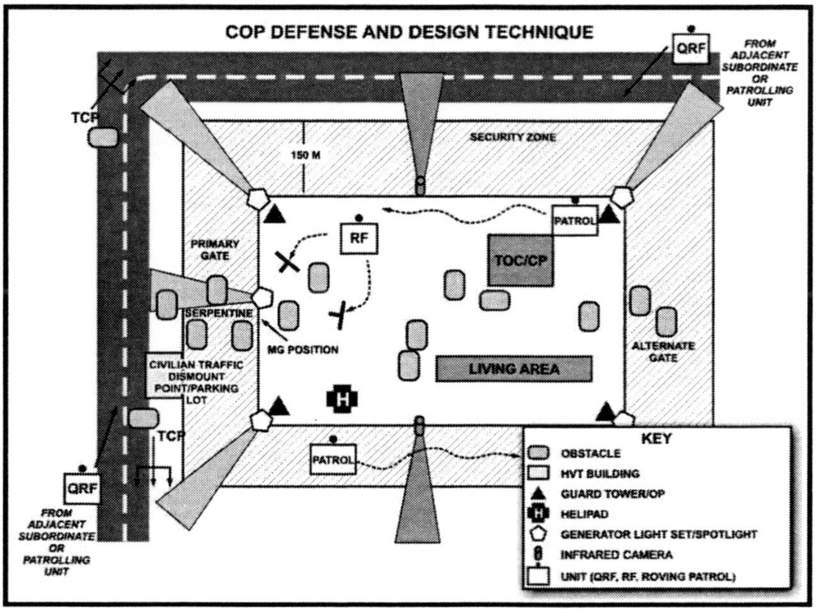


Figure 6-5. Typical US combat outpost design.

*Security*

- Position concrete block guard towers at each corner and reinforce with sandbags.
- Use chain-link screens to protect positions from rocket-propelled grenades and hand grenades.

**Work and Sleep Areas**

- Separate work areas from sleeping and eating areas for sanitation and health concerns.
- Develop a basing strategy that—
    - Projects where facilities will be located.
    - Identifies areas for expansion is required.

**Combat Outpost Equipment**

Standard equipment for the base includes—

- Kitchen sets.
- Motor pool assets.
- Gym sets.
- Power generators.
- Earth-moving equipment.

**Storage**

Create storage space by using military van containers (MILVANs).
Use MILVANs for bunkers, portable housing, and work units.

**Electrical**

When initially developing combat outposts—

- Envision the proper wiring and layout of zone power grids.
- Consider using generators for backup.
- Consider hiring contract electricians and construction workers, which can assist greatly in the development of this system.

**Plumbing**

- Field sanitation is paramount to the health and safety of personnel on the combat outpost.
- Bulk water from locally drilled wells is typically the main source of water.

**Fuel**

- Fuel stands are required for storing and distributing bulk fuel assets.

## Counterinfiltration and Early Warning

The key to an effective combat outpost defense lies in identifying the threats. Among the most dangerous threats are infiltrators. The best defense against these threats is the population that surrounds the combat outpost. As described in Chapter 1, internally displaced people, merchants, or shopkeepers are potential sources of intelligence about insurgent attacks on bases. Soldiers from the combat outpost must talk to their neighbors.

### *Base Defense*

This includes both normal and emergency military measures taken to nullify or reduce the effectiveness of enemy attacks or sabotage. The base commander is responsible for the local defense of the base and provision of resources for other activities within the overall area of operations. The base defense force commander executes base defense operations. Operations are conducted to ensure the continued effectiveness of base facilities and units to fulfill their missions. The area commander's responsibilities include protecting the resources of his area from interruptions caused by enemy activities. Figure 6-6 shows a possible organization of a base command.

## Base Commander

The mission of the base commander is to exercise command, control, and administration of the base and to exercise necessary control of resident and transient units not a part of the base command. A base commander may also be the area commander. The base commander's responsibilities include establishing the overall defense organization as well as planning, preparing, and executing all defense measures.

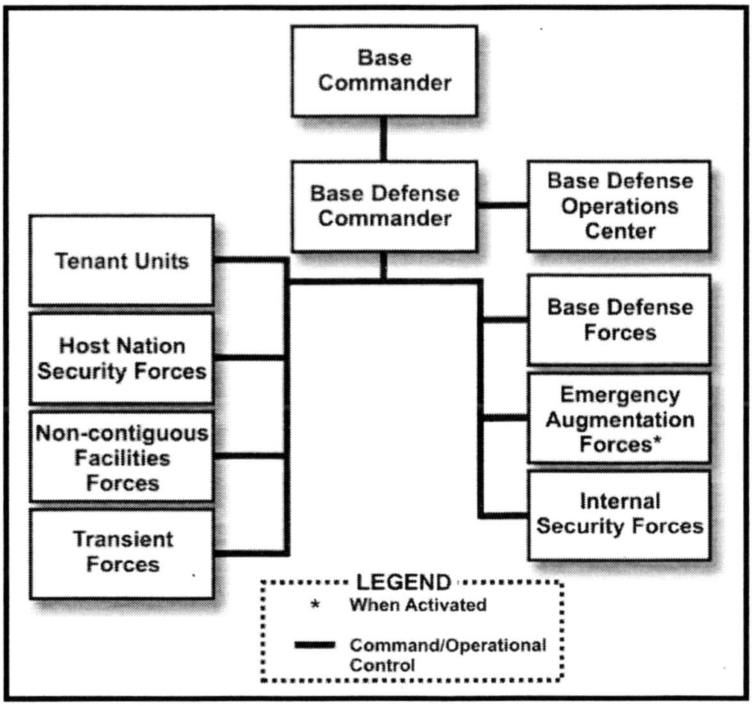


Figure 6-6. Organization of base command.

| | |
|---|---|
| *Permanent* | Includes elements assigned to the base permanently, and whose primary responsibility is for the defense or security of the base. Sustainment units often operate in this role. |
| *As Required* | Includes units or their elements of which normally occupy or operate in or from the base. |
| *As Available* | Includes units or their elements which of normally occupy or operate in or from the base. |

Figure 6-7. Base defense elements forces.

| | |
|---|---|
| **Bunkers** | These should be manned in daylight, darkness, reduced visibility, and increased enemy threat. Night and day vision devices, automatic weapons, grenade launchers, and hand grenades are commonly used in bunkers. |
| **Towers** | When coupled with the use of night and day vision aids, elevated platforms, either standard military or locally contracted, enhance the capability to detect perimeter infiltration and to determine the locations of insurgent mortar or rocket firing positions. |
| **Shelters** | Personnel shelters may be constructed throughout the billeting, administrative, and maintenance areas to provide individual protection against standoff attacks. |
| **Revetments** | Construction of revetments for critical resources provides additional protection against mortar and rocket fragmentation. |
| **Wire** | Tactical wire barriers should be used outside and within the perimeter to limit and canalize penetrations by enemy groups or individuals. |
| **Camouflage Netting** | Netting should be placed on top of barriers or towers for countersniper and counterobservation. |

Figure 6-8. Key base defense structures.

### Base Defense Commander

A base defense commander is appointed to supervise the preparation of detailed defense plans including establishing defense sectors, conducting required training, providing for or coordinating logistical support, and controlling base defense operations through an operations center. As the base commander's special representative, the base defense commander coordinates the planning efforts and operations of all elements that are to participate in the base defense.

### Base Defense Elements (Forces)

Three categories of elements that conduct base defenses are shown in Figure 6-7. The term *elements* is used in place of the term *units*, because base operations are often executed *ad hoc*:

### Key Base Defense Structures

The key base defense positions consist primarily of bunkers and towers in the base perimeter area. The positioning of bunkers and towers affords maximum observation and mutually supporting fires over the area forward of the perimeter to include the perimeter barrier and sensor system. See FM 3-34.300 for specifics on the construction of defensive positions associated with base defense (Figure 6-8).

### Base Defense Exercises

All base defense plans must be rehearsed, and should include the testing of the base defense alarm and communication systems. Diverse elements of the defense force must be trained to act in a coordinated effort. These exercises familiarize all elements of the defense force and base tenant units with their assignments in base defense. Exercises must be conducted frequently under various weather conditions during daylight and darkness. Defense exercises include the following:

- Defense of sectors of responsibility, to include—
  - Rehearsing counterattacks.
  - Manning defense positions.
- Employment of the reserve for counterattacking and for reinforcing the defense positions.
- Coordination of supporting fires and other means of support.
- Mass casualty exercises.
- Coordination with other forces assigned to base that may be used in a ground defense role.
- Command post exercises.

In 1965, the Marine Corps created the Marine Combined Action Program (CAP) in Vietnam. The plan was to incorporate a Marine Squad with a Vietnamese Popular Defense Force platoon on the Fort Page firebase.

On June 10, 1965, 12 Marines from C Company, 1st Battalion, 7th Marines were placed with a Vietnamese Popular Defense Force platoon of about 30 men in a village. Binh Nghia had long been a "hotbed" of Viet Cong activity. Aggressive patrols and intelligence gained by the Popular Defense Force soldiers immediately disrupted Viet Cong Operations, and the sector grew quiet. The Viet Cong requested support from the 409 North Vietnamese Army Battalion. They planned to attack Fort Page and destroy the CAP. This would free the Viet Cong to resume regular operations.

On September 14, 1965, the CAP was conducting normal night patrols, leaving Fort Page with only 6 Marines and 12 Vietnamese soldiers. Forgetting that the CAP's presence was responsible for the lull in Viet Cong activity, the squad leader lowered American security from 2 Marines to 1 Marine per guard shift. This left Vietnamese Soldiers almost completely in charge of Fort Page security. Viet Cong Sappers infiltrated that night and killed 6 Vietnamese Soldiers and 5 Marines. A reserve squad from a nearby Marine company firebase arrived on scene after the NVA battalion withdrew. The Viet Cong thought they had won.

The next morning, the Marine Division Commander asked remaining Marines if they wanted to stay. To a soldier, they did. At night, the Marines and Vietnamese soldiers doubled their security, and Viet Cong operations came to a complete stop.

In March 1967, the Viet Cong and the North Vietnamese Army planned another attack on Fort Page. The CAP had reliable intelligence about the attack, and sent out two patrols to provide early warning. Fearing a repeat of the 1965 attack, Marine commanders ordered the CAP to retire under threat of court martial. The Marine squad leader asked his squad what they thought.

One said, "*They're not getting this fort. They're not getting this ville. I'm not leaving here, no matter what.*"

The Marines stayed. When one of the CAP patrols killed the NVA battalion scout, the rest of the battalion withdrew.

Fort Page was never again threatened. By the time the Marines withdrew from Binh Nghia, the Vietnamese soldiers were patrolling in buddy teams on their own.

### *Counter Ambush*

Ambushes are a basic tactic commonly used by guerrilla forces. As such, counterinsurgent forces must continually operate under the understanding that they will encounter an ambush. Counterinsurgent forces must therefore develop specific local procedures to both react to and to counter ambushes. Typical complex insurgent ambushes involve IEDs, small arms fire, mortars, and sniper fire.

### Considerations during Counterinsurgencies

The goal for tactical units is to identify and attack or disrupt the ambush force before it can initiate the ambush. Additionally, if the insurgents' kill zone can be identified before the insurgents are in position, then the ambushers can be attacked as they move into position. Any counterinsurgency unit that is ambushed must immediately return fire and assault the ambushers. Specific considerations include—

- Avoid massing personnel or vehicles.
- Prepare a reaction plan that includes assaulting the ambushers.
- Ensure only part of the element is in a kill zone at the same time (usually means dispersion).
- Maintain control of your immediate tactical space (within 100 meters).
- Where insurgents are likely to establish ambushes, establish counter ambushes.
- Emplace sniper teams or other forces to overwatch likely insurgent routes to ambush sites.

### Mounted Procedures

The following procedures may be used when reacting to an ambush while mounted:

- Immediately return fire and assume a covered position if possible.
- If you are in the kill zone, leave rapidly.
- If you are *not* in the kill zone, use fire and maneuver to destroy the enemy if possible.
- Scan your area and prepare for additional attacks, especially—
    - An IED.
    - An enemy moving to flank.
    - An enemy moving to engage you with RPGs or antitank weapons.
- Report contact to higher HQ.
- Follow directions of the vehicle or convoy commander.

### *Countering Ieds*

An improvised explosive device (IED) is "a device placed or fabricated in an improvised manner incorporating destructive, lethal, noxious, pyrotechnic, or incendiary chemicals and designed to destroy, incapacitate, harass, or distract. It may incorporate military stores, but it is normally made from nonmilitary components" (JP 3-07.2). IEDs are key components of insurgent ambushes. (See also FM 3-90.119.)

### Considerations during Counterinsurgency

Between 2001 and 2007, US forces were attacked by over 81,000 IEDs. IEDs are the most dangerous and effective weapon system faced by Coalition Forces in Iraq. They have inflicted more casualties than all other weapon systems combined.

### Techniques

Wide usage, a destructive nature, and the resulting overall impact on military operations makes the IED a significant factor for tactical units. For this reason, leaders at all levels must consider IEDs. At a tactical level, a unit has three major techniques for defeating IEDs. They are—

- Attack the network.
- Defeat the device.
- Train the force.

*Attack the Network*

This method is used to defeat the complex network of IED makers, financiers, emplacers, suppliers, and others before the IED is emplaced. It includes actions and activities against networks designed to interrupt the enemy's chain of IED activities by identifying and exploiting vulnerabilities and enabling offensive operations. This effort is accomplished through intelligence, surveillance, reconnaissance, counter bomber targeting, device technical and forensic exploitation, disposal of unexploded and captured ordnance, persistent surveillance directed toward defeat of the enemy's capabilities, and the execution of information tasks which includes command and control warfare, military deception, and, especially, information engagements. Search operations and operations to kill or capture network members provide the final, critical step in the process. It is better to find 100 detonators in a cache than to deal with 100 IEDs on the battlefield.

Several analytical tools can be used by tactical units to detect members of an IED network and synchronize all information gained through various sources. They are part of the targeting process, which is discussed in Chapter 4. By correctly synchronizing intelligence at all levels, leaders can better apply combat power to attacking the IED network. The three major tools are imagery and geospatial intelligence analysis, pattern analysis, and link analysis.

***Imagery and Geospatial Intelligence Analysis***

Analysis of an AO where IEDs are employed would be incomplete without the use of imagery and geospatial intelligence. Imagery products include both aerial photography and satellite imagery. In many cases, tasked aerial reconnaissance platforms, including UASs, respond directly to the commander, thus ensuring timely and focused data collection.

***Pattern Analysis***

Units use pattern analysis to determine many of these patterns and predict potential hazards. Pattern analysis is the ability to observe a selection of events or actions over a period of time in a defined location or area. It is used to discover likely patterns or similarities that lead to a logical conclusion that the action or event will occur again in the same location. The two most common forms are coordinates register and pattern-analysis plot sheet.

### *Considerations*

For IED analysis, commanders and staffs use pattern analysis to determine the—

- Types of IEDs.
- Locations where IED incidents have occurred.
    - Characteristics of the IED sites.
    - Sizes of the IEDs.
    - Distance from road, mosque, other natural or man-made objects.
    - Type of terrain the threat prefers.
    - Other obstacles used in conjunction with IEDs.
    - Routes to and from the IED sites (worn paths, other).
- Distance from the IED site to the possible cache.
- Types of munitions and whether one or more types were used.
- Whether the IEDs used explosively formed penetrators (EFPs).
- Whether the IEDs used shape charges.
- Other IED main components such as—
    - Main charge (explosive)
    - Casing (materials around the explosives)
    - Initiators.

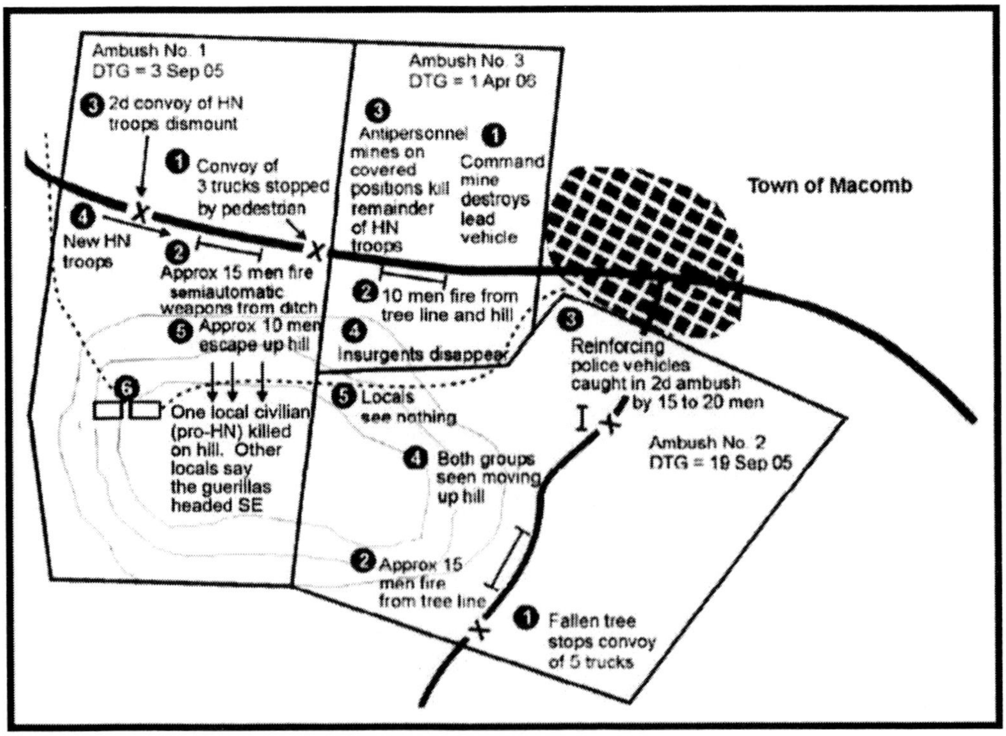


Figure 6-9. Coordinates register.

- Methods of initiation, such as—
  - Command wired.
  - Radio controlled.
  - Victim operated.
  - Timed.

***Coordinates Register***

A coordinates register shows cumulative events that have occurred within the AO and focuses on the "where" of an event. Figure 6-9, which is an example of a coordinates register, is also known as an incident map.

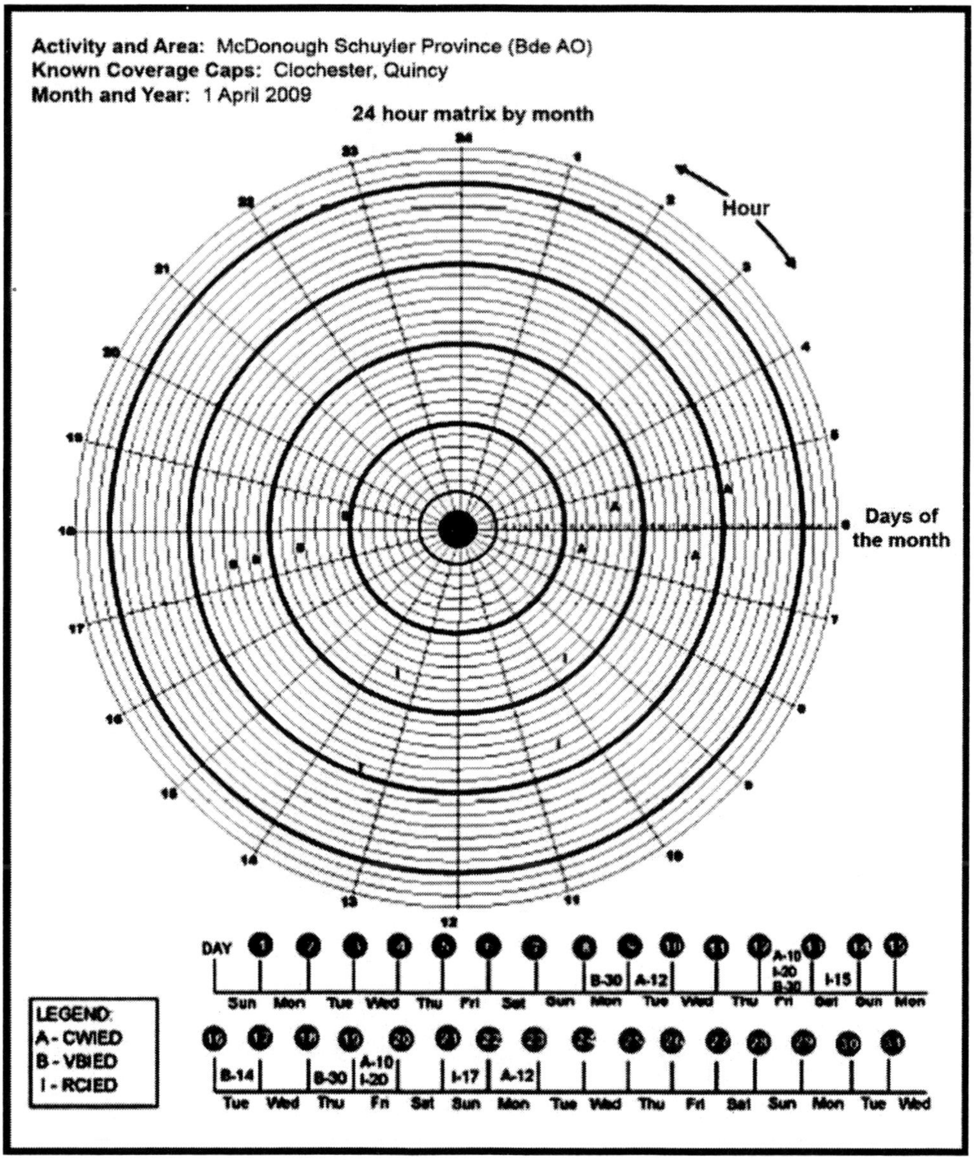


Figure 6-10. Example pattern-analysis plot sheet.

### *Pattern-Analysis Plot Sheet*

A pattern-analysis plot sheet focuses on the time and date of each serious incident that occurs within the AO. The rings show days of the month; the segments show the hours of the day. As shown in the plot sheet's legend, the chart shows the actual events; it identifies each by using an alphanumeric designation that directly corresponds to the legend used on the coordinates register. The legend icons can be changed to show other types of incidents such as vehicle-borne IEDs (VBIEDs), command-wired IEDs (CWIEDs), and radio-controlled IEDs (RCIEDs). Figure 6-10 shows an example pattern-analysis plot sheet.

*Note:* List daily entry number on calendar. In journal, cross-reference each incident to the incident overlay.

### *Link Analysis*

Link analysis is used to show contacts, associations, and relationships between persons, events, activities, organizations. The two most effective types of link analysis tools are the activities and association matrixes and the link diagram.

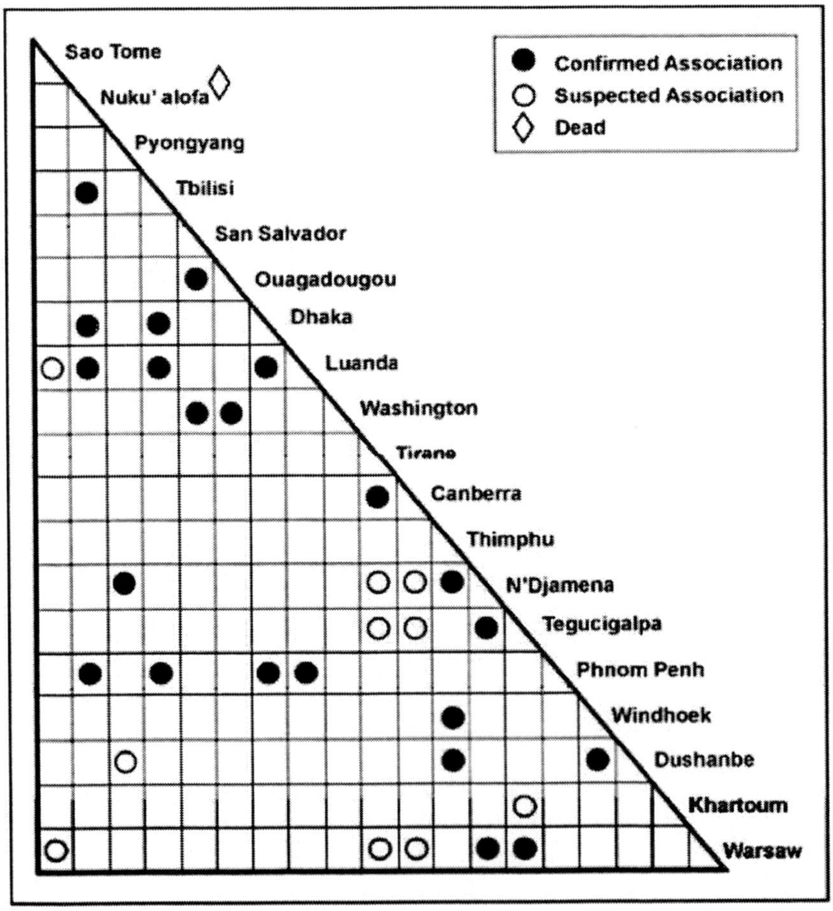


Figure 6-11. Example activities matrix.

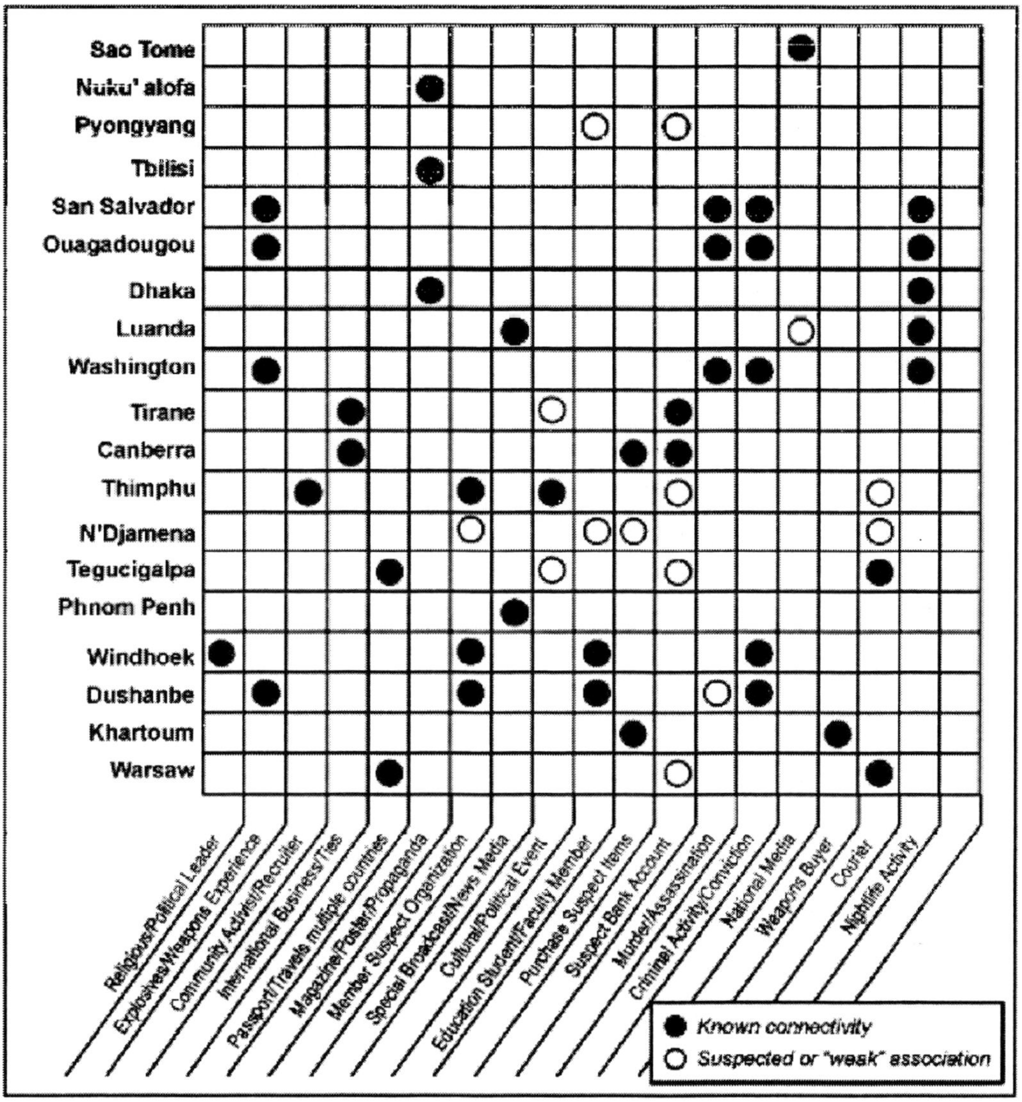


Figure 6-12. Example association matrix.

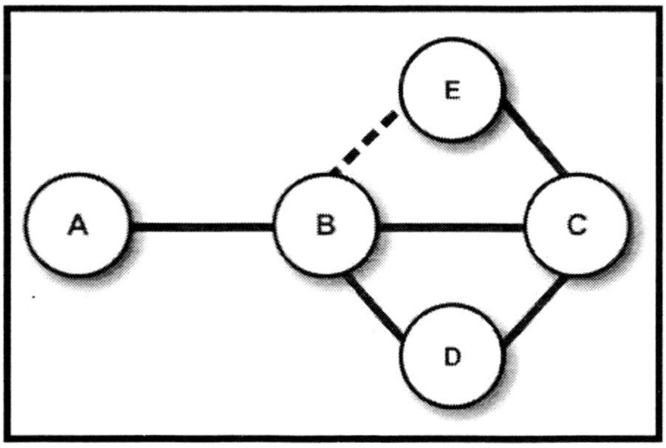


Figure 6-13. Example link diagram.

***Activities and Association Matrixes***

Using the activities and association matrixes, the analyst can pinpoint the optimal targets for further intelligence collection, identify key personalities within an organization, and considerably increase the analyst's understanding of an organization and its structure. The activities matrix is used to determine connectivity between individuals and any organization, event, address, activity, or any other nonpersonal entity. The association matrix is used to determine existence of relationships between individuals. Figure 6-11 shows an example activities matrix, and Figure 6-12 shows an example association matrix.

***Link Diagrams***

Often, the link analysis is the most effective method to show the connections between people, groups, or activities. An example of a link diagram is shown in Figure 6-13. The analyst can easily determine from the diagram that A knows B, and that B knows C and D. B is suspected of knowing E, and C knows D, B, and E. Although the same information could be shown on a matrix, it may be easier to understand when shown on a link analysis diagram. As situations or investigations become more complex, the ease in understanding a link analysis diagram becomes more apparent. In almost all cases, the available information is first shown and analyzed on both types of matrixes, which are then used to construct a link analysis diagram for further analysis.

***Defeat the Device***

In order to enhance commanders' freedom of maneuver for safe operations, actions taken to discover IEDs and activities reduce the effects of IED detonation at the point of attack by defeating the device. They include rapid identification, development, acquisition, and delivery of capabilities for route clearance, device neutralization, explosive detection, military explosive ordnance disposal, and vehicle and personnel protection. Additional considerations should address new technologies to include electronic warfare capabilities. Two techniques for defeating the device are dismounted patrols and route clearance.

***Dismounted Patrols***

Dismounted patrols are especially useful in countering IEDs. Use of dedicated counter IED patrols, especially along main supply routes, is an important technique to reduce the threat of IEDs along those routes. A great way to counter IEDs is to eliminate the popular support or acquiescence to the IED cell through regular engagements. If that is not possible, then identify, and neutralize the kill zones.

***Route Clearance***

Route clearance teams (RCTs) are used to thwart ambushes, clear natural or man-made obstacles, and detect IEDs. RCTs are comprised of engineers, EOD personnel, mechanical devices, and specialized robotics to increase standoff from the IED threat. The purpose of route clearance is to eliminate concealment for IEDs and munitions caches and to conduct the systemic detection and deterrence sweeps along the cleared routes.

Route clearance missions consist of the following two phases: right-of-way clearance and route maintenance and sweep operations. Right of way clearance is the removal of rubble, debris, berms, holes, trenches, vegetation, and trash from the medians and shoulders of routes. Right-of-way clearance can be conducted with special equipment such as an up-armored bulldozer. Cleaning the right of way in this manner serves multiple purposes: IED-detection devices become more effective; it is more difficult for the enemy to emplace IEDs; and all counterinsurgent forces can more easily identify IEDs should the enemy attempt to emplace them again. Another technique is to use by local national labor, which employs the population. Units should then conduct a deliberate route reconnaissance, identify and record the location of man-made objects (buried pipes and cables), and investigate suspicious areas. Route maintenance and sweep operations are when units conduct systemic, random detection sweeps of the cleared areas and progress to detection and deterrence sweeps along the cleared route. A visual detection sweep should focus on changed conditions.

***Train the Force***

Mitigating the effects of enemy IED employment through comprehensive training of our forces puts troops in the field who are situationally aware and who know their gear. This includes, but is not limited to, multiechelon and multicomponent training, training on new gear, information management and dissemination, strategic communications, doctrinal and institutional training changes, and unit mission rehearsals at Service combat training centers. All soldiers need to be trained in basic techniques, actions while mounted, and dismounted actions.

***Individual Techniques***

Counter-IED operations at the soldier level revolve around several principles. (See GTA 90-10-046 for information on combined arms improvised explosive device defeat operations.) Units need to know, control, and identify when something is out of place in their area of operations. Insurgents adapt their techniques based off of counterinsurgent unit actions. These principles are the following:

- Maintain an offensive mindset.
- Develop and maintain situational awareness.
- Stay observant.
- Avoid setting patterns.
- Maintain 360-degree security.
- Maintain tactical dispersion.
- Use blast and fragmentation protection.
- Use technology, such as electronic warfare capabilities, thermals, or other devices.
- Look for the triggerman.

***Mounted Actions***

Mounted action includes actions before encountering an IED and actions after encountering an IED.

***Precautionary Actions While Mounted***

Consider the following while conducting mounted operations:

- Alert all crewmembers of possible IED and its location.
- Never stop a vehicle within 50 meters of a suspected IEDs. The driver should speed up or back away to quickly move through the danger area. Minimum open area safe distance is 300 meters.

- Watch for approaching vehicle borne IEDs (VBIEDs). All personnel should be alert and constantly aware of any vehicle approaching or parked along the route
- Threats should be addressed with a direction, a short description, and an estimated distance. When time is limited, only the direction need be given: RIGHT FRONT or RIGHT FRONT MERGING TRUCK 200 METERS.
- Vehicle commands to the driver should be short, giving only a direction and a desired action: TURN RIGHT, SPEED UP, SLOW DOWN or STOP.
- Gunners should respond by scanning the sector (weapon ready) indicated in the threat alert.

***Reaction to IED while Mounted***

The following procedures may be used when reacting to an IED while mounted:

- Immediately establish security and scan for secondary IEDs.
- Scan your area and be ready for any follow on attacks, especially any additional IED or enemy moving to engage you with small arms, RPG, or antitank weapons. If necessary, use obscurants.
- Look for the triggerman or for anyone trying to escape the area.
- Conduct five and twenty-five meter vehicle perimeter checks.
- Report IED to higher HQ. If found before detonation, use the 9-line IED/UXO report.
- Follow directions of the vehicle/convoy commander.

***Dismounted Actions***

Dismounted actions include actions before encountering an IED and actions after encountering an IED.

***Precautionary Actions while Dismounted***

Consider the following while conducting dismounted operations:

- Alert patrol members of possible IED and its location.
- Never stop within 50 meters of a suspected IED. Move out of the danger area. Minimum open area safe distance is 300 meters.
- Watch for approaching VBIEDs or PBIEDs. All personnel should be alert and constantly aware of any vehicle approaching or parked along the route
- Threats should be addressed with a direction, a short description and estimated distance. When time is limited, only the direction need be given: RIGHT FRONT or RIGHT FRONT ONCOMING TRUCK 200 METERS.
- All Soldiers should respond by scanning the sector (weapon ready) indicated in the threat alert.

***Reaction to IED While Dismounted***

The following procedures may be used when reacting to an IED while dismounted:

- Establish security and scan for secondary IEDs (5m and 25m sweeps).
- Report IED to higher HQ. If found before detonation, use a 9-line IED/UXO report.

- Maintain security in case of an ambush.
- Follow the directions of the team/squad leader.

***React to a Suspected Person-Borne Improvised Explosive Device***

The following procedures may be used to react to a suspected person-borne IED (PBIED):

- Alert unit members of suspect and evacuate area if possible.
- Issue a verbal command in a loud and firm voice to the suspect to stop. Weapon should be trained on the suspect.

*Note:* After these actions, there is a high probability that the suicide bomber will attempt to detonate the explosive device. Use deadly force in accordance with the Theater ROE. Physically restraining the suspect from detonating the device requires coordination with other Soldiers and is inherently very dangerous.

- Direct the suspect to show hands palms up. The suspect must show palms and fingers spread, so it can be verified if the suspect is palming a detonator. Be advised that there may be a remote detonator.
- Tell the suspect to ground all carried items and step two paces away from them.
- Direct the suspect to remove outer clothing and place garments on the ground.
- Direct the suspect to raise or remove any undershirt, and to hold it up while turning a complete circle.
- Direct the suspect to lie face down, with arms outstretched palms up, and face turned away from you. DO NOT approach, even if the suspect is injured. Maintain cover and wait for the arrival of explosive ordnance disposal (EOD) personnel.
- Report suicide bomber to higher HQ using the 10-line explosive hazard spot report.
- If the suspect is noncompliant, deadly force will be used in accordance with the theater ROE.
- Evacuate the area around the suspect (minimum 300 meters or METT-TC) to any hard cover available.
- Establish security and scan for secondary PBIEDs.
- Maintain security in case of possible ambush.
- Follow the directions of the vehicle/squad leader.

## *Countersniper or Sniper Defeat*

The first step in countering snipers is for commanders, leaders, and Soldiers at all levels to be aware of the sniper threat. Plans to counter the sniper threat and protect friendly forces from insurgent snipers must be integrated into all counterinsurgent operations during the early stages of planning. Enemy snipers in a counterinsurgency environment vary from well-trained sniper teams to individuals taking shots of opportunity. US counterinsurgency forces must be prepared for both.

Figure 6-14 shows the three types of snipers an insurgent can employ. Identifying the type that is on the battlefield can aid a commander in deciding what countermeasure to use.

## Considerations during Counterinsurgencies

Sniper operations have the same effect in insurgencies as they do in conventional warfare, but the purpose is different. For example, in conventional warfare, a sniper targets leaders in order to cause confusion on the battlefield and disrupt operations. In an insurgency, a sniper will still target leaders, but more for use in propaganda that can be used after a successful attack. Additionally, he can create the attitude that the area of his operations is not secure and the Host Nation government is powerless to stop him. Since the sniper has the initiative, leaders must enforce compliance with the countermeasures. Implementing countermeasures halfheartedly invites casualties from snipers who can wait hours for the moment when a unit's guard is down. US forces have two categories of sniper countermeasures: active countermeasures and passive countermeasures. Each has its place, depending on the METT-TC mission variables.

### Active Countermeasures

Active countermeasures seek to detect and destroy the sniper before he can fire, or engage and neutralize him after he fires.

#### *Observation Posts and Observers*

Observation posts should have access to powerful spotting telescopes, medium power binoculars, night observation devices (thermal, if possible), and remote control closed circuit cameras. Additionally, laser countermeasure should be employed, such as laser protective glasses, binoculars with laser filters, and indirect-view optics to protect observers. Additionally, many electronic detection devices are available to aid in the detection of snipers such as acoustic devices. Once detected, snipers can then be neutralized or forced to withdraw. Observers can maintain a constant surveillance over potential sniper positions and detect snipers as they attempt to move into position for a shot. Constantly scanning an area for the minute movements made by a well-trained sniper is exhausting. Therefore, personnel on static OP duty should rotate frequently. Unmanned aerial systems (UAS) and rotary wing aircraft should be integrated into an OP plan to give it depth.

#### *Patrols*

Constant reconnaissance and security patrols around a unit's position hinder a sniper from getting into a firing position undetected. Small patrols are usually more effective than large ones. A moving sniper who has been discovered by a patrol is at a great disadvantage since he lacks the firepower to fight a long engagement. Small night reconnaissance patrols using night vision devices can be very effective in disrupting and interdicting a sniper. Reconnaissance patrols should move by covered and concealed routes to good observation points; should stop and observe; and then should move to another position. The patrol routes and times must vary, and a reaction force or supporting weapons must be ready if the patrol makes contact. A variation of the ambush patrol is the stay-behind ambush. A small ambush element moves as part of a larger patrol and occupies its position without being observed. It then observes its kill zone, which may be very large if the element has a sniper team with it, and engages enemy snipers as they attempt to move into position. Currently, SKTs, SCTs, or small observation teams (SOTs) conduct ambushes of likely sniper positions.

***United States Sniper Teams***

US sniper teams can be a very effective counter to enemy snipers. Not only do they have expert knowledge of sniping and likely enemy hiding places, they can normally engage enemy snipers or marksmen at a greater range than the enemy sniper can engage US forces. Their precision fires are also much less likely to cause civilian casualties than dozens of rounds from other weapons.

***Return Fire and Maneuver***

Before a unit encounters an enemy sniper, the unit must not only understand the rules of engagement but also war-game the rules of engagement to ensure the authorized level of responses are sufficient. Although the ROE and the tactical situation determine the response, ideally authorization should include the ability to respond with fire from all the units light weapons. If a unit can determine the general location of a sniper, it should return suppressive fire while maneuvering to engage the sniper from close range. This is not always successful because a well-trained sniper often has a route of withdrawal already chosen. Fire without maneuver will not be successful in defeating snipers.

***Obscurants***

Projected smoke that builds quickly is a good response to protect a unit from further casualties if engaged by an enemy sniper. It greatly limits his ability to acquire targets. The closer the smoke is placed to the sniper's location, the more effective it is. If the location of the sniper is unknown or cannot be reached by projected smoke, a smoke cloud established near the unit can be effective in reducing the sniper's chances of hitting a target and allowing a unit to conduct first aid and CASEVAC.

**Passive Countermeasures**

Passive countermeasures prevent the sniper from acquiring a clear target and prevent his fires from causing casualties. Passive countersniper measures are rarely successful by themselves. If passive measure are the only measures enacted, they may also create a siege mentality and pass the initiative over to the sniper. They include—

| | |
|---|---|
| **LEVEL I**<br>**Specially Trained Sniper** | The most dangerous sniper is the one who has been specially selected, trained, and equipped with an accurate sniper rifle equipped with a modern scope, night vision device, and even a thermal imager. |
| **LEVEL II**<br>**Trained Marksman** | This sniper is a trained Soldier, equipped with a standard issue weapon, who is an above average shot. |
| **LEVEL III**<br>**Armed Irregular** | This Soldier may have little or no formal military training but may have experience in urban combat or skill as a hunter. |

Figure 6-14. Types of snipers.

***Limiting Sniper Exposure***

If Soldiers limit their sniper exposure, they can marginalize snipers operations. Some examples include using covered and concealed routes, avoiding lighted areas at night, moving tactically while using traveling or bounding overwatch, and staying away from doors and windows.

#### *Wearing Protective Equipment*

Other sniper protective measures include wear of the Kevlar helmet, protective eyewear, and body armor systems. These should be worn any time Soldiers are exposed to potential sniper fire. This decreases the snipers' casualty-producing target area, which limits his effectiveness.

#### *Using Armored Vehicles*

Whenever possible, move around in the urban area in a protected vehicle with as little exposure as possible. Avoid open-sided cargo vehicles. Requisition or improvise vehicular armor against small-arms fire for all unarmored vehicles. This technique limits a unit's ability to interact with the population, and should be used as a last resort.

#### *Erecting Screens and Shields*

Use simple canvas or plastic screens to make a dangerous alleyway or street crossing safer for foot traffic. Adapt screens on windows to allow vision out while hiding personnel inside. Use moveable concrete barriers to provide protection for personnel at static positions. Use common items, such as rubble-filled 55-gallon drums and sandbags, to provide cover.

#### *Denying Enemy Use of Overwatching Terrain*

Either occupy overwatching terrain with friendly forces, or modify it to make it less useful to an enemy sniper. Pull down likely hiding places. Clear bushes and rubble. Board or brick up windows. Pile up earth and rubble in front of buildings to block lines of sight for snipers. Ensure all actions are in accordance with the laws of war.

### Technique Considerations

Countersniper actions have two parts: find and eliminate the sniper.

#### *Find the Sniper*

Once a unit determines it has a sniper or a sniper team in its AO, the unit should implement countersniper immediate action drills. Allowing the enemy sniper to fight another day is mission failure. First, the unit must find the sniper or the sniper team. Techniques for detecting a sniper consist of identifying likely sniper locations, focusing observation on sniper movement, and seeking audible and visual signs. *Backtracking*, which involves integrating the following techniques, is used to find the sniper:

#### *Identify Likely Sniper Locations*

A sniper is a line-of-sight asset. The larger the sniper kill zone, the closer or the higher he must position himself to cover it. Smaller sniper kill zones create for a narrow, if not linear, line of sight. In such cases, the sniper will always choose a position that offers that offers a line of sight, such as a road or alley, or maybe inside a building, shooting through a window, doorway, or other opening.

#### *Focus Observation on Sniper Movement*

Snipers always plan infiltration and exfiltration routes. After identifying possible sniper locations, a unit commander can focus observation on these routes and deny a sniper the ability to reposition. Snipers operate slowly, with calculated movements, to avoid detection.

The movement the sniper makes may not be easily detected. Units should look for movement that does not "fit" the environment.

***Seek Audible/Visible Signs***

Audible/visual indicators help find snipers based on the sniper's firing a shot. Actually, firing a weapon is a sniper's greatest moment of risk; it offers the best opportunity to locate him or his team. The acoustic, light and heat signatures associated with the projection of the bullet from his rifle are his greatest vulnerabilities. The muzzle flash can be detected with infrared sensors out to a kilometer. Often the muzzle flash and the blast can throw dust into the air. Acoustic sensors can detect a muzzle blast from several hundred meters to more than a kilometer. The bullet's shock wave is a mini-sonic boom. This noise, caused by the bullet travelling faster than the speed of sound, can sometimes be heard more than a kilometer away.

***Backtrack Sounds, Sights, and Angles***

In a city, the backtracking process is complicated by buildings that obstruct the view of the sniper's position. Backtracking is challenging, but it is how you relate shot angles, kill zones, movement, and audible and visible signs to help you find the sniper. Experience and training play a major role in accurate backtracking.

***Eliminate Sniper***

Once spotted, the enemy sniper must be killed or forced to surrender. To be successful in this goal, the unit must rapidly locate, fix, and finish an active enemy sniper to ultimately kill or capture him.

***Avoid the Kill Zone***

Figure out his kill zone and stay out of it. A unit may also want to eliminate or at least reduce his targets. If a unit removes its targets, the unit curtails mission success; the sniper will need to move, or at least adjust his position, to remain effective.

***Maneuver Rapidly***

When contact is made, the on-scene commander needs to rapidly calculate the disposition of his forces, the quick reaction force (QRF), and adjacent friendly forces and factor in time/space considerations to determine how to neutralize the threat. Because of the fleeting nature of the sniper, it is often preferable for the on-scene commander to begin to maneuver against the hostile sniper with an unengaged portion of his own force rather than wait for a QRF.

***Consider Sniper Hunting with Another Sniper***

Throughout history, many units engaged in urban combat often employed direct and indirect supporting arms, close air support, or large numbers of ground forces to counter urban snipers. However, the Russian and German forces of World War II found that the best solution to a sniper threat was the employment of their own snipers in a countersniper role. Advantages to using a trained sniper to counter enemy snipers include—

- The relative ease of movement over the urban battlefield of a sniper team versus that of a larger ground formation.

- Countersnipers who "know the habits and modus operandi" of their enemy counterparts.
- The ability to deliver "precision fire" and the resulting reduction in collateral damage or civilian and friendly casualties.

#### *Confirm that Enemy Sniper Is Neutralized*

In this context, neutralization means that the sniper is killed or captured. A dead or captured enemy sniper, besides ending that threat, can be a valuable source of intelligence.

### *Countering Drive-by Shootings*

A drive-by shooting is the firing of small arms by an occupant(s) of a vehicle as [the vehicle] drives past an area, followed by the vehicle's immediate departure of the area. This is an effective technique for an insurgent, because it allows rapid infiltration and exfiltration, requires little logistical support, has a chance of success, and has little chance of resulting in a decisive engagement. Any structure or location near a roadway is a target for a drive-by shooting. Usual targets for drive-by shootings are vulnerable, easily escapable spots. Attackers avoid well-defended locations.

#### Considerations

Establishing static checkpoints or closing roads near a high threat area often prevents further drive-by attacks along a specific avenue of approach. However, the insurgents are not eliminated and they will then look for new or different ways to attack. Timely and accurate reporting by subordinate units can allow higher headquarters to integrate UAS and rotary wing aircraft to track vehicles after a drive-by attack. This will open insurgents up to targeting counterinsurgent forces.

#### Techniques

One technique is to conduct mobile checkpoints. If effective, they can disrupt the ability of insurgents to transport weapons in vehicles. Random snap checkpoints may encourage drive-by shooters to look elsewhere for targets.

## Section III—Other Lines of Effort

In a counterinsurgency, defensive operations are conducted to prevent insurgents from attacking security forces, attacking the population, or disrupting actions along the seven COIN lines of effort. This section discusses special defensive operation considerations for the remaining five lines of effort.

### *Support Host Nation Security Forces*

The success or failure of a counterinsurgency falls heavily on the competence and capability of Host Nation security forces. When insurgent forces are stronger than the Host Nation security forces, insurgents focus on the destruction of weak Host Nation units to control terrain and consolidate gains through deliberate attacks on a Host Nation base, combined combat outpost, combined patrol base, or combined command post. In this situation, extra defensive efforts need to be applied to give time for the Host Nation security forces to develop. When Host Nation forces are stronger than insurgent forces, insurgents attempt to delegitimize Host Nation security force operations through attacks focused on lessening security. Then, defensive considerations need to be applied to enable Host Nation forces to increase the efficiency of their operations.

***Support to Governance***

This line of effort relates to the Host Nation governments' ability to gather and distribute resources while providing direction and control for society. By disrupting the Host Nation government's ability to conduct these functions, an insurgency can delegitimize the government. An example is the local security of a key judge who has been instrumental in issuing verdicts against criminal activities of insurgents in order to prevent his assassination. Defensive efforts may need to be applied so that a government can conduct its basic functions which include—

- Controlling military and police activities.
- Establishing and enforcing the rule of law.
- Public administration.
- Justice (a judiciary system, prosecutor/defense representation, and corrections).
- Property records and control.
- Civil information.
- Historical, cultural, and recreational services.
- An electoral process.
- Disaster preparedness and response.

***Restore Essential Services***

A common insurgent technique is to create a dissatisfied population by preventing the Host Nation from meeting the population's basic needs. An easy way for an insurgent to do this is to disrupt the population from receiving sewage, water, electrical power, and medical services. Units should apply defensive considerations for any attempt by combined or Host Nation security forces to build or restore those services. For example, a poorly planned generator drop allows an insurgent to attack and destroy the generator. Now, instead of restoring an essential service and reinforcing the legitimacy of the Host Nation government as intended, the insurgent has demonstrated the weakness of the Host Nation government. Instead, US and Host Nation security forces need to secure, protect, and prevent insurgents from influencing projects that restore essential services.

***Support to Economic and Infrastructure Development***

A poor and unemployed population is naturally dissatisfied. The major pool of insurgent recruits are unemployed, young, adult males. The primary motivation for a young male to join an insurgency is often the wage that it provides. Sometimes insurgencies foster the conditions to keep an economy stagnant. Naturally, the Host Nation and combined forces will conduct stability operations to create situations where businesses can thrive. An example is conducting route security, so that insurgents cannot interfere with the transportation of products to a market. Economic infrastructure that might need defensive considerations include—

- Natural resources.
- Industries such as manufacturing and transportation.
- Agriculture and livestock—crops, products, and storage facilities.
- Local merchants and markets.
- Credit associations.

***Conduct Information Engagement***

Information engagement is deliberately integrated into defensive operations to compliment and reinforce the success of operations. Exploiting insurgent violent tactics, behaving harshly towards the HN population, and failing to achieve their goals are matters of precise timing and coordination. Some units have chosen to prepare an information engagement plan to provide an immediate response to insurgent attacks or propaganda such as sending a field-grade officer to the site of an insurgent attack to talk with the local media and get the correct story out. These efforts may go far to influence the populace to provide information, either for reward or anonymously, that may break apart insurgent networks. Specifically, information supplied by the populace is critical in countering ambushes, IEDs, snipers, and drive-by shootings. It does this by interrupting the insurgent decision cycle.

## SUMMARY

General Clutterbuck in the Malayan emergency said, "(A Village police post)... was the only thing that could provide security against the threat that really mattered in the villages—the man with the knife, who lived in the village and prowled the streets at night seeking out those people who had actively supported the government or betrayed the guerrillas during the day." This chapter identified defensive operations as they apply specifically to establishing civil security and civil control, and the other LOEs. Defensive operations in COIN must secure the population, Host Nation security forces and government, and US forces from the actions of insurgents.

# 7. STABILITY OPERATIONS CONSIDERATIONS IN COUNTERINSURGENCY

*"Internal Warfare within a population, particularly in cities, generally involves an extensive police operation. There is also an intensive propaganda effort, destined primarily to make the steps that are taken understood [by the population]. A broad social program follows, the objective of which is to give the people the material and moral assistance necessary to permit them to resume their normal activities quickly after operations are over."*

Trinquier, Roger. Modern Warfare—A French View of Counterinsurgency, 1964.

*Stability operations* encompass various military missions, tasks, and activities conducted outside the United States in coordination with other instruments of national power to maintain or reestablish a safe and secure environment, provide essential governmental services, emergency infrastructure reconstruction, and humanitarian relief (JP 3-0). They leverage the coercive and constructive capabilities of the military force to establish a safe and secure environment; facilitate reconciliation among local or regional adversaries; establish political, legal, social, and economic institutions; and facilitate the transition of responsibility to a legitimate civilian authority (FM 3-07). In the absence of HN civil government, military forces will have to fulfill this role until HN, UN, or other US Government agencies assume these responsibilities. FM 3-07 describes in detail the requirements that may fall upon the Soldiers and leaders to perform.

## Section I—Overview

During an insurgency, stability operations are executed simultaneously with offensive and defensive operations. They compliment and reinforce offensive and defensive operations. Because they begin to address the root causes that lead to the insurgency, stability operations are often the most critical for defeating an insurgency. There are five primary tasks within stability operations—*establish civil security, establish civil control, support to governance, restore essential services*, and *support to economic and infrastructure development*. Two additional tasks— *support Host Nation security forces* and *conduct information engagement* are added to produce the seven COIN lines of effort. HN security forces are discussed in Chapter 8, while information engagements are in Chapter 4. These seven COIN LOEs are a means for tactical units to manage a stability operation's numerous tasks, achieve unity of effort, and restore the legitimacy of a Host Nation government. The stability tasks fall into three categories. This chapter will focus on the first two tasks:

- Military forces retain primary responsibility.
- Civilian agencies/organizations likely retain responsibility, but military forces prepared to execute.
- Tasks for which civilian agencies or organizations retain primary responsibility.

### *Nature of Stability Operations*

The stability tasks and their interrelation with the COIN LOEs are not sequential in nature. They must be continually assessed and re-evaluated. This iterative process occurs because units do not have the manpower to apply the same effort along each stability task. As in the rheostat, discussed in Chapter 3, success in the establish civil security LOE may cause units to apply more pressure along the restore essential services LOE. Figure 7-1 shows an example of a brigade that incorporated the five primary stability tasks into the seven COIN LOEs. At the commander's discretion, this brigade has chosen to combine the establish civil security and establish civil control LOEs.

### *Clear-Hold-Build Operations*

In a *clear-hold-build* operation, stability operations are conducted in the "*clear*" phase, "*hold*" phase, and the "*build*" phase. In the "*build*" phase, which focuses on building HN capacity and capabilities, stability operations are primary, although offensive operations and defensive operations occur simultaneously. Stability operations address multiple COIN LOEs.

## Section II—Five Primary Stability Tasks

All five primary stability tasks are used during counterinsurgency operations, normally as individual lines of effort. However, since each counterinsurgency, each insurgency, and even individual units and leaders are different, how these five stability tasks are applied and managed as individual lines of efforts may vary considerably. The five stability tasks follow:

- Establish civil security.
- Establish civil control.
- Support to governance.
- Restore essential services.
- Support to economic and infrastructure development.

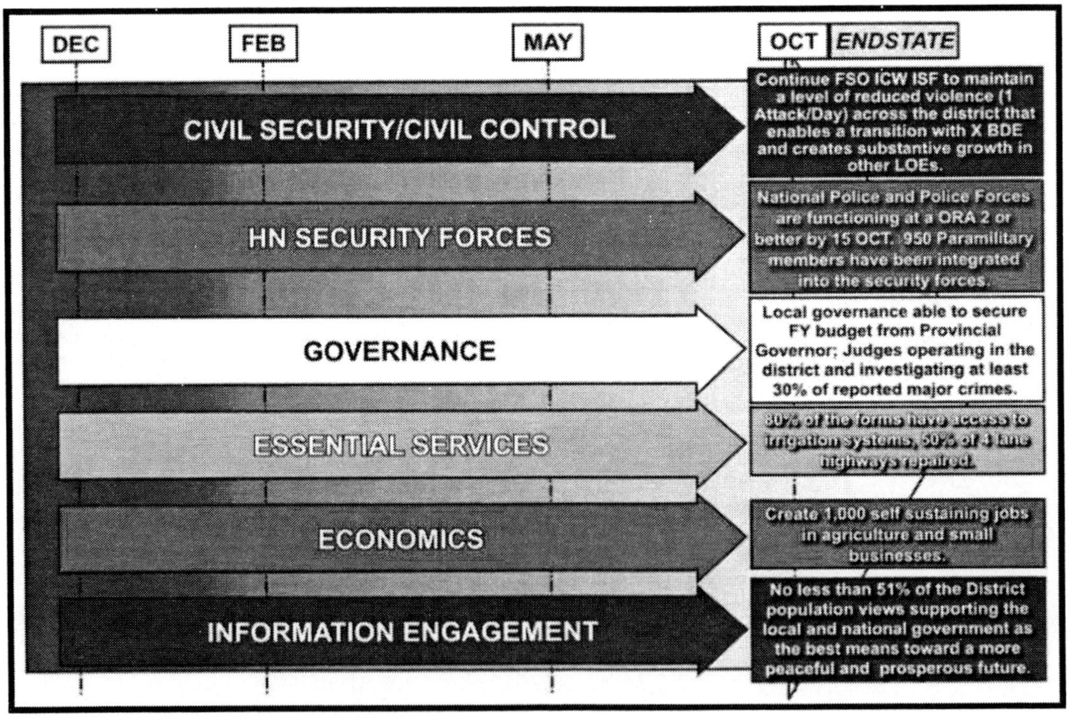


Figure 7-1. Example BCT using LOEs.

### *Establish Civil Security*

Counterinsurgent forces seek to create a safe, secure, and stable environment for the local populace. Key to this is the development of security institutions, or in their absence performing the duties normally associated with those institutions until those capabilities are developed or transitioned to HN, UN or other US agencies. Ultimately, the Host Nation must secure its own people. Civil security involves protecting areas, resources, and the populace from both external and internal threats. Ideally, Army forces focus on the external threats while police and security elements address internal security threats by criminals and small, hostile groups. However, during an insurgency, the Army must now also address the internal security threats by criminals and small, hostile groups—notably the armed insurgents.

**Subordinate Tasks**

Typical civil security subordinate tasks a counterinsurgent unit may perform are as follows (combat-oriented techniques for *establish civil security* are covered in Chapter 6):

- Enforce cessation of hostilities.
- Enforce peace agreements and other arrangements.
- Conduct disarmament, demobilization, and reintegration
- Conduct border control, boundary security, and freedom of movement.
- Support identification.
- Protect key personnel and facilities.
- Clear explosive and CBRN hazards.

**Enforce Cessation of Hostilities, Peace Agreements, and Other Arrangements**

These tasks aid in providing security and stability after an armed conflict, while at the same time setting the conditions needed to start disarming, demobilization, and reintegration. These tasks are critical to providing effective security for the local populace by reducing their exposure to the threat of violent conflict. The tasks help military forces establish a sustained peace by focusing on processes and activities fundamental to conflict transformation. In COIN, most efforts focus on identifying and neutralizing potential adversaries. For instance, in Iraq from 2003 to 2005, US forces fought a Sunni insurgency comprised of many members of the former regime's security apparatus. The list of essential tasks may include—

- Enforce cease fires.
- Supervise disengagement of belligerent forces
- Identify and neutralize potential adversaries.
- Provide security for negotiations.

*Conduct Disarmament, Demobilization, and Reintegration*

These tasks are fundamental to establishing stability and lasting peace. It includes physically disbanding guerrillas and reintegrating them into society. Other elements do not carry weapons; however, they support the guerrilla forces. Their communities and families perceive them as part of the insurgency. Together, the tasks of disarmament, demobilization, and reintegration reduce a potential resurgence of insurgency. These tasks provide a means for these individuals and groups to reenter society as contributing members. Some of these tasks could also be used for the demobilization of HN paramilitary forces. For more on HN paramilitary forces, see Chapter 8. Essential tasks may include:

- Negotiate arrangements with insurgents.
- Provide reassurances and incentives for disarmed groups.

- Establish and enforce weapons control programs, including collection and destruction.
- Establish a monitoring program.
- Ensure adequate health, food, and security for former insurgents.
- Disarm former insurgents or HN paramilitary forces.
- Reduce availability of unauthorized weapons.
- Reintegrate former insurgents and HN paramilitary forces and dislocated civilians into society.
- Secure, store, and dispose of weapons.
- Develop HN arms control capacity.

*Establish Border Control, Border Security, and Freedom of Movement*

A central component of civil security is the ability of the HN to monitor and regulate its borders. Generally, border and coast guard forces secure national boundaries while customs officials regulate the flow of people, animals, and goods across state borders. These border controls are necessary to regulate immigration, control the movements of the local populace, collect excise taxes or duties, limit smuggling, and control the spread of disease vectors through quarantine. In COIN, the ability for the counterinsurgent to control the borders reduces the effectiveness of the external support for the insurgency. Essential tasks may include—

- Establish border control and boundary security.
- Establish and disseminate rules relevant to movement.
- Dismantle roadblocks and establish checkpoints.
- Ensure freedom of movement.

*Support Identification*

This task complements efforts to vet Host Nation personnel, encourage participation in representative government, resolve property disputes, and validate professional credentials. Although vital to other programs for rebuilding a functioning civil society, identification programs are equally important to civil security. After the collapse of an authoritarian or hostile regime, these programs ensure that potential adversaries do not inadvertently reintegrate into society. Thus, they are deprived of the ability to sow the seeds of a future insurgency. It is a key part of populace and resource control operations. Tactical units often do this by conducting a census with biometric assets. Essential tasks may include—

- Secure documents relating to personal identification, property ownership, court records, voter registries, professional certificates, birth records, and driving licenses.
- Establish identification program.
- Ensure individuals have personal forms of identification.

*Protect Key Personnel and Facilities*

In COIN, military forces may extend protection and support to the HN population to ensure their continued contribution to the overall stability operation. In the interest of transparency, military forces specifically request and carefully negotiate this protection. Similarly, the long-term success of any counterinsurgency often relies on the ability of the security force to protect and maintain critical infrastructure until the HN government can resume that responsibility. Essential tasks may include identifying, securing, protecting, and coordinating disposition for stockpiles of munitions and CBRN materiel and precursors; facilities; and adversaries with technical expertise as well as the following:

- Protect—
  - Government-sponsored civilian reconstruction and stabilization personnel.
  - Contractor and civilian reconstruction and stabilization personnel and resources.
- Protect and secure—
  - Places of religious worship and cultural sites.
  - Critical infrastructure, natural resources, civil registries, and property ownership documents.
  - Strategically important institutions (such as government buildings; medical and public health infrastructure; the central bank, national treasury, and integral commercial banks; museums; and religious sites).
  - Military depots, equipment, ammunition dumps, and means of communications.
- Build Host Nation capacity to protect—
  - Civilian reconstruction and stabilization personnel.
  - Infrastructure and public institutions.
  - Military infrastructure.

*Clear Explosive and CBRN Hazards*

To an HN combatting an insurgency, the presence of explosive hazards (including minefields, IEDs, and unexploded ordnance) and CBRN hazards (resulting from intentional or accidental release) inflicts stress that the surviving institutions cannot bear. These hazards restrict freedom of movement, hinder international trade, and detract from the ability of an HN government to secure its population and borders. Military forces may clear unexploded ordnance and other explosive hazards to facilitate capacity-building activities. Removing these hazards ensures the safety, security, and well-being of the local populace. Essential tasks may include—

- Establish an explosive hazards coordination cell.
- Conduct emergency clearing of mines, IEDs, unexploded ordnance, and other explosive hazards.
- Map, survey, and mark mined areas, unexploded ordnance, and other explosive hazards.
- Remediate hazards remaining from the release of CBRN hazards and radiological fallout, as well as provide decontamination support.
- Create Host Nation capacity to conduct demining.
- Build Host Nation capability to export demining expertise.

***Establish Civil Control***

This task regulates selected behavior and activities of individuals and groups. This control reduces risk to individuals or groups and promotes security. Civil control channels the population's activities to allow provision of security and essential services while coexisting with a military force conducting operations (FM 3-0). A top priority for military forces conducting stability operations during a counterinsurgency will be the re-establishment of legal and justice systems to help aid with security sector reform. Units may find themselves involved with corrections activities, establishing public order and safety, resolving property disputes, and supporting reconciliation efforts. Most military efforts focus on building temporary or interim capabilities until more permanent capabilities are put in place by either the Host Nation or US and international agencies.

**Subordinate Tasks**

Typical civil control subordinate tasks a counterinsurgent unit may perform are—

- Establish public order and safety.
- Establish interim criminal justice system.
- Support law enforcement and police reform.
- Support judicial reform.
- Support property dispute resolution processes.
- Support corrections reform.
- Support public outreach and community rebuilding programs.

*Establish Public Order and Safety*

These tasks provide a broad range of activities to protect the civilian populace, provide interim policing and crowd control, and secure critical infrastructure. These essential tasks represent actions that must occur during and after an insurgency to ensure the long-term sustainability of any reform efforts. The speed and effectiveness in performing these tasks directly correlates with the length of time to defeat the insurgency. Executing these tasks as soon as practical after intervening reduces the time required for related efforts and allows the mission to be accomplished far sooner. However, the military's legal authorities for all activities in the justice sector, particularly involving enforcement and adjudication of the law, must be clear. Essential tasks may include—

- Secure the population.
- Ensure humanitarian aid and security forces reach endangered populations and refugee camps.
- Perform civilian police functions, including investigating crimes and making arrests.
- Locate and safeguard key witnesses, documents, and other evidence related to key ongoing or potential investigations and prosecutions.
- Control crowds, prevent looting, and manage civil disturbances.
- Secure facilities, records, storage equipment, and funds related to criminal justice and security institutions.
- Build Host Nation capacity to protect military infrastructure.
- Build Host Nation capacity to protect infrastructure and public institutions.

- Build Host Nation capacity for emergency response.
- Fostering sustainability where military forces identify modernization needs and the means to achieve them.

*Establish Interim Criminal Justice System*

Often in COIN, establishing or reestablishing an interim justice system is a prerequisite. This restoration requires a wide range of skilled professionals working under a clearly defined legal authority: judges, prosecutors, court administrators, defense lawyers, corrections personnel, law enforcement, and investigators. These personnel—and the institutions they represent—provide a temporary respite that allows the Host Nation to restore its legal system. Essential tasks may include an initial response in which military forces—

- Assess current laws and need for modifications or adoption of internationally accepted codes.
- Assess Host Nation capacity to combat crime.
- Deploy interim justice personnel to complement Host Nation criminal justice system.
- Establish mechanisms to review the legality of detentions and minor cases to minimize pretrial detention.
- Enact interim legal codes and procedures permitted by international law.

*Support Law Enforcement and Police Reform*

US military forces provide support to law enforcement and policing operations, which is integral to establishing civil control. HN police may provide this capability if the security environment permits. Usually in insurgencies, the HN police may have become corrupt or failed altogether. In failed states, especially during and immediately after conflict, military police forces are the only organizations able to fill this void. At times, HN police augment military forces, rather than the other way around. The preferred providers of civilian law enforcement services are HN police, augmented as required by military and paramilitary police units with policing capabilities. Civilian agencies typically provide training and capacity-building support for law enforcement services. However, US and HN military forces may be required to perform these services in the interim, until the situation permits transition of this function to civilian agencies or organizations. Essential tasks may include the following:

- Identify, secure, and preserve evidence of—
  - War crimes.
  - Crimes against humanity.
  - Corruption.
  - Transnational crime such as terrorism, organized crime, human trafficking, and narcotics.
- Identify and detain perpetrators of these offenses.
- Support vetting, checking credentials, and accounting for HN police forces.
- Inventory and assess police facilities and systems.
- Train and advise HN police forces.

- Rehabilitate or construct necessary facilities.
- Establish police academies.

*Support Judicial Reform*

The reform of judicial bodies is integral to rule of law and provides the necessary framework for broader security sector reform. The support provided to judicial institutions parallels efforts with police and security forces to enhance the state's capability to maintain civil control and security. Under most circumstances, other agencies organizations typically support the development of the judicial branch of government. In a failed state, however, military forces may initially perform these functions until they can be transitioned to an appropriate civilian agency or organization. Essential tasks may include—

- Identify Host Nation legal professionals.
- Educate criminal justice personnel on interim legal codes and international human rights standards.
- Inventory and assess courts, law schools, legal libraries, and bar associations.
- Deploy judicial advisors and liaisons.
- Rehabilitate or construct necessary facilities.
- Support vetting of Host Nation legal professionals.

*Support Property Dispute Resolution*

A vital service of the judiciary branch is to resolve property disputes. One of the causes of an insurgency might be old disputes over ownership and control of property. Authorities must implement dispute resolution mechanisms. This prevents the escalation of violence that can occur in the absence of law order as people seek resolution on their own terms. Typically, the military's role in resolving disputes is limited to transitional military authority where these mechanisms are implemented in the absence of a functioning Host Nation government. Essential tasks may include an initial response in which military forces—

- Implement mechanisms to prevent unauthorized occupation or seizure of property.
- Publicize dispute resolution process.
- Coordinate dispute resolution process to deter violence and retribution.

*Support Corrections Reform*

When the goal in COIN is to criminalize the insurgency and prosecute insurgents in the HN court system, corrections reform is an integral component of broader security sector reform. Corrections reform tasks focus on building HN capacity in the penal system, restoring the institutional infrastructure, and providing oversight of the incarceration process. Tasks also include a comprehensive assessment of the prisoner population to help reintegrate political prisoners and others unjustly detained or held without due process. Essential tasks may include—

- Identify and register all detention, correction, or rehabilitative facilities.
- Preserve and secure penal administrative records and reports.
- Inventory and assess prison populations and conditions.

- Implement humanitarian standards in prisons.
- Provide emergency detention facilities.
- Vet corrections personnel.
- Deploy penal trainers and advisors.
- Refurbish prison facilities at key sites.
- Coordinate jurisdiction and handover.
- Facilitate international monitoring.
- Rebuild corrections institutions.
- Train and advise corrections personnel to internationally accepted standards
- Develop reconciliation, parole, and reintegration mechanisms.

*Support Public Outreach and Community Rebuilding Programs*

These programs are central to the reconciliation process in a counterinsurgency, and to promoting public respect for the rule of law. They provide the HN populace with a means to form a cohesive society. While these programs generally do not involve substantial military involvement, some activities require the force's support to achieve success. Essential tasks may include an initial response in which military forces—

- Establish broad public information programs to promote reconciliation efforts.
- Assess needs of vulnerable populations.

## Techniques

Most combat-oriented techniques for performing establish civil control-type tasks are covered in Chapter 5 and Chapter 6. Checkpoint operations and vehicle searches help establish public order and safety by limiting the insurgent's freedom of movement.

*Establish Checkpoint (Control Traffic)*

A checkpoint is a manned position, usually at a chokepoint on a main thoroughfare, which can control all vehicular and pedestrian traffic passing along that road. A checkpoint may stop and check all vehicles or only a selected few. A checkpoint is not a roadblock. A roadblock prevents all vehicular or pedestrian traffic along a thoroughfare.

Checkpoints may be temporary, semipermanent, or permanent. A common discriminator between a temporary and a semipermanent checkpoint is its manning requirements. A temporary checkpoint has no shift requirements—the same individuals can establish, operate, and then break down the checkpoint. A semipermanent checkpoint requires at least one additional 'shift' of personnel to operate it for the length of time required. Over time, a semipermanent checkpoint may turn into a permanent checkpoint. A permanent checkpoint usually has no set end time. Most nations have permanent checkpoints, especially at international border crossings or at key access control points. Common reasons for establishing a checkpoint during counterinsurgency operations include—

- Demonstrating the presence and authority of the host government.
- Checking vehicles for explosive devices.
- Maintaining control of vehicular and pedestrian traffic.
- Apprehending suspects.

- Preventing smuggling of controlled items.
- Preventing infiltration of unauthorized civilians into or through a controlled area.
- Serving as an observation post.

*Considerations*

Checkpoints cause considerable inconvenience and sometimes fear to the population. Therefore, it is important that the citizens understand that checkpoints are a preventive and not a punitive measure. The use of a TPT to broadcast instructions and explanations and disseminate handbills to civilians waiting at a checkpoint helps minimize negative reactions to the control measure and to ensure compliance.

The rapid establishment of a temporary checkpoint can support a host of counterinsurgency operations. The two types of checkpoints follow:

***Snap Checkpoint***

This is a rapidly established, temporary checkpoint. If time allows, a designated, mobile unit is trained and resourced for this mission. Then, with little warning, this unit can establish a snap checkpoint almost anywhere.

***Enduring Checkpoint***

This is often established initially as a snap checkpoint. As the site is further developed, an enduring checkpoint becomes semipermanent.

Insurgent forces often target traffic checkpoints with harassing sniper or indirect fire. A plan to quickly locate and react to these threats often reaps great benefit. Additionally, checkpoints are often targeted by insurgent snipers or by a scripted traffic incident. The insurgent's goal is to induce the counterinsurgent to overact and then to exploit the overreaction in the media.

***Locations***

The precise location and degree of visibility of a checkpoint reflect its purpose. A highly visible checkpoint is often located where traffic can avoid it. This works well for entry-control points and border crossings. A concealed checkpoint is often located where traffic cannot avoid it by bypassing, turning around or getting off the road without being observed and, if ordered, detained. Figure 7-2 and Figure 7-3 show possible setups for snap and enduring checkpoints.

*Conduct Vehicle Searches*

Any unit may be tasked to conduct the routine search of vehicles. If there is a high probability that vehicles may be wired with explosives, then an EOD team should lead the search effort. All vehicles should be screened before entering the search area. Usually screening is done at the initial barrier and IAW local screening criteria. Vehicles are then directed either through the search area or to one of the search sites. Normally passenger vehicles are segregated from commercial vehicles as part of the screening process.

## Plans

Determine the specific purpose of the checkpoint. The purpose influences the location, degree of individual checks, length of time, resources, and manpower required.

- Define the parameters of searches—this includes whether to search all pedestrians and all vehicles. For vehicles, determine whether to search the vehicle, the occupants, or a combination. Determine the search method and level of search detail. Distribute a "Be on the look out" (BOLO) list, if available.
- Establish the screening criteria for searches and the method of screening. This can reduce time and manpower requirements.
- Determine the in-effect time. Checkpoints lasting longer than 12 hours will require shifts. Consider placing rest areas for shifts near the search area so they can be used as a reserve force.
- Establish both near and far security. Near security protects the checkpoint force. Far security, often concealed, prevents the escape of any vehicle or person attempting to turn back upon sighting the checkpoint. Plans must include the actions taken when this happens.
- Establish and brief the ROE and EOF procedures including warning signs.

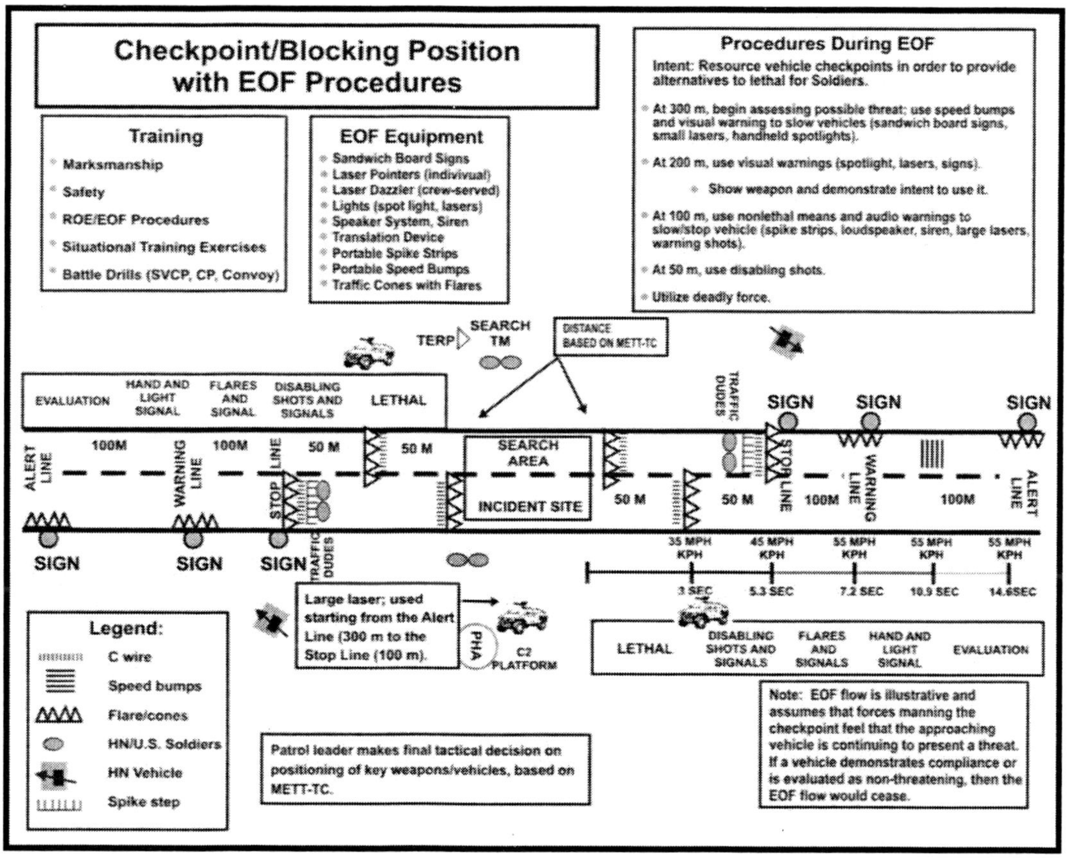


Figure 7-2. Well-equipped snap checkpoint layout.

- Rehearse checkpoint operations, especially EOF procedures.
- Design the checkpoint layout and gather the needed materials before arrival.
- Determine the personnel requirements.
- Emplace in positions where traffic cannot easily bypass the checkpoint.
- After it can operate on its own, determine if the HN security force should man the checkpoint.

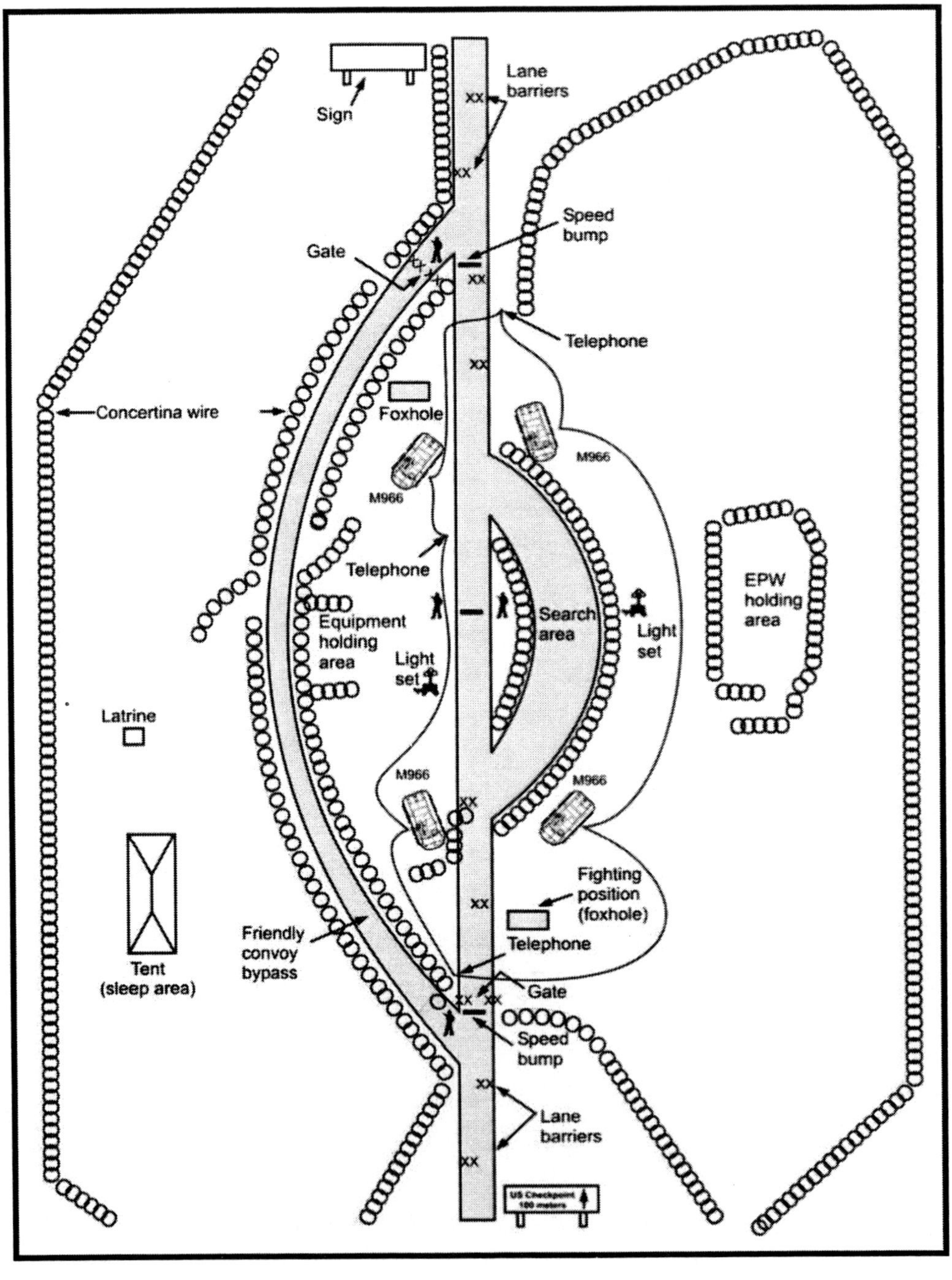


Figure 7-3. Typical enduring checkpoint layout.

**Material Considerations**

Consider equipment, for example:

- Equipment, such as speed bumps or filled barrels, to slow and canalize vehicles.
- Protection items, which may include both lethal and nonlethal munitions.
- Concertina wire to control movement around the checkpoint.
- Sandbags and wood for defensive positions or bunkers.
- Binoculars, night vision devices, and flashlights.
- Long-handled mirrors (for inspecting vehicle undercarriages)
- Logs and checklists for inspectors, supervisors, and so on.
- Caution flags and unit signs.

**Personnel Considerations**

Consider each element:

- Command element.
- Security element.
- Search element.
- Linguists and Host Nation representation.
- Military police.
- Medical personnel.
- Communications personnel.
- Tactical PSYOP team.
- HCTs.
- Military working dog teams.
- HN security force personnel.

**Establishment**

Position a combat vehicle off the road, but within sight, to deter resistance.

- Keep this vehicle in a hull-down position and protected by local security. It must be able to engage vehicles attempting to break through or bypass the checkpoint.
- Place obstacles in the road to slow or canalize traffic into the search area.
- Establish a reserve, if applicable.
- Establish a bypass lane for approved convoy traffic.
- Designate the vehicle search area.
- Establish a parking area adjacent to the search area.
- Designate male and female search areas near the vehicle search area. Normally, search women using a metal detector, searched by a female Soldier or searched by a Host Nation female.
- Consider a controlled bypass lane for HN civilians with medical emergencies, HN civilian officials, or other HN population concerns
- Consider the effect on the local population to include market times, sporting events, holidays, and collateral damage

### Operation

Screen all vehicles IAW with established, preferably written, criteria. Establish observation for vehicles or occupants attempting to avoid the checkpoint. This includes—

- Exiting a vehicle and walking away or around the checkpoint.
- Having a lead vehicle signal a following vehicle to avoid the area.
- Creating a distraction to allow the vehicle or occupants to slip away.

### Screening Considerations

In addition to having a basic published list of criteria—

- Note the number of occupants.
- Note the type or color of the vehicle.
- Devise a method of managing important changes or alerts.
- Observe all vehicles and all occupants.
- Always maintain some element of randomness.

#### *Basic Levels*

The three basic levels of vehicle searches follow:

#### *Initial*

This is a cursory search that is the same for all stopped vehicles. This relatively quick check should take no more than two minutes. It may be combined with initial vehicle screening. Basic vehicle initial search procedures include—

- Ensure at least two people conduct the search:
  - One guard (driver and occupants)
  - One searcher (vehicle)
- Stop the vehicle at the search site.
- Have all occupants exit the vehicle and move them to one location.
- Instruct the driver to turn off the engine, and then open the hood, trunk, and all doors.
- Continue the search once the above is complete.
- Question or search of the driver and occupants is optional.
- Begin and end the exterior search at the front of the vehicle, moving in a clockwise direction.
- Visually search of the exterior and top of the vehicle.
- Visually search of the underside of the vehicle.
- Search the interior starting at the front compartment, then the right side, then the rear compartment, then the left side, ending at the front.
- Complete the search.
- Instruct the driver and occupants to continue.

#### *Primary*

This is a full search of select vehicles. Vehicles may be selected randomly, IAW screening guidelines or due to similarities to vehicles associated with previous events (a be-on-the-look-out, or BOLO list.) This search usually entails checking the interior, exterior, engine and trunk compartments and mirror checks of the vehicle underside. It should take two to five minutes. Primary search include those of those of the initial search and may add the following—

- Question or search the driver and occupants.
- Check the inside of the front hood.
- Check the rear compartment.

***Secondary***

This is a detailed search of a suspect vehicle. A vehicle becomes suspect usually due to screening or discovery of items during a primary search. This thorough search is manpower and time intensive. It may include disassembling panels, checking interiors of wheels, and so on. At the basic level, to conduct a secondary search of a vehicle—

- Ensure at least four people conduct the search:
  - Two for the driver and occupants.
  - Two more for the vehicle.
- Assume that the driver and occupants or the vehicle itself is suspect.
- If possible, have the vehicle stop in a safe and secure location.
- Instruct the driver to turn off the engine, open the hood, open the trunk, and open all doors.
- Obtain the keys from the driver.
- Move the driver and occupants to a separate location.
- Detain and search driver and occupants while searching the vehicle.
- Begin the vehicle search at the front and end at the rear.
  - One searcher searches the right front, right side, and right rear.
  - The other searcher searches the left front, left side, and left rear.
- Conduct a visual search of the exterior and top.
- Conduct a visual search of the underside of the vehicle.
- Without entering, conduct a visual search of the interior of the vehicle.
- Ensure the engine is off, and then ensure the hood, trunk, and all doors are open.
- Check the inside of the front hood.
- Check the interior left and right sides.
- Check the rear compartment.
- Complete the search.
- Return keys to driver.
- Instruct driver and occupants to continue.

***Vehicle Types***

Vehicles are generally categorized into four types: automobiles (cars), motorcycles and bicycles, buses, and heavy goods vehicles (trucks). Search considerations for the car are covered above as part of basic vehicle considerations. Search considerations for the three additional vehicles follow:

***Motorcycles and Bicycles***

Basic motorcycle and bicycle search procedures include—

- Ensure at least two people conduct the search.
  - One for the driver and passenger.
  - One for the vehicle.
- Have the driver or a passenger turn off the engine and dismount.
- Visually inspect the vehicle and any compartments on the vehicle.
- Complete the search.
- Instruct driver and passenger to mount up and continue.

***Buses***

Basic bus search procedures include—

- Ensure at least four people conduct this search.
  - Two for the driver and occupants.
  - Two for the bus itself.
- Set aside an area to search buses due to their length, potentially large number of occupants, and lots of luggage.
- Instruct only the driver to turn off the engine and open all exterior compartments.
- Initially, board the bus, check all occupants' identification, note seating arrangements, and check carry-on baggage.
- If a full search is required, have all occupants exit the bus with their baggage, and then detain them while you search the bus and baggage.
- Complete the search.
- Instruct driver and occupants to reload and continue.

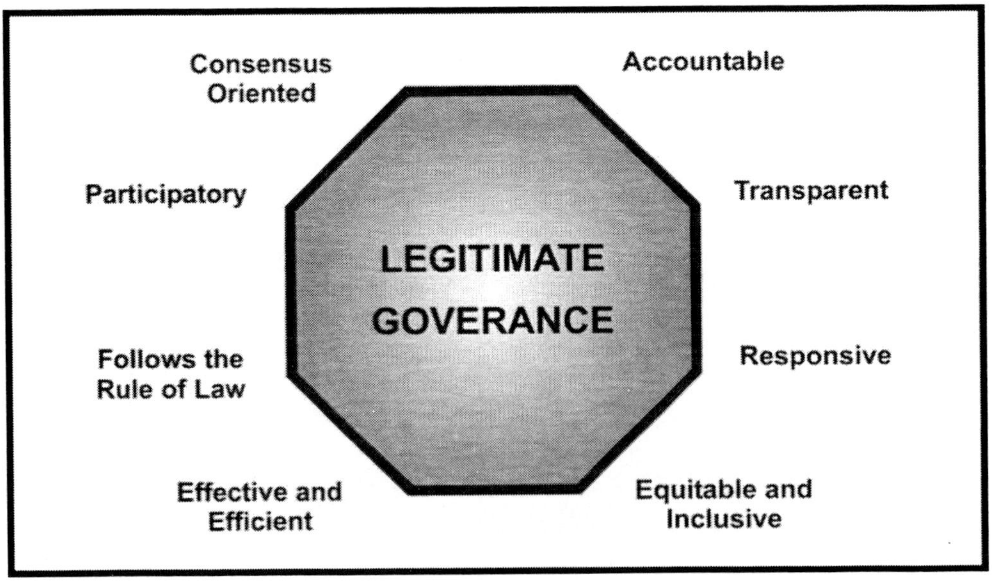


Figure 7-4. Legitimate governance.

*Heavy Goods Vehicles (Trucks)*

Basic heavy goods (truck) vehicle search procedures include—

- Ensure at least three personnel conduct the search.
  - One for the driver and occupants.
  - Two for the vehicle.
- Set aside an area to search these vehicles due to their length and amount of cargo.
- Instruct the driver to turn off the engine, open the hood, and open all doors.
- Check the driver's credentials and cargo manifests.
- Inspect to the degree required for the cargo.
- Complete the search.
- Instruct driver and occupants to secure load and continue.

***Support to Governance***

This helps build toward effective, legitimate governance. Specifically, it focuses on restoring public administration and resuming public services, while fostering long-term efforts to establish a functional, effective system of political governance. In every case, for the counterinsurgent to develop a long-term solution, unit actions must strengthen the Host Nation government and reinforce its legitimacy with the people (Figure 7-4). The formation of an effective local government is critical to the success of any counterinsurgency operation because they provide the foundation for legitimate governance at the national level. Additionally, effective local governments promote social stability when people have a say in their own government. They also increase financial transparency, which helps fight corruption. At the local level, the formation of local neighborhood councils serves as an excellent starting point for supporting governance. Units may have to initiate these efforts without an external agency support. Neighborhood and district councils are effective because they empower the population on many levels. They help the populace devise local solutions to local problems and help citizens and community leaders build skills in community decision making. All involved parties learn to resolve conflicts peacefully and in a transparent fashion. Local councils also help leaders at the local level develop skills that can help them serve at higher levels of government.

**Subordinate Tasks**

Typical tasks include—

- Support transitional administrations.
- Support development of local governance.
- Support anticorruption initiatives.
- Support elections.

## THE CHALLENGES OF GOVERNANCE IN IRAQ

In 2003, perhaps the most volatile site for this ethnic conflict was the city of Kirkuk, located in northern Iraq about 150 miles north of Baghdad. The city was located in what was historically Kurdish territory, but its population had long been a mix of Kurds, Arabs, Assyrian Christians, and Turkoman. Situated in the heart of Iraq's oil fields, Kirkuk had strategic importance for much of the 20th century. Partly for that reason, in the 1970s and 1980s Saddam attempted to "Arabize" Kirkuk by forcing Kurds to leave and moving Arab groups in. The arrival of Coalition troops in the city in May 2003 opened up the possibility for the Kurdish population to take control once again over an important cultural and economic center.

The 173d Airborne Brigade (173d ABN) took responsibility for the city soon after the toppling of the Saddam regime. In May 2003, Soldiers of the brigade found themselves attempting to mediate between groups of armed Arabs moving north to ensure the Kurds did not overwhelm the city and the Kurdish groups that had begun flexing their muscles by forcibly evicting some Arabs. On 17 May 2003, this conflict became violent with firefights erupting in the streets of Kirkuk.

Colonel William Mayville, the brigade commander, recalled that this event served as the "really big first lesson into, or insight into what some of the social dynamics in this community at play were." He added, "If you did not address [these dynamics], the consequences could be very, very violent."

Instead of finding citizens eager for democracy, the Soldiers of the 173d ABN discovered a multiethnic populace interested in removing all vestiges of Baathist power and solving the problems of Saddam's Arabization policies, while also ensuring their ethnic group retained its social, economic, and political position in the city. This presented a complex problem to the Soldiers of the brigade, the large majority of whom had no experience in politics of any type.

One of the first recommendations made by brigade officers was the establishment of a multiethnic city council that could help redress the grievances of the various groups and begin moving the city forward. By the end of May 2003, less than 6 weeks after the brigade arrived in Kirkuk, Mayville and his governance team selected 300 delegates from the city, who elected a 30-seat council, which included 6 representatives from each of the 4 ethnic groups.

Working with civilian contract consultants who had partnered with USAID, brigade officers convinced the city council to establish a new structure that included five directorates: employment, public safety, public works, budget office, and resettlement. The employment directorate would play a direct role in enforcing the de-Baathification process and US officers hoped the resettlement office could work with both Arabs and Kurds to defuse the tensions caused by land disputes. RTI consultants and CA officers also helped the new Kirkuk budget office prepare the city's budget for 2004 and established a citizen's bureau to help handle complaints from the public.

Adapted from "*On Point II: Transition to the New Campaign: The United States Army in Operation IRAQI FREEDOM*," May 2003—January 2005

*Support Transitional Administrations*

When the HN government has collapsed or been deposed, counterinsurgency efforts focus on immediately filling the void in governance. In either situation, the reliability and trustworthiness of local officials is suspect; due care and prudence are necessary to avoid empowering officials whose interests and loyalties are inconsistent with those of the occupying force. For example, a tactical unit may struggle with identifying and emplacing a good neighborhood council leader. Tasks may include—

- Vet Host Nation officials.
- Reconstitute leadership at local levels of government.
- Establish interim legislative processes.

*Support Development of Local Governance*

Establishing effective governance at the local level is necessary before developing governance institutions and processes throughout the HN. Initially, effective local governance almost depends entirely on the ability to provide essential civil services to the people. Most stability tasks require an integrated and synchronized effort across all sectors to achieve the desired end state. Tactical Units have often had success when working with a city, district or tribal council, or city manager. Essential tasks may include—

- Establish mechanisms for local-level participation.
- Identify, secure, rehabilitate, and maintain basic facilities for the local government.
- Restore essential local public services.
- Provide resources to maintain essential local public services.

***Support Anti-Corruption Initiatives***

Corruption of government official could be a root cause that led to the insurgency. Providing legal guidance and assistance to the transitional government mitigates the near-term effects of corruption. Long-term measures ensure lasting success. Corruption and graft can hinder efforts to establish governance, restore rule of law, or institute economic recovery. While some level of corruption is common to many cultures, its existence can unhinge reform efforts and put the entire mission at risk. Essential tasks may include an initial response in which military forces create mechanisms to curtail corruption across government institutions. Units must decide if there is an acceptable level of corruption in the HN society. Essential tasks may include an initial response to—

- Disseminate ethical standards for civil servants.
- Ensure transparency in the dispersal of government resources.
- Implement reporting procedures for corruption and intimidation.

*Support Elections*

The ability of the HN government and its local subdivisions to stage fair and secure elections is a significant milestone toward establishing legitimate, effective governance. While civilian agencies that maintain strict transparency guide the elections process, military forces provide the support that enables broad participation by the local populace. Essential tasks may include:

- Provide security for polling places, voters, and ballots.
- Determine identification requirements for voter registration.
- Establish or verify voter registry.

### Techniques

Two critical governance techniques are negotiating and mediating, and leader and meeting security. See FM 3-07 for additional technique considerations.

*Negotiations and Mediations*

In a counterinsurgency, situations may arise that require counterinsurgency personnel to negotiate, mediate, and perhaps arbitrate disputes. These usually involve minor points of contention between the belligerents or disagreements about the daily routines of the counterinsurgent force. Unit leaders must remain aware of their limitations. Larger points of contention or issues beyond their ability to resolve should be referred to a higher authority. If possible, arbitration should be referred to the existing legal system. Negotiations are made from a position of strength while mediation is made from a position of impartiality. In the joint, combined, and interagency environment, negotiations and mediations can be complex. Nonetheless, all negotiations and mediations require tact, diplomacy, honesty, patience, fairness, effective communications, cross-cultural sensitivity, and careful planning. For the purposes of this manual, the following definitions are used:

***Negotiation***

This is a discussion between at least two parties, with the intention to produce an agreement.

***Mediation***

This is an attempt to bring about a peaceful settlement or compromise between disputants through the objective intervention of a neutral party.

***Arbitration***

This is the process by which disputants submit their differences to the judgment of an impartial person or group appointed by mutual consent.

***Usage in a Counterinsurgency***

Unit leaders must be prepared to conduct negotiations or mediations with the leaders or members of various groups. These groups may be political, ethnic, religious, tribal, military, or even family. Understanding the composition of groups in the AO is vital for effective negotiations and mediations. Analyzing civil considerations using ASCOPE of various groups can be extremely helpful in preparation for these discussions. Furthermore, tensions or hostilities between groups may destabilize a society and provide opportunities for insurgents. Negotiations and mediations may be broken down into two categories: situational and preplanned.

***Situational Negotiations and Mediations***

These both allow immediate discussion and resolution of an issue or problem. For example, members of a ground patrol might encounter two groups arguing at an intersection. To negotiate a resolution to this problem, the patrol must thoroughly understand the ROE and rules of interaction (ROI). Soldiers apply this working knowledge to the process of discussing and, when possible, resolving issues and problems between opposing parties, which might include the unit itself. The leader on the ground must know when he has exhausted his options under the ROE and ROI, and turn over the discussion to a higher authority.

***Preplanned Negotiations***

These allow discussion and resolution of an upcoming, specific issue or problem. For example, a company commander may conduct a work coordination meeting between leaders of the belligerents to determine mine-clearing responsibilities. As with situational negotiations, preplanned negotiations require leaders to know and understand the ROE and ROI. Leaders must also know as much as possible about every aspect of the dispute or issue. The negotiator's goal is to reach an agreement that is acceptable to both sides, and that reduces antagonism and the threat of renewed hostilities.

***Considerations during Counterinsurgencies***

Commanders should identify powerful groups both inside and outside their AOs, and obtain key information about them to better prepare for negotiating and mediating. Analyzing civil considerations using ASCOPE is a useful tool. This prior preparation can be especially beneficial for the day-to-day situational negotiations and mediations that can easily arise during daily operations. This key information includes—

- Formal relationships such as treaties or alliances between groups.
- Informal relationships such as tolerance or friction between groups.
- Divisions and cleavages between groups.
- Cross-cutting relationships between groups such as religious beliefs, political parties, and business ventures.

***Guidelines for Negotiations and Mediations***

Credibility is key to success. As a rule, only communicate necessary information and those actions that authorities and commanders intend to do or have the authority to do. Successful negotiating often involves communicating a clear position on all issues. This includes visualizing several possible end states, while maintaining a clear idea of the best end state and conveying a willingness to bargain on negotiable points. Use tact to justify standing firm on nonnegotiable points while still communicating respect for other participants and consider having points that are disposable to use for concessions.

***Steps***

Several steps are common to negotiations and mediations, especially preplanned negotiations:

***Determine the Purpose***

Before contacting leaders of the belligerent parties to initiate the negotiation or mediation process, participants must know what they are trying to do, as well as the limits of their authority. Ensure that participants can recommend that their superiors ratify the agreement.

***Prepare***

Thorough preparation is needed to ensure the success of the negotiation or mediation process. Commanders must familiarize themselves with both the situation and the area in which their unit will operate. They should consider selecting one person who understands conflict dynamics and cross-cultural issues to observe a negotiation or mediation and give advice. This individual can watch for body language and other indicators of how the process is working. In turn, this person may be able to coach more effective techniques to the negotiators or mediators. Leaders should also develop an agenda for the meeting and consider the conduct, customs, and actions expected at the meeting

***Communicate***

Participants must establish an effective means of communicating with all parties. They should not assume that a certain leader or element is opposed to their efforts without careful investigation. Instead, they must ensure that facts are correct before forming any opinions. The commander must earn the trust and confidence of any opposing party. This includes establishing an atmosphere (and a physical setting) that participants will judge to be both fair and safe.

***Execute***

Always strive to maintain control of the session; be firm, yet even-handed, in leading the discussion. At the same time, be flexible, with a willingness to accept recommendations from the opposing parties and from assistants and advisors. Settle the easy issues first. Be prepared to precede issue-by-issue in a predetermined order. Actions can have different connotations to individuals of other cultures. Culture shapes how people reason, what they accept as fact, and what principles they apply to decision making. Also, nonverbal behavior, such as the symbolic rituals or protocols of the arrangement for a meeting, is important. If participants cannot reach agreement, they must keep the dialogue going. At a minimum, they must seek agreement on when the parties will meet again.

***Leader and Meeting Security***

Key leaders and important meetings, even in established nonviolent peaceful areas of the world, need security. This security allows leaders to focus on the meeting's purpose without distractions. During a counterinsurgency, security becomes an absolute requirement as all key leaders and important meetings are potential targets for insurgents. Successful counterinsurgency operations require active interaction with the population. This interaction varies from formal meetings and negotiations to informal interactions with local citizens. Formal meetings and negotiations are significant mission events and should be thoroughly planned and properly resourced. Informal interactions happen with little or no planning but must still be prepared for thorough training and SOPs. In most situations, unit leaders, from team leaders up are the individuals who will be involved in regular interactions and meetings. These personnel are often placed in exposed positions as they execute their mission. At the tactical level, a personal security detachment, MP platoon, or squad will typically secure a leader, VIP, or meeting. Leaders and units consider several things when preparing for meetings or negotiations:

#### *Individual Soldier Interactions*

In a counterinsurgency environment, individual Soldiers interact daily with the local population. When doing so, they must maintain situational awareness and be prepared to react immediately to a range of situations. While one Soldier interacts, another should provide security for the other. This is easy if the buddy system is used.

#### *Considerations for Leader Security*

At least one person should be ready to provide close-in, direct physical protection for the leader. However, this should not interfere with the leader's mission accomplishment, should not irritate to other participants, and should not include behavior that could be seen as a threat. Planning must also determine any restrictions on people approaching the protected leader, and should restrict the circulation of people around the leader. If armed at all, the leader should carry a sidearm. This leaves the leader's hands free to react to close-in threats. Security personnel should take positions where they can observe everything and everyone near the leader.

#### *Considerations for Meeting Security*

Before a meeting, crowd-control measures should be planned and prepared, even if crowds are not expected. Evacuation routes and EOF procedures should be planned for key personnel. Planning for outer security should include coverage of inner and outer cordons.

- The inner cordon protects the meeting participants, controls access to the meeting, and controls the immediate crowd.
- The outer cordon isolates and controls access to the meeting. This cordon may be close in and visible or out of sight of the actual meeting, focusing on outside elements trying to gain access.

If available, use special units or weapons such as snipers and shotguns or nonlethal weapons. Use at least one interpreter (two are preferred). One actively interprets while the other observes body language and listens to side conversations in the audience. All of these steps should be rehearsed before execution.

A typical technique for conducting a meeting includes reconnoitering the meeting site and also conducting a pre-site clearance before the meeting. Forces also establish outer and inner cordons, secure the meeting, collapse both cordons, and then conduct an orderly movement back to their base or combat outpost. All of these actions should be rehearsed extensively prior to execution.

### *Restore Essential Services*

Army forces establish or restore the most basic services and protect them until a civil authority or the Host Nation can provide them. Here, the counterinsurgent force works toward meeting the population's basic needs. Ideally, the military is simply providing the security for the humanitarian assistance. Normally, Army forces support other government, intergovernmental, and Host Nation agencies. When the Host Nation or other agency cannot perform its role, Army forces may provide the basics directly (FM 3-0). Tactical units focus primarily on addressing the immediate needs of a population and fostering Host Nation efforts at restoring essential services. However, in the absence of Host Nation capability, military forces may directly perform these tasks as the Host Nation develops the ability to do them on its own. Tactical units will generally provide for the basic needs of the population, such as food, water, shelter, and basic medical care. Should other agencies be present, close coordination between military forces and those agencies will become paramount. (See FM 3-07 for technique considerations.)

### Subordinate Tasks

Typical subordinate tasks to restore essential services follow:

- Provide essential civil services.
- Tasks related to civilian dislocation.
- Support famine prevention and emergency food relief programs.
- Support public health programs.
- Support education programs.

#### *Provide Essential Civil Services*

Although closely related to establishing and supporting effective local governance, efforts to provide essential civil services to the HN population involve developing the capacity to operate, maintain, and improve those services. This broader focus involves a societal component that encompasses long-range education and training, employment programs, and economic investment and development. At the tactical level, activities of military forces to provide essential civil services are often defined in terms of the immediate humanitarian needs of the people: providing the food, water, shelter, and medical support necessary to sustain the population until local civil services are restored. Once their immediate needs are satisfied, efforts to restore basic services and transition control to civil authorities typically progress using lines of effort based on the memory aid, SWEAT-MSO (sewage, water, electricity, academics, trash, medical, safety, and other considerations). These lines of effort are vital to integrating efforts to reestablish local HN services with similar, related actions to establish a safe, secure environment. Tactical units may support the effort to provide essential civil services by conducting detailed infrastructure reconnaissance or security for those types of reconstruction projects. Essential tasks may include:

- Provide for immediate humanitarian needs (food, water, shelter, and medical support).
- Ensure proper sanitation, purification, and distribution of drinking water.
- Provide interim sanitation, wastewater, and waste disposal services.

#### *Tasks Related to Civilian Dislocation*

In the fluid and uncertain nature of an insurgency, the population is often left homeless. The presence and uncontrolled flow of dislocated civilians can threaten the success of any stability operation. The treatment of displaced populations either fosters trust and confidence—laying the foundation for stabilization and reconstruction among a traumatized population—or creates resentment and further chaos. Local and international aid organizations are most often best equipped to deal with the needs of the local populace but require a secure environment in which to operate. Through close cooperation, tactical units can enable the success of these organizations by providing critical assistance to the populace. Nearly 80 percent of all dislocated civilians are women or children. Most suffer from some form of posttraumatic stress disorder, and all require food, shelter, and medical care. Following a major disaster, humanitarian crisis, or conflict, providing adequate support to dislocated civilians often presents a challenge beyond the capability of available military forces. Therefore, military forces offer vital support—coordinated with the efforts of other agencies organizations—to provide humanitarian assistance to the general population. The list of essential tasks includes:

- Assist dislocated civilians.
- Support assistance to dislocated civilians.
- Support security to dislocated civilian camps.

*Support Famine Prevention and Emergency Food Relief Programs*

Famine vulnerability may cause an insurgency. An insurgency could certainly lead to a food scarcity. The combination of weak institutions, poor policies, and environmental change often results in famine. Famine may result in food insecurity, increased poverty, morbidity, malnutrition, and mortality. Government agencies, such as the US Agency for International Development (USAID), numerous nongovernmental organizations, and the United Nations, are instrumental to response efforts in famine-prone states. They oversee the major relief programs that provide emergency food aid to suffering populations. Tactical unit of these efforts is vital to the overall success of the operation. Essential tasks may include—

- Monitor and analyze food security and market prices.
- Predict the effects of conflict on access to food.
- Estimate total food needs.
- Assess the adequacy of local physical transport, distribution, and storage of food.
- Deliver emergency food aid to most vulnerable populations.

*Support Public Health Programs*

This enables the complementary efforts of local and international aid organizations. The initial efforts of military forces aim to stabilize the public health situation within the operational area. These efforts may include assessments of the civilian medical and public health system such as infrastructure, medical staff, training and education, medical logistics, and public health programs. Achieving measurable progress requires early coordination and constant dialog with other actors; ultimately, this also facilitates a successful transition from military-led efforts to civilian organizations or the Host Nation. Essential tasks may include—

- Assess public health hazards within their AO including malnutrition, water sources, and sewer and other sanitation services.

- Assess existing medical infrastructure including preventative and veterinary services, health— physical and psychological—care systems, and medical logistics.
- Evaluate the need for additional medical capabilities.
- Repair existing civilian clinics and hospitals.
- Operate or augment the operations of existing civilian medical facilities.
- Prevent epidemics through immediate vaccinations.
- Support improvements to local waste and wastewater management capacity.
- Promote and enhance the HN medical infrastructure.

*Support Education Programs*

- Military activities to support education programs generally focus on repairing or building physical infrastructure such as classrooms, schools, or universities. In some cases, trained personnel with appropriate civilian backgrounds provide additional services such as administrative or educational expertise.

***Support to Economic and Infrastructure Development***

This helps a Host Nation develop capability and capacity. It includes both short—and long-term aspects. The short-term aspect concerns immediate problems, such as large-scale unemployment and reestablishing an economy at all levels. The long-term aspect involves stimulating indigenous, robust, and broad economic activity. The stability a nation enjoys is often related to its people's economic situation and its adherence to the rule of law. However, a nation's economic health also depends on its government's ability to continuously secure its population.

**Subordinate Tasks**

Typical subordinate tasks for support to economic and infrastructure development include—

- Support economic generation and enterprise creation.
- Support public sector investment programs.
- Support private sector development.
- Protect natural resources and environment.
- Support agricultural development programs.
- Restore transportation infrastructure.
- Restore telecommunications infrastructure.
- Support general infrastructure reconstruction programs.
- Use money as a weapon.

**Support Economic Generation and Enterprise Creation**

Economic recovery begins with an actively engaged labor force. Insurgencies often gain recruits by offering the unemployed a wage. When a tactical unit occupies its AO, the demand for local goods, services, and labor creates employment opportunities for the local populace. Local projects, such as restoring public services, rebuilding schools, or clearing roads, offer additional opportunities for the local labor pool. Drawing on local goods, services, and labor presents the force with the first opportunity to infuse cash into the local economy, which in turn stimulates market activity. However, this initial economic infusion must be translated into consistent capital availability and sustainable jobs programs. Thus, short-term actions are taken with an eye towards enabling financial self-reliance and the creation of a durable enterprise and job market.

The local economy requires this stimulus to sustain economic generation and enterprise creation. It includes efforts to execute contracting duties; identify, prioritize, and manage local projects; and implement employment programs. Often, such programs reinforce efforts to establish security and civil order by providing meaningful employment and compensation for the local populace. The assessment of the economic sector must include developing knowledge and understanding of local pay scales; this is essential to establishing jobs programs with appropriate wages. Inflated pay scales may divert critical professionals from their chosen field in pursuit of short-term financial gains from new jobs created by the force. Establishing appropriate pay scales is also significant when the environment includes illicit actors willing to pay for actions or services in direct conflict with the aims of the force. Adversaries can easily exploit relatively low pay scales and quickly undermine efforts to build positive perceptions among the people.

HN enterprise creation is an essential activity whereby the local people organize themselves to provide valuable goods and services. In doing so, they create jobs for themselves, their families, and neighbors that are inherently sustainable after the departure of other actors. Host nation enterprises may provide various goods and services, including essential services such as small-scale sewerage, water, electricity, transportation, health care, and communications. The availability of financing through banking or microfinance institutions is essential to enterprise creation. Essential tasks may include—

- Implement initiatives to provide immediate employment.
- Create employment opportunities for all ages and genders.
- Assess the labor force for critical skills requirements and shortfalls.
- Assess market sector for manpower requirements and pay norms.
- Implement public works projects.
- Support establishment of a business registry to register lawful business activity at the local or provincial level.
- Provide start-up capital for small businesses through small-scale enterprise grants.
- Encourage the creation of small lending institutions.
- Enable the development of financial institutions.

*Support Public Sector Investment Programs*

Organizations such as the US Agency for International Development usually manage public sector investment in a fragile state. However, the military force can also influence success in these programs. Public sector investment ensures the long-term viability of public education, health care, and mass transit. It also provides for development in industries—such as mining, oil, and natural gas—and hydroelectricity. At the tactical level, units may spur investment through grant programs or direct public investment projects. Essential tasks may include:

- Identify projects that require large amounts of labor.
- Prioritize public investment needs.
- Develop plans to allocate available resources.

*Support Private Sector Development*

Developing the private sector typically begins with employing large portions of the labor force. In addition to acquiring goods and services from the local economy, the tasks that support private sector development infuse much-needed cash into local markets and initiate additional public investment and development. Essential tasks may include:

- Identify projects that require large amounts of labor.
- Assess the depth of the private sector and enterprise creation.
- Identify obstacles to private sector development.
- Facilitate access to markets.
- Strengthen the private sector through contracting and outsourcing.
- Provide investors with protection and incentives.

*Support Agricultural Development Programs*

The agricultural sector is a cornerstone of a viable market economy, providing crops and livestock vital to local markets and international trade. The development of this sector may be hindered by property disputes, difficulty accessing nearby markets, poor irrigation, animal disease, minefields, or unexploded ordnance. Therefore, development agencies prioritize and integrate projects with related tasks in other stability sectors to establish and institutionalize practical solutions to the long-term growth of the agricultural sector. The military contribution to agricultural development parallels related efforts to spur economic growth in local communities. Together, they draw on local labor pools to help reestablish basic services central to the agricultural sector. Essential tasks may include—

- Assess the state of agricultural sector.
- Secure and protect post-harvest storage facilities.
- Rebuild small-scale irrigation systems.
- Establish work programs to support agricultural development.
- Protect water sources.
- Identify constraints to production.
- Assess health, diversity, and numbers of animals.
- Establish transportation and distribution networks.
- Encourage Host Nation enterprise creation to provide goods and services to the agricultural sector.
- Ensure open transit and access to local markets.

*Restore Transportation Infrastructure*

Even at the local level, this is central to economic recovery. An underdeveloped or incapacitated transportation infrastructure limits freedom of movement, trade, social interaction, and development. Goods must make it to the markets. Military forces often initiate immediate improvement to the transportation and distribution networks of the Host Nation. These networks enable freedom of maneuver, logistic support, and the movement of personnel and materiel to support ongoing operations. They also may reduce the risk of IED attacks on security forces. These improvements facilitate the vital assistance efforts of civilian agencies organizations that follow in the wake of military forces. Essential tasks may include an initial response in which military forces:

- Assess overall condition of local transportation infrastructure (airports, roads, bridges, railways, coastal and inland ports, harbors, and waterways), including facilities and equipment.
- Determine and prioritize essential infrastructure programs and projects.
- Conduct expedient repairs or build new facilities to facilitate commercial trade.

*Restore Telecommunications Infrastructure*

This exists to support every element of a society, from the government to the financial sector, and from the media to the HN populace. The failure of this infrastructure accelerates the collapse of the HN, isolates the HN government and populace from the outside world, and hampers development efforts. The military contribution to reconstruction efforts in the telecommunications infrastructure is limited; normally, few essential tasks exist in this area. Essential tasks may include military forces:

- Assess overall condition of the national telecommunications infrastructure.
- Determine and prioritize essential infrastructure programs and projects.

*Support General Infrastructure Reconstruction Programs*

General infrastructure reconstruction programs focus on rehabilitating the state's ability to produce and distribute fossil fuels, generate electrical power, exercise engineering and construction support, and provide municipal and other services to the populace. Such capacity building spurs rehabilitation efforts that establish the foundation for long-term development. As with the restoration of essential services, support to general infrastructure programs requires a thorough understanding of the civil considerations using ASCOPE. Civil affairs (CA) personnel support this information collection to help prioritize programs and projects. Essential tasks may include an initial response in which military forces:

- Assess overall condition of local energy infrastructure.
- Determine and prioritize essential infrastructure programs and projects.
- Assess condition of existing power generation and distribution facilities.
- Assess condition of existing natural resources conversion and distribution facilities.
- Assess condition of existing facilities needed to effectively execute essential tasks in other sectors.
- Assess conditions of existing municipal facilities that provide essential services.
- Conduct expedient repairs or build new facilities to support the local populace.

*Use Money as a Weapon System*

Recent experiences have shown the effectiveness of using money to win popular support and further the interests and goals of units conducting counterinsurgency operations. Money should be used carefully. In most cases, higher authority will tightly control funds through strict accountability measures that ensure the money is used properly. When used effectively, and with an end state in mind, money can be an effective means to mobilize public support for the counterinsurgent's cause and further alienate the insurgents from the population. A counterinsurgency force can use money to:

- Fund civic cleanup and other sanitation projects, and the equipment to complete those projects.
- Fund small scale infrastructure improvements.
- Fund agricultural projects to improve farming practices and livestock health, or help implement cooperative farming programs.
- Repair civic and cultural sites and facilities.
- Repair institutions and infrastructure critical to governance and rule of law such as prisons, courthouses, and police stations.
- Purchase education supplies or repair infrastructure critical for educating the local populace.
- Pay rewards to citizens who provide information on enemy activities and locations
- Support the creation, training, and operation of Host Nation security forces.
- Fund events and activities that build relationships with Host Nation officials and citizens.
- Repair damage resulting from combined and coalition operations.
- Provide condolence payments to civilians for casualties from combined and coalition operations.

## Techniques

Two techniques include the USAID principles of reconstruction and development, which can be used by tactical units for project selection and small-scale building projects. (See FM 3-07 for additional considerations.)

### *Principles of Project Selection*

The following principals apply to project selection and the corresponding use of resources, especially financial resources, in support of counterinsurgency operations:

#### ***Host Nation Ownership***

This principle holds that units must ensure that there is counterinsurgent and Host Nation ownership of any project. The local population and the government officials who serve them should view any project as their own and not one that has been imposed by outside agencies. A project that has been locally conceived, funded, and constructed legitimizes the government and contributes to stability.

#### ***Capacity Building***

This principle involves the transfer of knowledge, techniques, and skills to the local people, institutions, and government so that they develop the requisite abilities to deliver essential services to the population. Ultimately, the local officials and institutions that gain capacity are better prepared to lead their regions through political, economic, and security-related issues.

### *Sustainability*

This principle says that commanders should design and select projects and services that will have a lasting effect on the local population. In other words, the impact of the projects under consideration must endure after the facility or service is handed off to local authorities and the unit (or contractor's) departure from the site. Sustainability also implies that the local government has the necessary resources to staff and maintain the project. There are examples where commanders have failed to conduct adequate analysis and built new schools or medical clinics only to discover that too few teachers or doctors were available to staff these facilities. Similarly, some commanders have purchased large generator systems to address electricity shortfalls for neighborhoods within their AOs. However, without addressing the fuel, maintenance, and service requirements of these systems, the machines eventually failed and were unable to serve as a sustainable solution for the local electrical deficit.

### *Selectivity*

The development community defines this principle as the allocation of resources based on need, local commitment, and foreign policy interests. These strategic characteristics are equally important tactically. Commanders seldom receive all of the financial resources they need when implementing their essential service, economic, and governance lines of effort. Therefore, they must "mass" available resources into select reconstruction projects that offer the following advantages:

- The local government strongly supports them.
- They will positively impact the most people
- They will achieve the commander's desired effects.

### *Assessment*

This principle advises the commander and his staff to carefully research nominations, adopt best practices, and design for local conditions in their proposed projects. Commanders should request or conduct an assessment of local conditions before investing financial resources into any potential relief or reconstruction program. Money should not be invested in an intelligence vacuum—all available information about local conditions should be considered such as the population's requirements, animosities, traditions, capabilities, economics, and so on. A detailed assessment of local conditions will best advise the commander on the project's potential to deliver its desired effects.

### *Results*

This principle relates directly to the principle of assessment and advises commanders to direct resources to achieve clearly defined, measurable, and long-term focused objectives. This development principle is analogous to the military principle of objective—direct every military operation toward a clearly defined, decisive, and attainable objective. It is equally critical that the commander and his staff also assess potential unintended results from their use of money. Many second or third order effects from a project may potentially outweigh the benefits from the original intent.

***Partnership***

This principle encourages close collaboration with local governments, communities, donors, nonprofit organizations, the private sector, international organizations, and universities. Partnership plays a central role in any relief, reconstruction, or development program as it supports each of the other principles of reconstruction and development. In the context of money as a weapon system, effective partnership will ensure the unit's financial resources are well invested and deconflicted with other programs from other agencies.

***Flexibility***

This principle states that units must adapt to changing conditions, take advantage of opportunities, and maximize effectiveness as part of their reconstruction and development program. Just as the conditions for offensive and defensive operations are often changing and uncertain, so are the relief and reconstruction tasks associated with stability operations. These conditions will often require commanders to change tactics to achieve desired objectives.

***Accountability***

Enforcing accountability, building transparency into systems, and emplacing effective checks and balances to guard against corruption are important components to any relief, reconstruction, or development program. Accountability in all actions, including the unit's use of money, reinforces the legitimacy of the commander and his operations, as well as the legitimacy of the local government, in the eyes of the population.

***Small-Scale Building Projects***

Small-scale building projects encompass almost all construction, both permanent and temporary, that is limited in scale. Examples include schoolhouses, clinics, simple irrigation works, farm-to-market roads, or small police stations. These projects also try to provide immediate, short-term economic stimuli and long term benefits to a local area. The HN should have the highest possible profile and participation in all phases of these projects. Before construction begins, leader's should understand the second and third order effects of the project, such as which group(s) will profit from the project or use the project and any conflicts of interest that arise from it. They should also consider the following:

- The local population should need or want the project.
- The project should support the Host Nation's plan.
- The project should support the unit's COIN mission and long-range plans.
- The project should support Host Nation civilian authority.
- The counterinsurgent unit should develop a sustainability plan.

***Basic Technique Considerations***

The following outlines a basic process to conduct a small-scale building project:

***Concept Development***

Identify needs and potential projects to fill those needs. Soldiers and leaders at every level of responsibility should help in identifying potential projects that will further the unit's and the Host Nation's interests in an AO. Leaders identify potential projects through meetings with local officials and through interaction with the local populace, along with their own observations of their area of operations. Concurrently, possible locations are identified, surveyed, and discussed with the local population and leaders to ensure suitability.

***Project Planning***

Once the site has been selected and the project is ready to move forward, a plan for security is developed and implemented. Whenever possible, Host Nation police forces should be used. The counterinsurgent unit should also continually assess the security situation and adjust the security plan accordingly based on the changing threat. Consideration should also be given to sustaining the security forces for the duration of the project.

- Create the design and obtain approval of design.
- Determine resource requirements.
    - Material.
    - Equipment.
    - Personnel—Labor (professional, skilled, and unskilled).
    - Wages and payment system.
    - Infrastructure needs (SWEAT-MSO).
- Conduct available resource assessment.
    - Availability of local HN civilian resources.
    - Availability of National/regional HN civilian resources.
    - US civilian and international resources.
    - US government and military resources
- Determine project participants and division of responsibilities.
    - HN national and regional.
    - Local HN.
    - US Government employees and contractors.
    - NGO.
- Contract the services.
    - Material support.
    - Equipment support
    - Personnel support (individual labor).
    - Subcontractor support (specific service).
    - Infrastructure support (warehouses, electrical, sewage, water, and so on).
- Identify, design and incorporate information engagement theme and message.

***Project Execution***

- Execute the project.
- Establish periodic progress meetings or reviews.

***Project Completion***

- Hand over ownership to HN authority—preferably local officials.
- Plan on a follow-up visit to resolve questions or issues and ensure sustainability.

**SUMMARY**

The seven counterinsurgency lines of effort (LOEs), establish civil security, establish civil control, support Host Nation security forces, support to governance, restore essential services, and support to economic and infrastructure development are critical to the success of any stability consideration during a counterinsurgency operation. Ultimate success in any counterinsurgency operation requires a combination of military and nonmilitary efforts. As President Bush stated in 2007 in reference to operations in Iraq:

"A successful strategy for Iraq goes beyond military operations. Ordinary Iraqi citizens must see that military operations are accompanied by visible improvements in their neighborhoods and communities."

Stability operations during a counterinsurgency seek to bring about those community improvements and create a safe, secure, and productive environment for the populace.

# 8. SUPPORT TO HOST NATION SECURITY FORCES

*"Do not try to do too much with your own hands. Better the Arabs do it tolerably than that you do it perfectly. It is their war, and you are to help them, not to win it for them. Actually, also, under the very odd conditions of Arabia, your practical work will not be as good as, perhaps, you think it is."*

T.E. Lawrence, *27 Articles*

This chapter addresses the working relationship between US forces and Host Nation security forces. It begins with a discussion of the benefits and challenges involved and resources required. It provides a framework for successful US and Host Nation operations that accomplish both HN and US objectives.

## Section I—Benefits, Challenges, and Goals

Success in counterinsurgency operations requires establishing a legitimate government, supported by the people, and able to address the root causes that insurgents use to gain support. Achieving these goals requires the Host Nation to secure the populace, defeat the insurgents, uphold the rule of law, and provide a basic level of essential services and security for the populace. Key to all these tasks is developing an effective Host Nation security force. A goal in any counterinsurgency operation is the eventual transition of the responsibility for security to the Host Nation security forces. Many factors influence the amount and type of assistance required in developing Host Nation security forces. These factors include—

- Existing HN security force capabilities.
- Character of the insurgency.
- Population and culture.
- Level of commitment and sovereignty of the Host Nation.
- Level of commitment from the United States and other nations.
- Impact of US forces on the local infrastructure and HN legitimacy.

***Benefits***

A Host Nation is a nation which permits, either by written agreement or official invitation, government representatives, agencies, forces or supplies of another nation to operate in or to transit through its territory under specified conditions (JP 1-02). A Host Nation naturally needs security forces. Therefore, a primary goal of partnering is to get the best that both sides have to offer. Figure 8-1 shows a typical flow of these benefits.

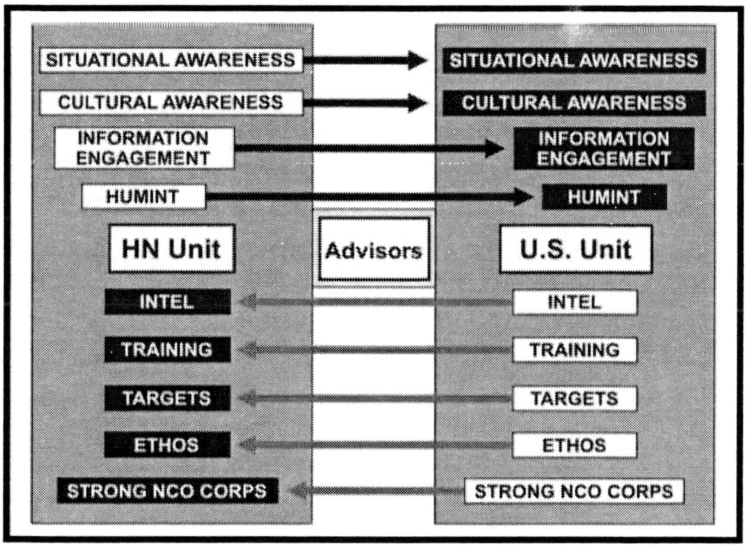


Figure 8-1. Partnership benefits.

| | |
|---|---|
| Understand the Operational Environment | The AO is their home and their culture; they know the language, the different groups, the political situation, educational levels, economic considerations, historical bad actors, and unofficial community leaders. |
| Provide Human Intelligence | The HN security force is better able to gather information that leads to HUMINT for a host of reasons, to include speaking the same language, understanding the important players in the area, and so on. |
| Put the Pieces Together | They can often better integrate the different fragments of intelligence into the context of the operational environment. |
| Determine Credibility of Intelligence Assets (Sources, Walk-Ins, Call-Ins) | They possess a vastly superior sense of cultural and situational awareness vital to managing and assigning credibility to sources. |
| Validate and Check Interpreters | They can confirm not only the interpreters' ability, but also the interpreters' loyalty. |
| Identify and Root Out Infiltrators | They can pick out minute differences between normal and abnormal behavior. |
| Gain Information Superiority | They can write messages that resonate with the local populace. |

Figure 8-2. Host nation contributions.

### US Army Assets

These bring numerous assets to the counterinsurgency fight. They bring a professional military with its training, leadership, targeting, technology, assets, resources, and expertise.

### HN Assets

These also bring many assets to the fight. If properly measured, organized, rebuilt, trained, equipped, advised, and mentored, an HN security force unit is uniquely able to protect the HN population, because they understand the operational environment far better than US Forces. Their cultural and situational awareness is a significant force multiplier during any counterinsurgency operation, especially in HUMINT, information engagement, negotiations, and targeting. In terms of specific COIN operations, HN security forces can aid or even take the lead on *clear-hold-build* operations. They should be completely integrated in the planning process, instrumental in the decision of what areas need to be cleared, included in the site selection process for the *hold* phase, and leveraged in the decisions on what stability operations need to be conducted during the *build* phase. Thus, a Host Nation unit potentially brings to a counterinsurgency fight the abilities shown in Figure 8-2.

### *Challenges*

Most nations have at least some cultural obstacles to developing a professional military that is responsive and accountable to the HN population. Part of the challenge is to design a professional military that minimizes these culture obstacles. Most challenges arise from the differing perspectives of the US and the Host Nation.

### Examples

Common cultural challenges include the following:

- Nepotism, such as, rank or positions based on who you know.
- Denial of negative results or errors in the interest of saving face.
- Corruption, such as unofficial or under the table entitlements.
- Influence from competing loyalties (ethnic, religious, tribal and political allegiances).

### Categories

These challenges fall under one of four broad categories of resources, leadership, abusing power organizational structures.

#### *Resources*

Governments must properly balance national resources to meet the people's expectations. Funding for services, education, and health care can limit resources available for security forces. HN spending priorities may result in a security force capable of protecting only the capital and key government facilities, leaving the rest of the country unsecured. HN security forces will typically not have the same resources, equipment, money, or salaries as US units. US units may not have the means to partially sustain the HN security forces to maximize their effectiveness. Conducting effective COIN operations requires allocating resources to ensure integration of efforts to develop all aspects of the security force.

### *Leadership*

Counterinsurgents may need to influence the existing HN approach to leadership. HN leaders may be appointed and promoted based on family ties or membership in a party or faction, rather than on demonstrated competence or performance. Leaders may not seek to develop their subordinates or feel the need to ensure the welfare of subordinates. In some cases, leaders enforce the subordinates' obedience by fear and use their leadership position to exploit them. Positions of power can lead to corruption, which can also be affected by local culture. Ethical climate and decision making in the HN security force leadership is often reflected in the HN security force organization's treatment of detainees. This is a proven barometer of the commander's ability to guide his unit and provide lasting security for the HN population.

US Forces tasked to develop HN security forces should take special interest in how the Host Nation appoints, promotes, and develops leaders. The best way US advisors and partner units can influence leaders is to provide positive leadership examples.

### *Abuse of Power*

The behavior of HN security force personnel is often a primary cause of public dissatisfaction. Corrupting influences of power must be guarded against. Cultural and ethnic differences within a population may lead to significant discrimination within the security forces and by security forces against minority groups. In more ideological struggles, discrimination may be against members of other political parties, whether in a minority cultural group or not. Security forces that abuse civilians do not win the populace's trust and confidence; they may even be a contributing factor of the insurgency. US partners and advisors must identify and address biases, as well as improper or corrupt practices.

Abuse of power, such as hoarding property, extortion, or reprisal attacks on a specific group, could be a root cause that lead to the insurgency or could worsen the insurgency. Units need to understand issues in the OE and define the problem before attempting to "fix" problems. A current example is how various groups may have a long cyclical history of conflict with other population groups. Ethnic, historical, political, racial, religious, social, and territorial turmoil may contribute to cycles of violence in the Host Nation so intractable that it will take establishing discipline, a code of honor, patriotic nationalism, and a merit based promotion system inherent in a professional military to break this societal norm. Even then, these efforts may not be sufficient to stop the abuse of power.

### *Organizational Structures*

Perhaps the biggest hurdle for US forces is accepting that the Host Nation can ensure security using organizational and operational structures that differ from US practices. The goal is not to make the HN unit a US unit, nor even to make it operate like a US unit. HN security force commanders must be given leeway in resolving their own security problems. Mirror imaging HN security force structure is often impractical and does not approach the problem from the perspective root causes that lead to the insurgency. The population, and subsequently their military, will have cultural aspects that will differ with US norms.

Commanders must recognize that the "American way is best" bias is counterproductive with respect to most Host Nation security forces. While relationships among US police, customs, and military organizations and individual unit structure work for the United States, those relationships may not exist in other nations that have developed differently. Units and advisors should develop innovative ways to maximize the effectiveness of the Host Nation's organization, units, and bureaucracy.

### *Goals*

Training HN security forces is a slow and painstaking process. It does not lend itself to a "quick fix." To ensure long-term success, commanders clarify their desired end state for training programs early. These goals consist of a set of military characteristics common to all militaries. Those characteristics have nuances in different countries. Figure 8-3 shows how well-trained HN security forces should be characterized.

**PROFICIENT**

Security forces can work well in close coordination to suppress lawlessness and insurgency.

Military units are—

- Tactically and technically proficient
- Able to perform their national security responsibilities
- Able to integrate their operations with those of multinational partners.

Nonmilitary security forces—

- Can maintain civil order, enforce laws, control borders, secure key infrastructure, and detain criminal suspects.
- Are well trained in modern police ethics.
- Understand police and legal procedures, including the basics of investigation and evidence collection.

**WELL LED**

Leaders at all levels—

- Have sound professional standards and appropriate military values.
- Are selected and promoted based on competence and merit.

**PROFESSIONAL**

Security forces—

- Are honest, impartial, and committed to protecting and serving the entire population, operating under the rule of law, and respecting human rights.
- Are loyal to the central government and serving national interests, recognizing their role as the people's servants and not their masters.
- Operate within a code of conduct and ethical behavior.
- Recognize the merits of a trained and capable NCO and junior officer corps.

**FLEXIBLE**

Security forces can accomplish the broad missions required by the Host Nation—not only to defeat insurgents or defend against outside aggression, but also to increase security in all areas. This requires an effective command organizational structure that makes sense for the Host Nation.

**INTEGRATED INTO SOCIETY**

Security forces represent the Host Nation's major ethnic groups. They are not seen as instruments of just one faction. Cultural sensitivities toward the incorporation of women must be observed, but efforts should also be made to include women in police and military organizations.

**SELF-SUSTAINED**

Security forces must be able to manage their own equipment throughout its life cycle. They must also be able to provide their own administrative and sustainment support, especially fuel, water, food, and ammunition.

Figure 8-3. Characteristics of well-trained HN security forces.

## Section II—The Framework

Developing HN security forces is a complex and challenging mission. The United States and multinational partners can only succeed if they approach the mission with the same deliberate planning and preparation, energetic execution, and appropriate resourcing as the combat aspects of the COIN operation. Accordingly, COIN force commanders and staffs need to consider the task of developing HN security forces during their initial mission analysis. They must make that task an integral part of all assessments, planning, coordination, and preparation. This section discusses developing a plan and developing Host Nation security forces.

### *Developing a Plan*

To defeat the insurgency and be able to sustain success, the Host Nation should develop a plan, with US assistance when necessary, to improve the unit's organization, training, material, leadership, personnel, and facilities. However, these elements are tightly linked, simultaneously pursued, and difficult to prioritize. Commanders monitor progress in all domains. The HN security force plan must be appropriate to HN capabilities and requirements. At the tactical level, the US advisors or partner units will outline goals, allocate resources, and schedule events as part of the plan under the COIN line of effort—support Host Nation security forces.

### Considerations

Competently trained and led Host Nation security forces are vital to winning a counterinsurgency and to sustaining a stable, secure, and just state. If local communities do not perceive US and HN forces as legitimate, or that they cannot provide for security, then the population will not risk providing the timely intelligence necessary to defeat the insurgency. In addition, if the populace does not see at least a gradual transition to Host Nation responsibility, they may begin to view the HN security forces as an instrument or puppet of US forces. Finally, these failures will result in HN security forces that are ill equipped to persevere in a protracted struggle after US/coalition forces withdraw.

US leaders must gain trust and form bonds with HN security force units and personnel as part of the partnership mission. Working towards cultural understanding improves the US forces relationship with the partner units. Commanders must treat the individuals in partner units as they treat their Soldiers by living, eating, sleeping, socializing, planning, and fighting side by side. The Host National security force must know that US forces care for them both individually and professionally. It is this bond, not the materials or support that commanders provide, that inspires that level of commitment and loyalty. To successfully train and support HN security forces, the US partner unit or advisor must—

- Ensure security forces understand that they support the HN government *and* the people.
- Maintain relevancy of security forces for their culture, their population, and their laws.
- Understand and define the security problem.
- Ensure credibility and legitimacy to all counterinsurgency operations.

- Provide a model for society by using military units of mixed ethnicity, religion, political affiliation, for example, who can work together to secure and protect all the people.
- Conduct multinational operations with each newly trained security force.
- Promote mutual respect between US and HN forces and between the military, police, and paramilitary.
- Train the trainers first, and then train the HN cadre.
- Support the HN cadre in training the whole force.
- Separate HN military and police forces, especially during their training.
- Place the HN cadre in charge as soon as possible.
- Recognize achievement, especially excellence.
- Train all security forces not to tolerate abuses or illegal activity outside of culturally acceptable levels.
- Develop methods to report violations.
- Enable HN to assume the lead in counterinsurgency operations to alleviate effects of a large US presence.
- Create, as needed, special elements in each force such as SWAT, waterway, border, or SOF.

| | |
|---|---|
| **Team Chief** | This Soldier is the principal advisor to the HN battalion commander and is the Advisor team commander. |
| **Team Sergeant** | This is the advisor team NCOIC, who also serves as principal advisor to the HN command sergeant major or equivalent. |
| **Executive Officer** | This Soldier is the principal advisor to the HN security forces executive officer in addition to performing the same functions that any unit executive officer would do in a traditional unit. |
| **Intelligence Officer** | This Soldier is the principal advisor to the HN Intelligence Staff and provides intelligence to the Advisor Team Chief. |
| **Operations Officer** | This Soldier is the principal advisor to HN Staff for Operations and Training (S3) and plans the advisor team operations. |
| **Operations NCOIC** | This Soldier is the training NCOIC for the Advisor Team and is the principal advisor to the Operations Staff NCO in the HN security force. |
| **Operations NCO** | This Soldier is the advisor team armorer and principal instructor for basic rifle marksmanship (small arms), short-range marksmanship (SRM)/close-quarters battle (CQB), urban operations, patrolling, checkpoints, and any other individual training deemed necessary based on the OE. |
| **Operations NCO** | This Soldier is the advisor team S1. He is also the principal instructor for fires and effects considerations (CAS and artillery support). |
| **Medical NCO** | This Soldier is the advisor team medical NCOIC and primary instructor to HN security forces concerning medical issues. |
| **Communications NCO** | This Soldier is the advisor team communications NCOIC and principal communications instructor. |
| **Logistics Officer** | This Soldier is the principal advisor to the HN staff for logistics and is the logistics Chief for the advisor team. |

Figure 8-4. Possible duties of the advisor team.

- Establish and use mobile training teams.
- Ensure infrastructure and pay is appropriate and managed by the Host Nation government.
- Promote professionalism that does not tolerate internal incompetence. Develop methods to redress.

**Organizing US Forces**

As planning unfolds, mission requirements should drive the initial organization for the unit charged with developing security forces. To achieve unity of effort, a single organization should receive this responsibility. Typically, these duties are undertaken by some form of an advisor team. Due to manpower constraints, some functions are best undertaken by partner units.

*Partner Units*

This is a unit that shares all or a portion of an area of operation with an HN security force unit. US forces operating as partner units to HN security forces need to be prepared to make some organizational changes. US forces should consider establishing combined staff cells for intelligence, operations, planning and logistics. These staff cells support transparent operations and unity of effort, enhance the relationship between the BCT and the HN force by demonstrating a degree of trust, and develop HN capacity in key staff areas by having HN personnel get intimately familiar with various staff procedures by performing them alongside their CF partners.

Additionally, before deploying, US forces should train in partnering with HN security forces. This will ease the transition to multinational operations. At a minimum, this should include cultural awareness training, basic language training, and basic soldier skills training such as marksmanship and first aid.

*Advisor Units*

An *advisor* is a military member who conducts operations that train Host Nation military individuals and units in tactical employment, sustainment, and integration of land, air, and maritime skills; provide advice and assistance to military leaders, and provide training on tactics, techniques, and procedures required to protect the HN from subversion, lawlessness, and insurgency, and develop indigenous individual, leader, organizational skills. Advisor units can be internally or externally resourced. Internally resourced advisor teams are created from the partner unit when they are required to establish them on their own. Externally resourced teams are usually DA resourced for the duration of the advisor team mission. One way of organizing for this role is to have an eleven person team with clearly divided responsibilities. Each person's duties should reflect his responsibilities to the internal team, and to advising their HN security force counterpart. In certain situations, the partner unit commander might need to provide a security element to support an advisor team. This security element might need to be large enough to guard a compound on an HN security force base and to crew several vehicles. A ten-person security element has proved be useful in Iraq and Afghanistan. Figure 8-4 shows possible duties on the advisor team.

### *Developing HN Security Forces*

The mission to develop HN security forces at all levels can be organized around seven tasks— measure (assess), organize, rebuild/build facilities, train, equip, advise, and mentor. The memory aid MORTEAM can help partner units and advisors. These tasks incorporate all doctrine, organization, training, material, leadership and education, personnel and facilities (DOTMLPF) requirements for developing the HN security forces. Although described sequentially, these tasks are normally performed concurrently. For example, training and equipping operations must be integrated and, as the operation progresses, assessments will lead to changes. If US forces are directly involved in operations against insurgents, the development program requires a transition period during which major COIN operations are handed over to HN security forces. Figure 8-5 shows the seven framework tasks as an iterative and cyclical process, in which both advisor teams and partner units measure and mentor HN security force units at each step.

### Measure (Assess) Host Nation Security Forces

Commanders of advisor units and partner units must measure and assess the HN security forces as part of the comprehensive program of analyzing the OE and the insurgency. The partner unit and advisor team must work closely in assessing the quality of HN security forces. From the assessment, planners develop short-, mid-, and long-range goals and programs. As circumstances change, so do the goals and programs. In measuring the state of the Host Nation's security forces, it may be determined that some existing security forces are so dysfunctional or corrupt that the organizations must be disbanded rather than rehabilitated. In some cases, commanders will need to replace some HN leaders before their units will become functional. While every situation is different, leaders of the development program should assess and measure the following HN security force factors throughout planning, preparation, and execution of the operation:

- Social structure, organization, demographics, interrelationships, and education levels of security force elements.
- Extent of acceptance of ethnic and religious minorities.
- Laws and regulations governing the security forces and their relationship to national leaders.
- Corruption, abuse of power, and nepotism.
- Equipment and facilities, with a priority placed on maintenance.
- Logistic and support structure, and its ability to meet the force's requirements.
- The unit's methods and proficiency at conducting COIN operations.
- State of training at all levels, and the specialties and education of leaders.

### *Assessment Techniques*

HN security forces must be trained and tactically proficient for some time before they are considered as ready to conduct operations on their own. US Partner units are responsible for the mentoring, training, and welfare of their HN counterparts. Advisors assigned to the unit provide the day-to-day connection, but are not sufficiently equipped to do everything required to build a well trained, well led, sustainable and professional force.

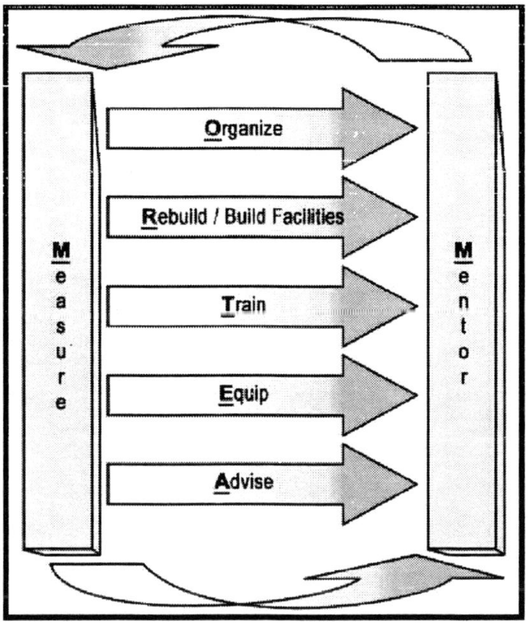


Figure 8-5. Seven framework tasks.

A nation's armed force that behaves unprofessionally can quickly lose legitimacy, which is needed to conduct counterinsurgency operations and draw quality recruits. When the US military mission fails to prepare HN security forces to take the lead—a key political objective—unity of effort suffers at many levels. The results are often corruption, nepotism, and bureaucracy, which generate obstacles to units who must rapidly adapt to the insurgent strategies and tactics. Due to these factors, the HN will eventually lose the ability to persevere against a steadfast insurgent fighting a protracted war.

US forces often choose to use a readiness assessment of an HN security force unit. Figure 8-6 shows an example format for a readiness assessment.

*Assessment Periods*

Assessing Host Nation security forces should be done for three distinct periods of time—short, mid, and long. Considerations for each include—

***Short***

The advisor and partner unit are involved in the training of the Host Nation security force unit. An example of a short term goal for an advisor unit would be ensuring that the HN operations officer is tracking all of his units conducting missions. An example for a partner unit would be training the HN Soldiers on marksmanship and room clearing.

***Mid***

The HN unit is more self-sufficient, but still not fully capable, and the advisor acts in a supervisory role. An example of a mid-range goal for an advisor unit would be ensuring that the HN staff plans for logistical support during missions. An example of a mid range goal for a partner unit would be training platoon size units to move tactically during patrols.

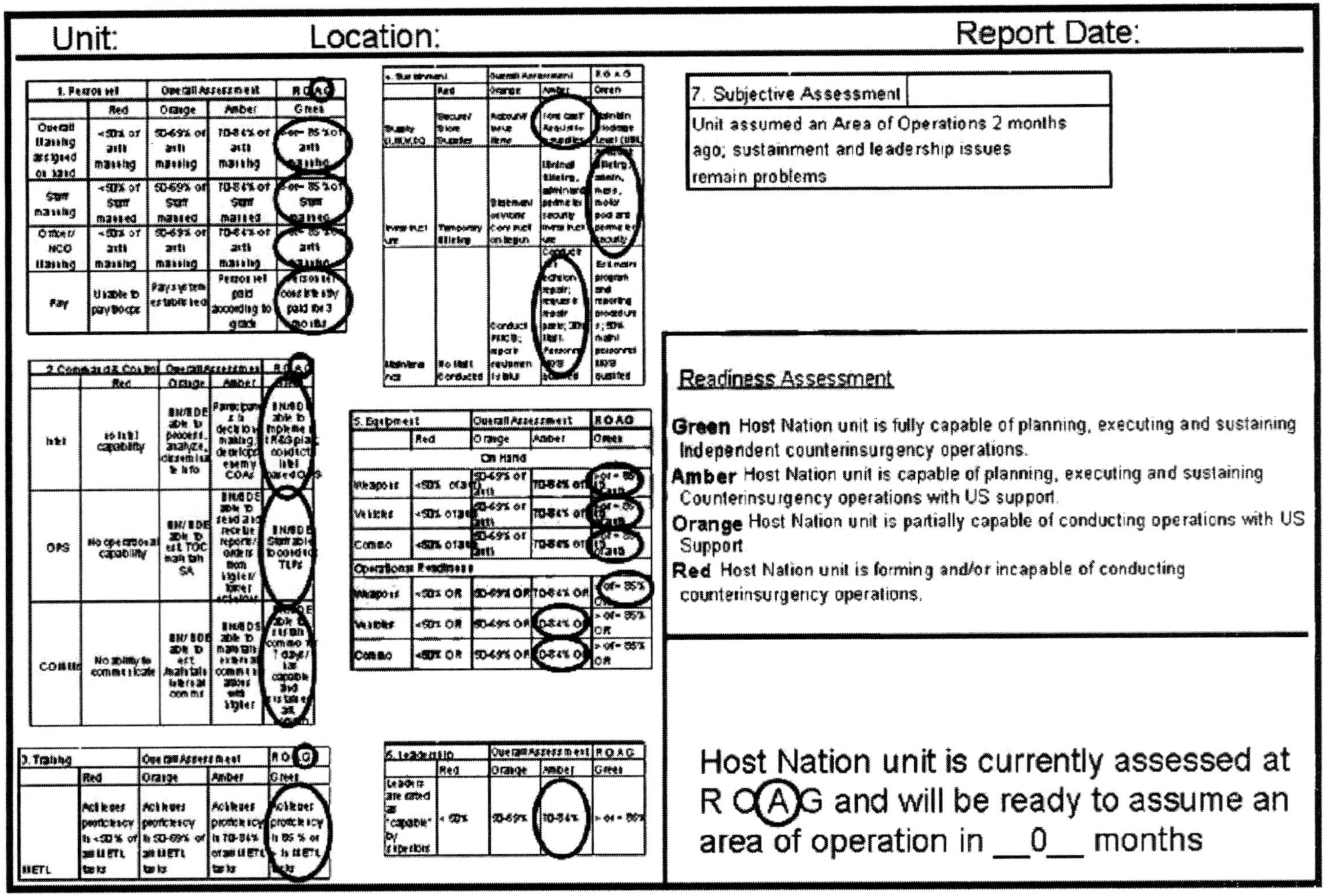


Figure 8-6. Example format for readiness assessment.

***Long***

The HN unit can conduct training, planning, sustainment, and operations with little guidance and the advisor provides oversight and mentorship. An example of advisor unit long-range goal would be to ensure that the HN intelligence officer gathers, analyzes, and disseminates intelligence and is fully integrated into the planning process. An example of a partner long-range goal would be to migrate to providing only the additional forces and quick-reaction force (QRF) capabilities to the Host Nation security force unit.

*Assessing Methods and Proficiency*

The training, leadership, equipment, logistics and operations of Host Nation security forces must be assessed. If US advisors are not on the objective with the Host Nation forces, US forces may not know their capabilities. Assessment of HN security forces may include their ability to—

- Plan and brief an operation.
- Perform correct actions on contact.
- Properly clear and mark rooms, halls, and stairwells.
- Perform other small unit tasks appropriate to the HN OE.
- Use NCOs.
- Employ fire control measures and maintain fire control.
- Call for air reserves or quick reaction force.
- Sustain themselves with Class I, III, and V supplies and provide medical support.
- Maintain vehicles and equipment.
- Maintain professionalism.

## Organize

Organizing HN forces depends on the Host Nation's social and economic conditions, cultural and historical factors, and security threat. The development program's aim is to create an efficient organization with a command, intelligence, logistic, and operations structure that makes sense for the Host Nation. Conventional forces with limited special-purpose teams, such as explosive ordnance disposal and special weapons and tactics (SWAT), are preferred. Elite units tend to divert a large share of the best leadership and remove critical talent from the regular forces. Doctrine should be standard across the force, as should unit structures. The organization must facilitate the collection, processing, and dissemination of intelligence across and throughout all security forces.

In many situations, simple personnel accountability is one of the greatest impediments to effective organizations. The advisor team should mentor and advise the Host Nation security forces on keeping better accountability. Formations, timelines, and uniform standards can be slow to be accepted, which can stem from lack of supplies. Also, it is important that the unit can manage its leave and pass system. Once achieved, this will further the capabilities of the Host Nation unit and allow them to focus on operations and training, not troops to task.

*Organizational Structure*

As much as possible, the Host Nation should determine the security force organization's structure. The Host Nation may be open to proposals from US forces, but should approve all organizational designs. As the HN government strengthens, US leaders and trainers should expect increasingly independent organizational decisions. These may include changing the numbers of forces, types of units, and internal organizational designs. Culture and conditions might result in security forces being given what may be considered nontraditional roles and missions. HN police may be more paramilitary than their US counterparts, and the military may have a role in internal security. Eventually, police and military roles should clearly differ. Police should counter crime, while the military should address external threats. However, the exact nature of these missions depends on the HN situation. In any event, police and military roles should be clearly delineated.

*Organizational Types*

The day-to-day troops-to-task requirements and priorities of effort requires a wide assortment of Host Nation security forces to conduct counterinsurgency operations. Although the police force should be the frontline of COIN security forces, other units are required to assist the police on a temporary basis until the police can conduct operations on their own. Police forces are not equipped or trained in operating in high intensity situations.

***Military***

Military forces might have to perform police duties at the start of an insurgency; however, it is best to establish police forces to assume these duties as soon as possible. Few military units can match a good police unit in developing an accurate human intelligence picture of their AO. Because of their frequent contact with populace, police often are the best force for countering small insurgent bands supported by the local populace.

Although the long term objectives in training Host Nation militaries need to focus on foreign defense, realities of counterinsurgencies often dictate that the military establish civil security and civil control. Commanders must ensure that military units coordinate with local police with respect to intelligence gathering, interacting with civilians, and civil affairs.

A Host Nation security force unit will be task-organized into elements that can accomplish one of three functions—conducting offensive operations, conducting defensive operations, and implementing civil control. These organizations' sizes vary based on the operational environment; the components, the elements and dynamics of the insurgency; and the phase of the counterinsurgency. A Host Nation security force must also have C2, a reserve, and some form of sustainment support.

***Offensive Force***

A HN unit needs the capability to conduct offensive operations as well as specific counterinsurgency operations, such as search and attacks, raids, cordon and searches, ambushes, and site exploitation (SE). Due to the elusive nature of the insurgent, this force may be the smallest organization in the unit.

***Defensive Force***

A HN unit needs to have the ability to conduct defensive operations. Typically, HN counterinsurgency units focus on executing area defense at the tactical level that includes securing important events, critical infrastructure, and bases. Depending on the phase of the counterinsurgency and insurgent capabilities, this force may be quite large.

***Civil Control or Constabulary Force***

A HN unit needs to have a force that contributes to national security, as well as performs police duties and civil functions. Typically, this type of force conducts tasks associated with the civil control LOE such as manning checkpoints, gathering intelligence, conducting investigations into crimes and attacks, and performing reconnaissance. Often the largest force, it has constant interaction with the population which increases security of both the unit and the population. Historical and current examples include the Philippine Constabulary, the Italian Carabinieri, the Iraqi National Police, and Haitian Constabulary.

***Police***

The primary frontline COIN force is often the police—not the military. However, the police are only a part of the rule of law. Common roles of the police forces are to—investigate crimes, provide traffic control, preserve the peace by resolving simple disputes and civil disturbance control. Police require support from a code of law, judicial courts, and a penal system. Such support provides a coherent and transparent system that imparts justice. Upholding the rule of law also requires other civil institutions and an HN ability to support the legal system. If parts of the rule of law do not work, then commanders must be prepared to meet detention requirements.

Countering an insurgency requires a police force that is visible day and night. The Host Nation will not gain legitimacy if the populace believes that insurgents and criminals control the streets. Well-sited and protected police stations can establish a presence in communities, which secures communities and builds support for the HN government. Daily contact with locals gives police a chance to collect information for counterinsurgents.

Police might be organized on a national or local basis. Whatever police organization is established, Soldiers must understand it and help the Host Nation effectively organize and use it. This often means dealing with several police organizations and developing plans for training and advising each one. A formal link or liaison channel must exist between the HN police, HN military forces, and US forces. This channel for coordination, deconfliction, and information sharing enables successful COIN operations. Police often consist of several independent but mutually supporting forces. These may include—

- Criminal and traffic police.
- Border police.
- Transport police for security of rail lines and public transport.
- Specialized police forces.
- Internal Affairs.

***Paramilitary***

Success in a counterinsurgency depends on isolating the insurgent from the population. Insurgents intimidate the population into passive support and prevent the population from providing information to counterinsurgent units. A technique to combat this is to establish a paramilitary organization. Paramilitary forces are designed to support the rule of law and stabilize the operating environment, particularly for when the HN police forces and military forces are standing-up organizing. Temporary in nature, the paramilitary unit's mandate should be two-fold.

- It needs to provide the population with a means of securing themselves.
- It should have the means to gather and report intelligence about the insurgency.
- It can sometimes be used covertly to target insurgents within higher headquarters guidance.

***Organization***

Once citizen leaders are identified, units can work with locals in organizing the paramilitary units. The units need to have recognizable uniforms, standard light weapons, and a salary. The salaries will take away from the insurgent recruitment base.

***Limitations***

Paramilitary organizations should be prevented from conducting offensive operations. They should also be monitored closely for insurgent infiltration or abuse of power.

***Demobilization***

Eventually, the best way to demobilize a paramilitary force might be to integrate part of it into the HN military or police force.

***Corrections***

If counterinsurgents seek to delegitimize an insurgency by criminalizing it, then the Host Nation must establish a robust corrections system. A counterinsurgency with this goal, should be provided with a more robust armament than in stable governments.

The situations in Iraq and Afghanistan are not the first time US forces have created Host Nation security forces from scratch and allowed them to win in a counterinsurgency. The Philippine-American War began in 1898 and ended officially in 1902, but hostilities remained until 1913. The war was not a popular one in the United States, but resulted in a stable and free Philippine Republic.

The Philippine Constabulary (PC) was established on August 8, 1901, to assist the United States military in combating the remaining Filipino revolutionaries. The PC was entrusted into the hands of Captain Henry T. Allen, named as the chief of the force and later dubbed as "the Father of the Philippine Constabulary." With the help of four other army officers, Captain Allen organized the force, trained, equipped, and armed the men as best as could be done under difficult conditions.

General Henry Adler said it best when he said, "For some time to come, the number of troops to be kept here should be a direct function of the number of guns put into the hands of the natives...It is unwise to ignore the great moral effect of a strong armed force above suspicion."

The Constabulary was instrumental in defeating the insurgency. Originally officered by Americans, the Filipinos slowly over took all operations of the unit. At a tactical level, they were initially used to augment US forces and moved into having their own area of operations. In July of 1901, the US Army garrisoned 491 positions and by December it only garrisoned 372 positions. Although poorly armed with shotguns and revolvers, the constabulary soon maintained ownership of entire provinces.

### Rebuild or Build Facilities

HN security forces need infrastructure support. Soldiers and police need buildings for storage, training, and shelter. Often, requirements include barracks, ranges, motor pools, and other military facilities. Construction takes time; the Host Nation needs to invest early in such facilities if they are to be available when needed. Protection must be considered in any infrastructure design, including headquarters facilities, since infrastructure is an attractive target for insurgents. (See FM 5-104 for information on hardening measures to increase infrastructure survivability and improve protection.)

During an insurgency, HN military and police forces often operate from local bases. Building training centers and unit garrisons requires a long-term force-basing plan. If possible, garrisons should include housing for the commissioned officers, NCOs, and families; government-provided medical care for the families; and other benefits that make national service attractive.

The easiest way for a partner unit to do this is to build a combined facility to house both itself and the HN unit. As the situation improves, the Host Nation security forces can occupy the entire compound once the partner unit leaves.

Advisor units can use funding that they have to build showers, tents, toilets, dining facility equipment, barriers and minor building renovations. Improving the military facilities can result in an increase in the morale and performance of the HN unit.

### Train

US and multinational training assistance should address shortfalls at every level with the purpose of establishing self-sustaining training systems.

#### US Trainers

Soldiers assigned to training missions should receive training on the specific requirements of developing HN forces. The course should emphasize the Host Nation's cultural background, introduce its language, and provide cultural tips for developing a good rapport with HN personnel. The course should also include protection training for troops working with HN forces. US trainees must become familiar with the HN organization and equipment, especially weapons not in the US inventory. This training must emphasize the following:

- Sustaining training and reinforcing individual and team skills.
- Using the smallest possible student-to-instructor ratio.
- Developing HN trainers.
- Training to standards—not to time.
- Providing immediate feedback and using after-action reviews.
- Respecting the HN culture, but learning to distinguish between cultural practices and excuses.
- Learning the HN language.
- Working with interpreters.

*Establish Standards*

The Host Nation unit and trainers must establish clear measures for evaluating the training of individuals, leaders, and units. Insurgent strategies and their corresponding responses from targeted governments vary widely. COIN operations require many of the same individual and collective skills performed in conventional military operations, but also include additional requirements for COIN. Small units execute most COIN operations; therefore, effective COIN forces require strong junior leaders. Soldiers and Marines know how to evaluate military training. Metrics for evaluating units should include subjective measures, such as loyalty to the HN government, as well as competence in military tasks. However, the acceptance of values, such as ethnic equality or the rejection of corruption, is far more difficult than evaluating task performance.

*Soldiers and Police*

Security force members must be developed through a systematic training program. The program first builds their basic skills, then teaches them to work together as a team, and finally allows them to function as a unit. Basic military training should focus first on Soldier skills, such as first aid, marksmanship, and fire discipline. Leaders must be trained in planning tactics, including patrolling, urban operations, or other skills necessary. Everyone must master the rules of engagement and the law of armed conflict. Required skills include the following:

- Conduct basic intelligence functions.
- Manage their security.
- Coordinate indirect fires.
- Conduct logistic (planning, maintenance, sustainment, and movement) operations.
- Provide for effective medical support.
- Provide effective personnel management.

***Advisor Teams***

Advisor teams are generally responsible for initial Host Nation unit training and then provide oversight as the Host Nation units implement their own training plan. Leader and ethics training should be integrated into all aspects of training.

***Partner Units***

Partner units play an important role in advising. Advisors cannot go out with every Host Nation operation, but, as almost all operations in COIN are combined, a partner unit can effectively advise a Host Nation unit. Partner units can be most effective at advising Host Nation units in the planning process, especially MDMP. The MDMP model may need to be modified to suit the HN security force. Staff sections should work closely with their HN peers as part of the planning process.

Partner units might have to provide limited support or supplies to their Host Nation unit. For example, a unit might provide its HN unit JP-8 or Class VIII, or it might help recover a disabled tank. It should never create a reliance on partner unit support. The commander allocates partner support, considering the realities of the situation.

HN security force staffs, when fully capable, should be able to achieve the following:

***Operations***

Plans are synchronized with consideration given to all Warfighting Functions. They are disseminated to appropriate personnel in a timely manner and operate within a short-, mid-, and long-term framework.

***Intelligence***

Intelligence is shared with other units and agencies. It is fully integrated into the planning process, is bottom fed, and uses multiple sensors.

***Sustainment***

Units have synchronized methods for requesting supplies and sustainment assets. Planners account for logistical capabilities during the planning process.

***Command and Control***

The command structure is unified. Leaders empower subordinate leaders within the commander's intent. Commanders visualize, describe, and direct their units.

***Soldiers***

Training can be divided into individual, unit, and staff training.

***Individual***

Individual skills training covers marksmanship, first aid, land navigation, and individual movement techniques. Host nation security forces do not always have the institutional military instruction that US forces have, so much of the individual skills training occurs at the unit. Marksmanship will be of particular concern, due to the need for precise fires in COIN.

***Unit***

Unit training is focused on getting Host Nation battalions, companies, platoons, and squads ready to conduct operations. This typically means training focused on three key tasks: checkpoint operations, combat patrols (mounted and dismounted), and cordon and search operations.

***Staff***

Staff training involves training staffs to use and implement systems focused on planning, logistical support, intelligence integration, and command and control.

***Police***

Police training is best conducted as an interagency and multinational operation. Ideally, leaders for police training are civilian police officers from the Departments of Justice and State, along with senior police officers from multinational partners. Civilian police forces have personnel with extensive experience in large city operations and in operating against organized crime groups. Experience countering organized crime is especially relevant to COIN; many insurgent groups are more similar to organized crime in their organizational structure and relations with the populace than they are to military units. US military police units serve best when operating as a support force for the professional civilian police trainers.

However, military police units may be assigned the primary responsibility for police training, and they must be prepared to assume that role if required.

Higher level police skills, such as civilian criminal investigation procedures, anti-organized crime operations, and police intelligence operations—are best taught by civilian experts.

Effective policing also requires an effective justice system that can process arrests, detentions, warrants, and other judicial records. Such a system includes trained judges, prosecutors, defense counsels, prison officials, and court personnel. These people are important to establishing the rule of law.

Military police or corrections personnel can also provide training for detention and corrections operations. HN personnel should be trained to handle and interrogate detainees and prisoners according to internationally recognized human rights norms by the appropriate US personnel. Prisoner and detainee management procedures should provide for the security and the fair and efficient processing of those detained.

Police forces, just like military forces, need quality support personnel to be effective. This requires training teams to ensure that training in support functions is established. Specially trained personnel required by police forces include the following:

- Armorers.
- Supply specialists.
- Communications personnel.
- Administrative personnel.
- Vehicle mechanics.

*Leaders*

The effectiveness of the HN security forces directly relates to the quality of their leadership. Building effective leaders requires a comprehensive program of officer, staff, and specialized training. The ultimate success of any US involvement in a COIN effort depends on creating viable HN leaders able to carry on the fight at all levels and build their nation on their own. One of the major challenges that partner units and advisor teams may face is the perceived low quality of leader, especially the junior leader ranks.

The leader training methodology must reinforce the different levels of authority within the HN security force. The roles and responsibilities of each commissioned officer and NCO rank must be firmly established so recruits understand what is expected of them. Their subordinate relationship to civilian authorities must also be reinforced to ensure civilian control. In addition, training should establish team dynamics. In some cultures, security forces may need training to understand the vital role of members not in primary leadership positions.

***Commissioned Officers***

Officer candidate standards should be high. Candidates should be in good health and pass an academic test with higher standards than the test for enlisted troops.

***Officer Candidates***

These should be carefully vetted to ensure that they do not have close ties to any radical or insurgent organizations.

### *Basic Officer Training*

Various models for basic officer training exist:

- One-year military college.
- Two-year military college.
- Four-year military college.
- Officer Candidate School.
- Military training at civilian universities.

### *Additional Training*

In addition to tactical skills, commissioned officers should be trained in accountability, decision-making, delegation authority, values, and ethics. Special COIN training should address—

- Intelligence collection and legal considerations.
- Day and night patrolling.
- Site security.
- Cordon and search operations.
- Operations with—
    - Other US Forces.
    - Other HN governmental agencies.
    - Intergovernmental organizations.
    - Nongovernmental organizations.
- Treatment of detainees and prisoners.
- Psychological operations.
- Civil military operations.
- Negotiations.
- Ethnic and religious sensitivity.

### *Noncommissioned Officers*

Professional and effective security forces all have a professional NCO Corps. NCOs need training in tactical skills, accountability, values, and ethics. Relations and responsibilities between the Officer and NCO Corps should be clearly defined, and should empower the NCO Corps. Units often create special NCO academies to train NCOs from partner Host National units.

### *Civilian Leaders*

Additionally, Host Nation civilian leaders may need specific training to improve their skills and performance. These key representatives might include—

- City mayor.
- Officials from public works, utilities, transportation, and communication.
- Local police chief.
- Fire-fighting officials.
- Superintendent of schools.

- Religious leaders.
- Health and medical officials and leaders.
- Judicial representatives.
- Editors of local news media.
- Business and commercial leaders.

*Augmenting*

This is an arrangement where the Host Nation provides either individuals or elements to US units or vice versa. Augmentation can occur at a number of levels and in many different forms. For example, a US squad can be augmented with HN individuals, a US company can be augmented with an HN platoon, and a US battalion can be augmented with an HN company. The benefit of this type of training strategy is that Host Nation security forces can emulate US forces in actual combat operations. In addition, US forces can gain valuable cultural, language, and intelligence-gathering skills. Typically a chain of command will be agreed upon prior to execution, however, command may be executed as a partnership. Figure 8-7 shows an example of how augmenting a unit can be implemented.

## Equip

The requirement to provide equipment may be as simple as assisting with maintenance of existing formations or as extensive as providing everything from shoes and clothing to vehicles, communications, and investigation kits. If insurgents use heavy machine guns and rocket-propelled grenades, HN security forces need comparable or better equipment. This especially applies to police forces, which are often lightly armed and vulnerable to well-armed insurgents.

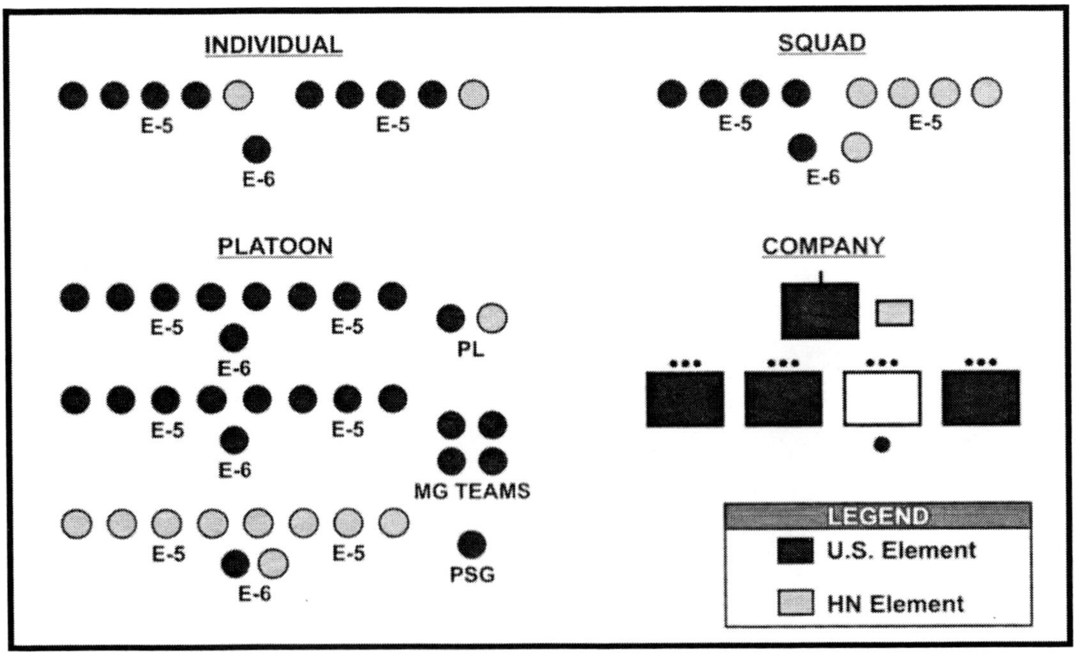


Figure 8-7. Augmentation of an example unit.

Primary considerations should include maintainability, ease of operation, and long-term sustainment costs. Few developing nations can support highly complex equipment. In COIN operations, having many versatile vehicles that require simple maintenance is often better than having a few highly capable armored vehicles or combat systems that require extensive maintenance. Developing an effective HN maintenance system may begins with major maintenance performed by contractors. The program then progresses to partnership arrangements with US forces as HN personnel are trained to perform the support mission.

### Advise

Advisors are the most prominent group of US personnel that serve with HN units. Advisors live, work, and fight with their HN units. Segregation is kept to an absolute minimum. The relationship between advisors and HN forces is vital. US partner unit commanders must remember that advisors are not liaison officers, nor do they command HN units. Additionally, partner units regularly advise their HN security force counterparts.

Effective advisors are an enormous force enhancer. The importance of the job means that the most capable individuals should fill these positions. Advisors should be Soldiers known to take the initiative and set the standards for others.

Professional knowledge and competence win the respect of HN troops. Effective advisors develop a healthy rapport with HN personnel but avoid the temptation to adopt HN positions contrary to US or multinational values or policy.

Advisors who understand the HN military culture understand that local politics have national effects. Effective advisors recognize and use cultural factors that support HN commitment and teamwork. A good advisor uses the culture's positive aspects to get the best performance from each security force member and leader.

#### *Guidelines*

Important guidelines for advisors include—

- Remain patient. Be subtle. In guiding Host Nation counterparts, explain the benefits of an action and convince them to accept the idea as their own. Respect the rank and positions of Host Nation counterparts.
- Exercise diplomacy in correcting Host Nation security forces. Praise each success and work to instill pride in the unit.
- Work to continually train and improve the unit, even in the combat zone. Help the commander develop unit SOPs.
- Know light infantry tactics and unit security procedures.
- Use "confidence" missions to validate training.
- Understand that an advisor is not the unit commander but an enabler. The Host Nation commander makes decisions and commands the unit. Advisors help with this task.
- Train Host Nation units to standard and fight alongside them. Consider Host Nation limitations and adjust.
- Flexibility is key. It is impossible to plan completely for everything in this type of operation.

- Constantly look forward to the next issue and be ready to develop solutions to problems that cannot be answered with a doctrinal solution.
- Remember that most actions have long-term strategic implications.
- Try to learn enough of the language for simple conversation, at a minimum, greetings.
- Keep Host Nation counterparts informed; try not to hide agendas.
- Remain prepared to act as a liaison to multinational assets, especially air support and logistics. Maintain liaison with civil affairs and humanitarian teams in the area of operations.
- Remain ready to advise on the maintenance of equipment and supplies.
- Stay integrated with the unit. Eat their food. Do not become isolated from them.
- Remain aware of the operations in the immediate area to prevent fratricide.
- Insist on Host Nation adherence to the recognized human rights standards concerning treatment
  of civilians, detainees, and captured insurgents. Report any violation to the chain of command.
- Remain objective in reports on Host Nation unit and leader proficiency. Report gross corruption or incompetence.
- Maintain a proper military bearing and professional manner.

*Advisor Teams*

Advisor teams link Host Nation units and US partner units to fill the gaps of the HN's supply system. They have the ability to bring attention to shortfalls in equipment both to their higher US chain of command and, through their chain of command, the Host Nation higher headquarters. The advisor should ensure that the Host Nation logistical system is being used properly first. In addition, the advisor team may assist by hand walking his HN counterpart through the HN system.

Advisor teams often have authority to provide tactical equipment such as radios through their appropriated funds. The advisor team must use this money since it is one of the few forms of leverage they posses.

It is important to note that advisors should not use bribery or coercion, since results achieved from these actions are only temporary. As soon as the "payment" is made, or the "force" is removed, the Host Nation counterpart has no reason to comply. In practice, these techniques are not efficient and will not achieve the long-term goal of developing proficiency, competence, and initiative in the counterpart.

*Advisor Art and Science*

The most important mission of an advisor is to enhance the military professionalism of his counterpart. Rapport, credibility, and legitimacy can only be established through time, proximity, and interaction. Advisors mentor HN leadership at every opportunity. Partner units are engaged in every aspect of the HN's development.

Advisors sleep and eat daily with HN security force units. They must "leave the wire." The closer they operate with the HN security force, the faster the unit will improve. The social and cultural aspects of the mission are just as or more important than patrolling with the unit.

Just as the credibility of US advisors is critical for influencing the leadership and men of the HN security force, the actions of these advisor teams are equally vital to winning the population's confidence. Advisors build their credibility by contact, visibility, technical and tactical proficiency, the ability to provide resources, success in battle, and respectful interaction with civilians.

Often, to be successful, advisors partner with their HN equivalents. For instance, US leaders partner with HN leaders, US staffs partner with HN staffs, US soldiers partner with HN soldiers.

Combat advisors go on patrol and operations in order to—

- Lead and advise by example.
- Share the risk.
- Improve legitimacy and unity of effort.
- Bond with the counterparts.
- Coordinate assets (CAS, QRF, EOD).
- Prevent fratricide.
- Provide accurate and specific AARs.
- Prevent, report, remedy human rights violations.
- Gather information for intelligence.
- Ensure site exploitation is thorough.
- Ensure evidence is correctly gathered and processed.
- Improve situational awareness.

*Challenges*

US advisors and units share the same challenges and often develop the same practices. Some of these practices are negative while others are positive. Some challenges are common to all US advisors and units. They include aspects often referred to as the American military culture.

Just as in counterinsurgency operations, advising is an iterative process. The advisor team and the HN unit will not get it right the first time. However, both the HN unit and the advisor team must "learn and adapt" faster than insurgents.

Some typical advisor challenges include—

- Adjusting to native cuisine can pose a problem for the advisor. Refusal to accept food and beverages when offered might be considered an insult.
- The advisor does not become discouraged. Not all advice will be accepted. Some will be implemented later.
- The advisor cannot forget that a careless word or action can cost the United States dearly in good will and cooperation that may have been established with great effort and at considerable cost.
- The advisor does not criticize HN policy in front of HN personnel. It is the advisor's obligation to support the incumbent government just as he does his own. This obligation is US national policy.
- The advisor studies his counterpart to determine his personality and background. He makes every effort to establish and maintain friendly relationships. He learns

something about his counterpart's personal life and demonstrates an interest in his likes and dislikes.

- The advisor recognizes and observes military courtesy and local customs and courtesies. He recognizes that in many cultures, observance of formal courtesies must take place before other business can be conducted. When in doubt, he leans toward the polite.
- The advisor does not get caught up in personality clashes between HN officers/personnel.
- The advisor keeps in mind that HN partners may consider person-to-person relationships more important than organizational frameworks.
- The advisor may have to deal with the HN norms regarding time and timeliness.

| | |
|---|---|
| **By, With, and Through** | Not counting immediate action battle drill responses, the mark of an effective advisory effort is the amount of stake the Host Nation security forces take in their own operations. |
| **Empathy Leads to Cultural Competence** | Truly understanding other human beings and where they come from allows honest relationships to develop. These relationships are critical factors of success. |
| **Success is Built on Personal Relationships** | This relationship is likely to be tested on numerous occasions and challenges; only one built on a solid relationship of mutual trust can survive and ensure mission success. |
| **Advisors are not 'Them'** | Increasing the advisors' level of frustration is the rapid realization that, when the dealing with partner units, advisors are not one of "them." The advisors are often alone navigating between two military systems and two cultures, never quite fitting in with either of them. |
| **You Will Never Win... Nor Should You** | The advisor attaining a tactical objective does not achieve success; success is achieved by the Host Nation forces achieving the objective. |
| **Advisors are not Commanders** | Advisors are not intended to lead Host Nation security forces in combat; they are ultimately responsible for command and control only of their own small TEAM of US combat advisors. |
| **Advisors are Honest Brokers** | Advisors are advocates for the Host Nation security forces with partner units. |
| **Living with Shades of Gray** | Advisors will likely find themselves isolated with great autonomy, often with no supervision and will encounter moral and ethical dilemmas on a daily basis. |
| **Talent is Everything, but Understand Rank** | The paradox lies in that in some Host Nation forces, recognized talent can take a back seat to rank. Advisors must understand that rank on the uniform is important to many armies, but it is skin deep; the ways around rank are the relationship and talent. |
| **Make Do** | Advisors will never have everything that they feel they need to succeed. Scrounging, bartering, and horse-trading are daily activities of the combat advisor. An enormous amount of energy must be devoted to these activities. These efforts will not only help the advisor achieve mission success but also endear him to his counterpart. |

Figure 8-8. Principles of advising.

*Advising Principles*

Advising principles can best be described as shown in Figure 8-8.

*Advisor Relationship with Host Nation Peers*

The advisor spends maximum time with the unit so the troops get to know and trust him. The advisor talks to and gets to know the troops, not just the unit leaders, so he receives excellent feedback through the common Soldier's candid comments. Such comments often reflect troop morale and operational effectiveness. He stays abreast of what is going on in the unit by staying in close contact with the commander and staff.

- The advisor encourages frequent command inspections by the commander. In some cultures, this action is a new concept or not a common practice. Many HN commanders are reluctant to inspect. They rely solely on correspondence and reports to evaluate unit effectiveness.
- The advisor continually stresses the obvious advantages of good military-civilian relations to avoid the idea of military arrogance, which irritates the civilian populace. The development of a proper Soldier-civilian relationship is a critical factor in counterinsurgency. Improper behavior by Soldiers toward civilians must be immediately corrected.
- The advisor keeps training standards high so that the unit is prepared for combat at all times.
- The advisor stresses human rights and the consequences of mistreating suspects and prisoners.
- The advisor constantly promotes unit esprit-de-corps to sustain the unit in the face of difficulties.
- The advisor persuades the HN personnel to pass information up, down, and laterally.

**Mentor**

Effective mentorship is based on mutual respect, building relationships and trust. Maintaining patience is the key in developing positive situations for Mentorship. All of which will go into the building of an effective team. Both advisor teams and partner units have equal ability to mentor, if the conditions have been set.

*Respect*

In order to get respect, mentors need to give it.

*Relationship*

Mentors need to have a good working relationship with Host Nation forces.

*Trust*

The best way to gain trust is through shared danger and hardship. Nothing builds trust faster than facing the enemy together.

*Team*

The end state should be a unity of effort for the advisor team, the HN unit, and the partner unit. If a recommendation is made to the Host Nation unit, mentors should move on. If every recommendation becomes a point of contention, the mutual respect and relationship that has been carefully cultivated will be damaged. Figure 8-9 shows the process that is needed to build a relationship that ultimately develops into an effective team. Giving respect sets the conditions for trust and, finally, a good team, where mentoring can occur.

***Host Nation AAR***

First, an advisor should never take credit for an idea; instead, they should empower their counterpart. Discuss the issue and come up with the solution together, even if he has the answer. The point is that advisors are mentoring them to come up with solutions of their own in the future. Second, by giving them the credit and making them look good the leader will more readily come to the advisors for advice and take advice more readily in the future.

Advisors should cover all activities and salient points (sustainment and improvement) for each phase of the operation, including information engagements. The AAR should conclude with a plan to follow-up and make improvements as necessary. Just as with US forces, an AAR with the HN security forces is not only potentially of great value but it is also potentially full of pitfalls. To evaluate the HN security forces, the advisors should assess at a minimum—

- Leadership
- Level of training demonstrated on operations
- Use and effectiveness of administrative/logistics (as applicable).
- Professionalism to include human rights and dealing with civilians.

Advisors need to bring up general positive points with the groups, but they should refrain from singling out a leader, staff, soldier, or unit to provide criticism in front of others. The US military culture of group AARs to discuss all mistakes is a technique we find valuable because our culture accepts it as important for improvement. Most other cultures do not accept pointing out criticisms of one person or group to another. The advisor should take that member or unit aside and discuss the issue. Let them talk about how to improve this point in a productive way. Figure 8-10 shows an example AAR for a Host Nation security force after and operation.

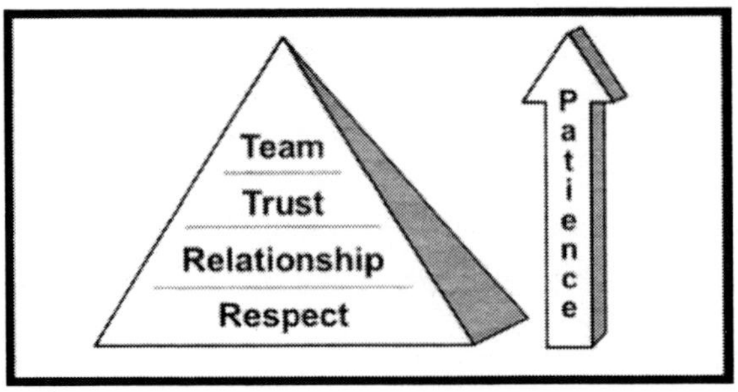


Figure 8-9. Team-building process.

### *Commander's One-on-One*

Remember some AAR points are best kept between the advisor and the Host Nation commander.

### *Internal Advisor AAR*

In order to effectively advise HN security forces, the US Advisor Team should also conduct an internal AAR, focusing on what the Team must sustain and improve. An internal AAR is for the advisors only and should cover three primary topics:

- Conduct of advisors and US partner
- Conduct of the HN security force
- Review ROE/EOF for effectiveness and compliance/training level

## *Employment of Newly Trained Forces in COIN*

Building the morale and confidence of HN security forces should be a long term objective. Committing poorly trained and badly led forces results in high casualties and invites tactical defeats. While defeat in a small operation may have little effect on the outcome of a conventional war, even a small tactical defeat of HN security forces can have serious strategic consequences, since COIN is largely about perceptions. Effective insurgent leaders can quickly turn minor wins into major propaganda victories. In short, the HN security forces must be prepared for operations so that they have every possible advantage. The decision to commit units to their first actions and their employment method requires careful consideration. As much as possible, HN security forces should begin with simpler missions. As their confidence and competence grows, these forces can assume more complex assignments.

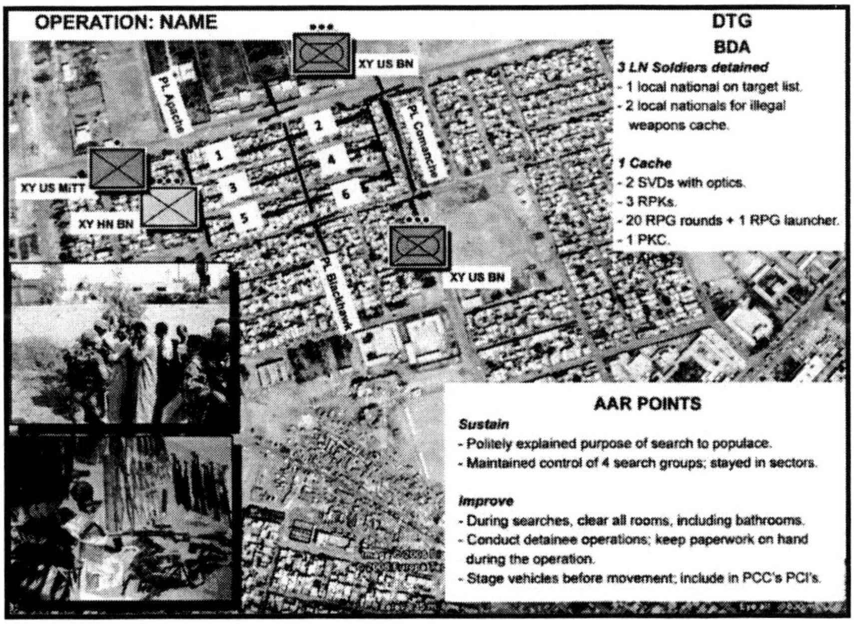


Figure 8-10. Example Host Nation security force AAR—operation summary.

Newly trained units should enter their first combat operation in support of more experienced HN, US, or multinational forces. Partner units and advisor teams need to closely monitor the situation and support each other and the Host Nation. It is paramount that the Host Nation security force not be defeated in the initial stage of their operations. Host nation units can be broken down into smaller elements and paired with US forces in order to put a Host National "face" on the operation and provide the Host Nation with support and training experience.

### SUMMARY

A successful COIN effort establishes HN institutions that can sustain government legitimacy. Developing effective HN security forces—including military, police, and paramilitary forces—is one of the highest priority COIN tasks. Soldiers and Marines can make vital contributions to this mission by training and advising the HN security forces. Effective tactical commanders must understand the importance of this mission and select the right personnel as trainers and advisors. Using the MORTEAM framework may enable tactical leaders to successfully train, mentor and conduct operations with Host Nation security force.

## APPENDIX A. INTELLIGENCE PREPARATION OF THE BATTLEFIELD

*"In no class of warfare is a well organized and well served intelligence department more essential than in against guerrillas"*

COL C. E. Callwell, Small Wars, 1896

The complexity of an insurgency and increased number of variables (and their infinite combinations) increases the difficulty of providing timely, relevant, and effective intelligence support to counterinsurgency (COIN) operations. Conducted effectively, however, the intelligence preparation of the battlefield (IPB) allows commanders to develop the situational understanding necessary to visualize, describe, and direct subordinates in successfully accomplishing the mission.

### Overview

IPB is the systematic process of analyzing the threat and environment in a specific geographic area—the area of operations (AO) and its associated area of interest (Figure A-1). It provides the basis for intelligence support to current and future operations, drives the military decision-making process, and supports targeting and battle damage assessment. The procedure (as well as each of its four steps) is performed continuously throughout the planning, preparation, and execution of a COIN operation.

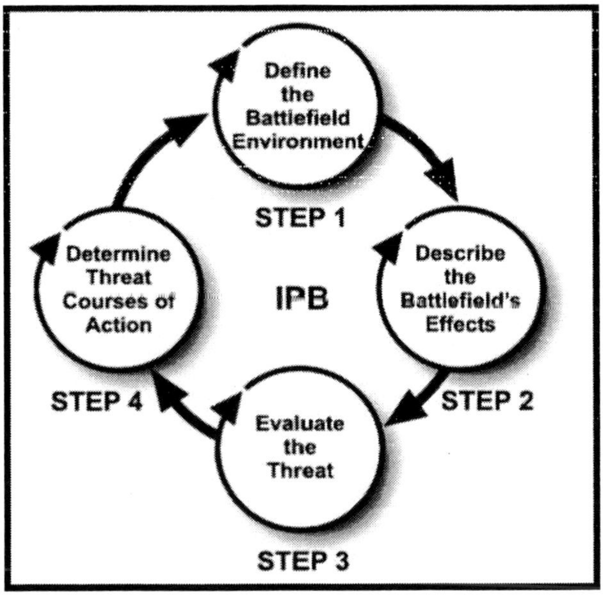


Figure A-1. The steps of IPB.

### *Increased Complexity*

Uncovering intricate relationships takes time, careful analysis, and constant refinement to determine actual effects on friendly and threat courses of action (COAs). These relationships exist among—

- Population groups.
- The infrastructure.
- The historical, cultural, political, or economic significance of the area in relation to surrounding urban and rural areas or the nation as a whole.
- The physical effects of the natural and man-made terrain.

A primary goal of any IPB is to accurately predict the threat's likely COA (step four—which may include political, social, religious, informational, economic, and military actions). Commanders can then develop their own COAs that maximize and apply combat power at decisive points. Understanding the decisive points in counterinsurgency operations allows commanders to select objectives that are clearly defined, decisive, and attainable.

### *Reduced Uncertainty and Its Effects*

Commanders and their staffs may be unfamiliar with the intricacies of a counterinsurgency and more adept at thinking and planning in other environments. Therefore, without detailed situational understanding, commanders may assign missions that their subordinate forces may not be able to achieve. As importantly, commanders and their staffs may miss critical opportunities because they appear overwhelming or impossible (and concede the initiative to the threat). They also may fail to anticipate potential threat COAs afforded by the distinctive operational environment. Commanders may fail to recognize that the least likely threat COA may be the one adopted precisely because it is least likely and, therefore, may be intended to maximize surprise. Misunderstanding the AO's effect on potential friendly and threat COAs may rapidly lead to mission failure and the unnecessary loss of Soldiers' lives and other resources. A thorough IPB of the AO can greatly reduce uncertainty and contribute to mission success.

### *Amplified Importance of Civil Considerations*

In COIN operations, the terrain and enemy are still essential considerations, but the societal component of the COIN environment is considered more closely and throughout the operational process. Moreover, the human or civil considerations gain importance in COIN or stability operations. As discussed in Chapter One, a key tool for the counterinsurgent unit is analyzing civil considerations using AS COPE. Overall, population effects are significant in how they impact the threat, Army forces, and overall accomplishment of strategic and operational goals.

Describing the battlefield's effects—step two in IPB—ascribes meaning to the characteristics analyzed. It helps commanders understand how the environment enhances or degrades friendly and insurgent forces and capabilities. It also helps commanders understand how the environment supports the population. It also explains how changes in the "normal" environment (intentional or unintentional and because of insurgent or friendly activities) may affect the population. Included in this assessment are matters of perception. At each step of the IPB process, commanders must try to determine the HN population's perceptions of ongoing activities to ensure Army operations are viewed as intended. Throughout this process, commanders, staffs, and analysts cannot allow their biases—cultural, organizational, personal, or cognitive—to markedly influence or alter their assessment (see FM 34-3). This particularly applies when they analyze the societal aspect of the operational environment. With so many potential groups and varied interests in such a limited area, misperception is always a risk.

## Signifigant Characteristics in COIN

COIN intelligence analysis must include consideration of the AO's distinguishing attributes— terrain, society, infrastructure, and the threat. Because the COIN environment is so complex, it is often useful to break it into categories. Then commanders can understand the intricacies of the environment that may affect their operations and assimilate this information into clear mental images. Commanders can then synthesize these images of the AO with the current status of friendly and threat forces, and develop a desired end state. Then they can determine the most decisive sequence of activities that will move their forces from the current state to the end state. Identifying and understanding the environment's characteristics (from a counterinsurgent, insurgent, and HN population's perspective) allows commanders to establish and maintain situational understanding. Then they can develop appropriate COAs and rules of engagement that will lead to decisive mission accomplishment.

Figures A-2 and A-3 are not all-encompassing lists of COIN characteristics. Instead, they provide a starting point or outline useful for conducting a COIN-focused IPB and analysis that can be modified to fit the specific operational environment and meet the commander's requirements. Commanders and staffs can compare the categories presented with those in the civil affairs area study and assessment format found in FM 3-05.40 and the IPB considerations for stability operations found in ST 2-91.1.

### *Interconnected Systems*

Since the COIN environment comprises an interconnected "system of systems," considerations among the key elements of the environment will overlap during a COIN intelligence analysis. For example, boundaries, regions, or areas relate to a physical location on the ground. Hence, they have terrain implications. These boundaries, regions, or areas often stem from some historical, religious, political, administrative, or social aspect that could also be considered a characteristic of the society. Overlaps can also occur in a specific category, such as infrastructure. For instance, dams are a consideration for their potential effects on transportation and distribution (mobility), administration and human services (water supply), and energy (hydroelectric).

This overlap recognition is a critical concern for commanders and their staffs. In "taking apart" the COIN environment and analyzing the pieces, commanders and staffs cannot lose perspective of how each piece interacts with any other and as part of the whole. Otherwise, their vision will be shortsighted, and they will fail to recognize the second-and third-order effects of their proposed COAs; the actual end state differing dramatically from the one envisioned by the commander. The increased density of combatants and noncombatants, infrastructure, and complex terrain means that a given action will likely have unintended consequences—positive or negative. Those consequences will be more widely felt and their impact will spread in less time than in other environments. These unintended results may have important strategic and operational consequences. The multiple ways these dynamic COIN elements and characteristics combine make it necessary to approach each COIN environment as a unique challenge for intelligence analysis.

### *Terrain and Weather*

Earlier admonitions that civil considerations are more closely considered in COIN do not necessarily mean that considerations for terrain and weather are deemphasized.

## Terrain

In every COIN operation, terrain and its effects on both threat and friendly forces must be assessed and understood. Then commanders can quickly choose and exploit the terrain (and weather conditions) that best supports their missions. Effective terrain analysis thoroughly assesses structures as well as the ground on which they stand (Figure A-2). An analysis of terrain first considers broader characteristics and effects and then progresses to a more detailed examination.

#### *Natural Terrain*

Natural terrain features significantly influence unit operations. They dictate where buildings can be constructed, the slopes and patterns of streets, and even the broad patterns that develop over longer periods, all of which influence a unit's scheme of maneuver. The military aspects of terrain—observation and fields of fire, avenues of approach, key terrain, obstacles, and cover and concealment (OAKOC)— remain critical to the analysis of natural terrain in, under, and around areas where COIN operations will be conducted. Fortunately, commanders and their staffs are normally accustomed to this type of analysis.

| **Forms and Functions** | **Construction and Placement** | **Military Aspects of Terrain: OAKOC** |
|---|---|---|
| Cores.<br>Industrial areas.<br>• Toxic industrial material production and storage facilities<br>• Standard signs and markings for toxic chemicals<br>Outlying high-rise areas.<br>Residential areas and shantytowns.<br>Commercial ribbon areas.<br>Forts and military bases.<br><br>**Broad Urban Patterns**<br>Types.<br>• Satellite<br>• Network<br>• Linear<br>• Segment<br>Dominant or central hub (if any).<br>Area covered (square miles).<br><br>**Street Patterns**<br>Basic types.<br>• Radial<br>• Grid<br>• Irregular (planned and unplanned).<br>Variations.<br>• Rayed<br>• Radial ring<br>• Contour forming<br>• Combined<br>Widths. | Construction.<br>• Mass or framed<br>• Light or heavy clad<br>• Material (dirt, wood, stone, brick, cinder block, concrete, steel, and glass)<br>• Density and thickness (roofs, floors, and interior and exterior walls)<br>• Load-bearing walls and columns<br>• Height (floors)<br>• Doors, windows, fire escapes, and other openings<br>• Interior floor plan (including crawl spaces, elevators, and stairs)<br>Placement.<br>• Random<br>• Close, orderly block<br>• Dispersed<br>Ownership. | Observation and fields of fire.<br>• Smoke (fire), dust (explosions), and flying debris<br>• Rubble<br>• Engagement ranges (including minimum safe distances and backblast factors) and obliquity/angles (ricochets)<br>• Elevation and depression considerations<br>• Lasers and reflective concerns<br>Avenues of approach (mobility corridors).<br>• Airspace<br>• Surface<br>• Supersurface<br>• Subsurface<br>Key terrain<br>• Landmarks<br>• Buildings of significant cultural, social, political, historical, or economic significance<br>Obstacles<br>• Rubble and vehicles<br>• Steep embankments<br>• Medians<br>• Inadequate bridges and overpasses (destroyed, weight-restricted, or narrow)<br>• Tunnels and underpasses (destroyed or narrow)<br>• Mines and roadside improvised explosive devices<br>• Masking of fires<br>• Burning buildings or other fire hazards<br>• Rivers and lakes<br>Cover and concealment.<br>• Building protection<br>• Weapon penetration (single shot and multiple rounds) considerations<br>• Rubble and vehicles |

Figure A-2. Significant terrain characteristics common to COIN operations.

### *Man-Made Terrain*

Building composition, frontages, placement, forms and functions, size, floor plans, and window and door locations affect maneuver, force positioning, and weapons deployment considerations. Angles, displacement, surface reflection, and antenna locations influence command and control. Structures also influence ISR operations. The increased density and volume created by man-made structures increases how much information commanders and their staffs collect and assess as well as the number of forces required. Building materials and construction will also influence force structures to include weapons and equipment required. Heavily constructed buildings combined with hot and cold extremes may affect target identification for thermal sights. Thick walls, for example, may make combat vehicle identification difficult by distorting hotspots. Additionally, the increased use of heaters and warming fires may clutter thermal sights with numerous hotspots. The ability to maneuver through the urban dimensions—airspace, supersurface, surface, and subsurface—and shoot through walls, ceilings, and floors also creates increased psychological stress. The physical characteristics of man-made terrain can also be analyzed using OAKOC.

## Weather

Weather and its effects are often considered when examining the military aspects of terrain. Military aspects of weather include temperature (heat and cold), light conditions, precipitation (cloud cover, rain, snow, fog, and smog), and wind. Their military effects during COIN are similar to any operational environment (see FM 34-81 and FM 34-8 1-1). Extremes of heat and cold affect weapon systems and the Soldiers that operate them. Precipitation affects mobility and visibility.

Commanders also analyze weather for its potential effect on civilians and civilian infrastructure as well as Soldiers and military equipment. Rain might create sewage overflow problems in areas with collapsed sewage infrastructure, increasing disease and even creating panic. Rain and flooding may also make some subsurface areas impassable or extremely hazardous to civilian and military forces alike. Other weather effects on COIN can include—

- Heavy snowfall may paralyze area transportation and distribution infrastructure, hindering the government's ability to provide vital human services (police, fire fighting, medical, and rescue). Heavy rains and flooding may have similar effects especially on poorly designed and constructed roads or roads that have been damaged by tracked vehicles.
- Extreme hot and cold weather climates, which increase the dependence (and military significance) of many elements of the infrastructure. For example, the energy infrastructure may be critical. Without it, civilians may be inadequately cooled or heated, or they may be unable to cook their food.
- In tropical areas, rain can occur at the same time each day during the wet season. Threat forces may attack during these periods knowing aircraft will have difficulty responding. Bad weather also reduces the effectiveness of surveillance, direct and indirect fire, and logistic support.
- Inclement weather may preclude demonstrations or rallies by threats. Good weather may mean a maximum turnout of civilians for events such as festivals, sporting events, and other social, cultural, or religious activities.

- Severe weather may affect psychological, civil-military and humanitarian assistance operations. Heavy rains and severe dust storms may disrupt leaflet drops, construction projects, food and water distribution, and medical and veterinary assistance programs.

**POPULATION DEMOGRAPHICS**

General population size.
- Village
- Town
- City
- Metropolis
- Megalopolis

Group size based on race, age, sex, political affiliation, economics, religion, tribe, clan, gang, criminal activities, or other significant grouping.
- Significant U.S. or coalition populations
- Distribution, densities, and physical boundaries and overlaps
- Majority, minority, and dominant groups

Increasing or decreasing migration trends.
- Dislocated civilians

Nongovernmental organizations (NGOs).
- Local
- National
- International

Languages (distribution dialects, relationship to social structure).

Educational levels and literacy rates.

Crime rates.

Birth and death rates.

Labor statistics and considerations.
- Skilled and unskilled
- Imported and exported
- Unemployment
- Standard wages and per capita income
- Workday and workweek norms

**HEALTH**

Diseases.

Nutritional deficiencies.

Local standards of care.

Pollution and environmental hazards (air, water, food, and soil).

Health workers (types, numbers, and degree of skill).

**HISTORY**

General and for a specific group.
- Internal or external
- Recent conflicts

Relationship with U.S. allies and other participating multinational forces.

Applicable international treaties.

Status-of-forces agreements.

Antagonists/protagonists.

Heroes.

Events, facts, and dates considered important or celebrated.

Urban area's historical importance.

**LEADERSHIP AND PROMINENT PERSONALITIES**

Identification, location, and prioritization of influential leaders (exploitation, evacuation, protection, etc.)

Affiliation (ethnic, religion, military, government, industry, criminal, or entertainment).

Education attained.

Organization and distribution of power.

Associations among different leaders and groups.

**ETHNICITY AND CULTURE**

Values, moral codes, taboos, and insults (verbal and nonverbal).

Attitudes towards age, sex, and race (including same-sex interaction).

Role of the clan, tribe, or family.

Biases between ethnic groups.

Privacy and individuality.

Recreation, entertainment, and humor.

Fatalism or self-determination.

Exchanges of gifts.

Displays of emotion.

Lines of authority.

Dating and marriage.

Greetings, leave-takings, and gestures.

Visiting practices.

Alcohol and drug use.

Important holidays, festivals, sporting, or entertainment events.

Eating and dietary practices.

Significance of animals and pets.

Urban-rural similarities and differences.

Driving habits.

Clothing.

**RELIGION**

Sects, divisions, and overlaps.

Religious biases and problems.

Relationship and influence on government, politics, economics, and education.

Impact on ethnic and cultural beliefs.

Key events or celebrations (daily, weekly, monthly, or annually).

Funeral and burial practices.

**GOVERNMENT AND POLITICS**

Present and past forms.

Organization and powers (executive, legislative, judicial, and administrative divisions).

Scheduled elections and historical turnouts.

Degree of control over the population
- Identification required
- Border-crossing procedures

Relations with U.S. or multinational governments, national government, and criminal elements.

Political factions and boundaries.

Political traditions.

Grievances.

Censorship.

Nepotism and other clan, tribal, or social ties.

Civil defense and disaster preparedness (organization, plans, training, equipment, and resources).

Legal system.
- System of laws
- Applicable treaties
- Courts and tribunals
- Procedures
- Records (birth and deeds).

Property control.

Monetary system (formal and informal).

Domestic and foreign trade.
- Taxation and tariffs
- Customs requirements
- Rationing and price controls
- Economic performance and contribution to gross national product

Economic aid.

Perception of relative deprivation.

Trade unions.

Competition with the black market and organized crime.

Figure A-3. Societal considerations.

***Society***

This manual shows that societal considerations take on added importance in COIN. Critical to operational success is knowing and understanding which groups live in an area, what relationships exist among them, and how each population group will respond to friendly and threat activities. Often determining any of this is very difficult. Cultural acuity is also essential in helping commanders and their staffs to view the area as the residents view it. The demographics presented show what conditions exist, while the other categories help to explain the root causes or why conditions exist (Figure A-3). Other categories besides basic demographics that are important to gain this understanding include health, history, leadership, ethnicity and culture, religion, and government and politics.

# APPENDIX B. READINGS FOR COIN TACTICAL LEADERS IN A TIME-CONSTRAINED ENVIRONMENT

## Articles

"Best Practices in Counterinsurgency." *Military Review,* May-Jun 2005, Kalev I. Sepp.

"COIN Cliff Notes: Techniques for the Conventional Rifle Platoon in Layman's Terms." Infantry Magazine, July-August 2008, Craig Coppock.

"Counterinsurgency *Redux*," *Survival,* Winter 2006-2007, David Kilcullen.

"The Decisive Weapon: A Brigade Combat Team Commander's Perspective on Conduct Information Tasks." *Military Review,* May-Jun 2006, Ralph O. Baker.

"Learning Counterinsurgency: Observations from Soldiering in Iraq." *Military Review,* Jan-Feb 2006, David Petraeus.

"'Twenty-Eight Articles': Fundamentals of Company-level Counterinsurgency." *Military Review,* May-Jun 2006, David Kilcullen.

"The 27 Articles of T.E. Lawrence.," *Military Review*, May-June 2006, Professional Forum

## Books

Galula, David. *Counterinsurgency Warfare—Theory and Practice.* London: Praeger, 1964.

Hammes, T.X. The *Sling and the Stone: On War in the 21st Century.* Osceola, WI: Zenith Press, 2004

Kitson, Frank. *Low Intensity Operations: Subversion, Insurgency and Peacekeeping.* London: Faber and Faber, 1971.

Trinquier, Roger. *Modern Warfare—A French View of Counterinsurgency.* New York: Praeger, 1964.

Zedong, Mao. *On Guerrilla Warfare.* London: Cassell, 1965.

# APPENDIX C. TWENTY-EIGHT ARTICLES: FUNDAMENTALS OF COMPANY-LEVEL COUNTERINSURGENCY

LTC David Kilcullen, PhD, originally submitted his essay, *28 Articles*, to *Military Review* for the CAC CG's *Special Topics Writing Competition: Countering Insurgency*. However, he was asked to publish it immediately to help Soldiers in the field. Even though doing so cost him the opportunity to compete—and possibly win—the competition, the Australian Army officer graciously agreed, and pulled the essay from the contest. The article is reprinted in this appendix, with permission, from the May-June 2006 issue of *Military Review*, and formatted to fit this publication.

## Introduction

Your company has just been warned about possible deployment for counterinsurgency operations in Iraq or Afghanistan. You have read David Galula, T.E. Lawrence, and Robert Thompson. You have studied FM 3—24, and now understand the history, philosophy, and theory of counterinsurgency. You have also watched *Black Hawk Down* and *The Battle of Algiers*, and you know this will be the most difficult challenge of your life.

But what does all that theory mean at the company level? How do the principles translate into action at night, with the GPS down, the media criticizing you, the locals complaining in a language you don't understand, and an unseen enemy killing your people by ones and twos? How does counterinsurgency actually happen?

There are no universal answers, and insurgents are among the most adaptive opponents you will ever face. Countering them will demand every ounce of your intellect. But be comforted: You are not the first to feel this way. There are tactical fundamentals you can apply to link the theory with the techniques and procedures you already know.

## What Is Counterinsurgency?

If you have not studied counterinsurgency theory, here it is in a nutshell: Counterinsurgency is a competition with the insurgent for the right to win the hearts, minds, and acquiescence of the population. You are being sent in because the insurgents, at their strongest, can defeat anything with less strength than you. But you have more combat power than you can or should use in most situations. Injudicious use of firepower creates blood feuds, homeless people, and societal disruption that fuel and perpetuate the insurgency. The most beneficial actions are often local politics, civic action, and beat-cop behaviors. For your side to win, the people don't have to like you but they must respect you, accept that your actions benefit them, and trust your integrity and ability to deliver on promises, particularly regarding their security. In this battlefield, popular perceptions and rumor are more influential than the facts and more powerful than a hundred tanks.

Within this context, what follows are observations from collective experience, the distilled essence of what those who went before you learned. They are expressed as commandments, for clarity, but are really more like folklore. Apply them judiciously and skeptically.

## Preparation

Time is short during predeployment, but you will never have more time to think than you have now. This is your chance to prepare yourself and your command.

1. Know your turf. Know the people, the topography, economy, history, religion, and culture. Know every village, road, field, population group, tribal leader, and ancient grievance. Your task is to become the world expert on your district. If you don't know precisely where you will be operating, study the general area. Read the map like a book: Study it every night before sleep and redraw it from memory every morning until you understand its patterns intuitively. Develop a mental model of your area, a framework in which to fit every new piece of knowledge you acquire. Study handover notes from predecessors; better still, get in touch with the unit in theater and pick their leaders' brains. In an ideal world, intelligence officers and area experts would brief you; however, this rarely happens, and even if it does, there is no substitute for personal mastery. Understand the broader area of influence, which can be a wide area, particularly when insurgents draw on global grievances. Share out aspects of the operational area among platoon leaders and noncommissioned officers; have each individual develop a personal specialization and brief the others. Neglect this knowledge, and it will kill you.
2. Diagnose the problem. Once you know your area and its people, you can begin to diagnose the problem. Who are the insurgents? What drives them? What makes local leaders tick? Counterinsurgency is fundamentally a competition between each side to mobilize the population in support of its agenda. So you must understand what motivates the people and how to mobilize them. You need to know why and how the insurgents are getting followers. This means you need to know your real enemy, not a cardboard cut-out. The enemy is adaptive, resourceful, and probably grew up in the region where you will be operating. The locals have known him since he was a boy; how long have they known you? Your worst opponent is not the psychopathic terrorist of Hollywood; it is the charismatic follow-me warrior who would make your best platoon leader. His followers are not misled or naïve; much of his success may be due to bad government policies or security forces that alienate the population. Work this problem collectively with your platoon and squad leaders. Discuss ideas, explore the problem, understand what you are facing, and seek a consensus. If this sounds unmilitary, get over it. Once you are in theater, situations will arise too quickly for orders or even commander's intent. Corporals and privates will have to make snap judgments with strategic impact. The only way to help them is to give them a shared understanding, then trust them to think for themselves on the day.
3. Organize for intelligence. In counterinsurgency, killing the enemy is easy. Finding him is often nearly impossible. Intelligence and operations are complementary. Your operations will be intelligence-driven, but intelligence will come mostly from your own operations, not as a product prepared and served up by higher headquarters. So

you must organize for intelligence. You will need a company S2 and an intelligence section (including analysts). You might need platoon S2s and S3s, and you will need a reconnaissance and surveillance (R&S) element. You will not have enough linguists—you never do—but carefully consider where best to use them. Linguists are a battle-winning asset, but like any other scarce resource, you must have a prioritized "bump plan" in case you lose them. Often during predeployment the best use of linguists is to train your command in basic language. You will probably not get augmentation for all this, but you must still do it. Put the smartest soldiers in the S2 section and the R&S squad. You will have one less rifle squad, but the intelligence section will pay for itself in lives and effort saved.

4. Organize for interagency operations. Almost everything in counterinsurgency is interagency. And everything important, from policing to intelligence to civil-military operations to trash collection, will involve your company working with civilian actors and local indigenous partners you cannot control, but whose success is essential for yours. Train the company in interagency operations: Get a briefing from the US Department of State, aid agencies, and the local police or fire brigade. Train point-men in each squad to deal with the interagency people. Realize that civilians find rifles, helmets, and body armor intimidating. Learn how not to scare them. Ask others who come from that country or culture about your ideas. See it through the eyes of a civilian who knows nothing about the military. How would you react if foreigners came to your neighborhood and conducted the operations you planned? What if somebody came to your mother's house and did that? Most importantly, know that your operations will create a temporary breathing space, but long-term development and stabilization by civilian agencies will ultimately win the war.
5. Travel light and harden your combat service support (CSS). You will be weighed down with body armor, rations, extra ammunition, communications gear, and a thousand other things. The enemy will carry a rifle or rocket-propelled grenade launcher, a shemagh (head scarf), and a water bottle if he is lucky. Unless you ruthlessly lighten your load and enforce a culture of speed and mobility, the insurgents will consistently out-run and out-maneuver you. But in lightening your load, make sure you can always reach back to call for firepower or heavy support if needed. Also, remember to harden your CSS. The enemy will attack your weakest points. Most attacks on Coalition forces in Iraq in 2004 and 2005, outside preplanned combat actions like the two battles of Fallujah or Operation Iron Horse, were against CSS installations and convoys. You do the math. Ensure your CSS assets are hardened, have communications, and are trained in combat operations. They may do more fighting than your rifle squads.
6. Find a political/cultural adviser. In a force optimized for counterinsurgency, you might receive a political-cultural adviser at company level, a diplomat or military foreign area officer able to speak the language and navigate the intricacies of local politics. Back on planet Earth, the corps and division commander will get a political advisor; you will not, so you must improvise. Find a POLAD (political-cultural adviser) from among your people— perhaps an officer, perhaps not (see article 8). Someone with people skills and a feel for the environment will do better than a political-science graduate. Don't try to be your own cultural adviser: You must be fully aware of the political and cultural dimension, but this is a different task. Also,

don't give one of your intelligence people this role. They can help, but their task is to understand the environment. The POLAD's job is to help shape it.

7. Train the squad leaders—then trust them. Counterinsurgency is a squad and platoon leader's war, and often a private soldier's war. Battles are won or lost in moments: Whoever can bring combat power to bear in seconds, on a street corner, will win. The commander on the spot controls the fight. You must train the squad leaders to act intelligently and independently without orders. If your squad leaders are competent, you can get away with average company or platoon staffs. The reverse is not the case. Training should focus on basic skills: marksmanship, patrolling, security on the move and at the halt, and basic drills. When in doubt, spend less time on company and platoon training, and more time on squads. Ruthlessly replace leaders who do not make the grade. But once people are trained and you have a shared operational diagnosis, you must trust them. We talk about this, but few company or platoon leaders really trust their people. In counterinsurgency, you have no choice.
8. Rank is nothing; talent is everything. Not everyone is good at counterinsurgency. Many people don't understand the concept, and some can't execute it. It is difficult, and in a conventional force only a few people will master it. Anyone can learn the basics, but a few naturals do exist. Learn how to spot these people, and put them into positions where they can make a difference. Rank matters far less than talent—a few good men led by a smart junior noncommissioned officer can succeed in counterinsurgency, where hundreds of well-armed soldiers under a mediocre senior officer will fail.
9. Have a game plan. The final preparation task is to develop a game plan, a mental picture of how you see the operation developing. You will be tempted to try and do this too early. But wait, as your knowledge improves, you will get a better idea of what needs to be done and a fuller understanding of your own limitations. Like any plan, this plan will change once you hit the ground, and it may need to be scrapped if there is a major shift in the environment. But you still need a plan, and the process of planning will give you a simple, robust idea of what to achieve, even if the methods change. This is sometimes called "operational design." One approach is to identify basic stages in your operation, for example "establish dominance, build local networks, marginalize the enemy." Make sure you can easily transition between phases, forward and backward, in case of setbacks. Just as the insurgent can adapt his activity to yours, so you must have a simple enough plan to survive setbacks without collapsing. This plan is the solution that matches the shared diagnosis you developed earlier. It must be simple, and known to everyone

## Golden Hour

You have deployed, completed reception and staging, and (if you are lucky) attended the in-country counterinsurgency school. Now it is time to enter your sector and start your tour. This is the golden hour. Mistakes made now will haunt you for the rest of your tour, while early successes will set the tone for victory. You will look back on your early actions and cringe at your clumsiness. So be it. But you must act.

10. Be there. The most fundamental rule of counterinsurgency is to be there. You can almost never outrun the enemy. If you are not present when an incident happens, there is usually little you can do about it. So your first order of business is to establish presence. If you can't do this throughout your sector, then do it wherever you can. This demands a residential approach: living in your sector, in close proximity to the population rather than raiding into the area from remote, secure bases. Movement on foot, sleeping in local villages, night patrolling—all these seem more dangerous than they are. They establish links with the locals, who see you as real people they can trust and do business with, not as aliens who descend from an armored box. Driving around in an armored convoy, day-tripping like a tourist in hell, degrades situational awareness, makes you a target, and is ultimately more dangerous.
11. Avoid knee-jerk responses to first impressions. Don't act rashly; get the facts first. The violence you see may be part of the insurgent strategy; it may be various interest groups fighting it out with each other or settling personal vendettas. Normality in Kandahar is not the same as in Seattle—you need time to learn what normality looks like. The insurgent commander wants to goad you into lashing out at the population or making a mistake. Unless you happen to be on the spot when an incident occurs, you will have only second-hand reports and may misunderstand the local context or interpretation. This fragmentation and "disaggregation" of the battlefield, particularly in urban areas, means that first impressions are often highly misleading. Of course, you can't avoid making judgments. But if possible, check them with an older or a trusted local. If you can, keep one or two officers from your predecessor unit for the first part of the tour. Try to avoid a rush to judgment.
12. Prepare for handover from day one. Believe it or not, you will not resolve the insurgency on your watch. Your tour will end, and your successors will need your corporate knowledge. Start handover folders, in every platoon and specialist squad, from day one. Ideally, you would have inherited these from your predecessors, but if not you must start them. The folders should include lessons learned, details about the population, village and patrol reports, updated maps, and photographs—anything that will help newcomers master the environment. Computerized databases are fine, but keep good back-ups and ensure you have hard copy of key artifacts and documents. This is and tedious, but essential. Over time, you will create a corporate memory that keeps your people alive.
13. Build trusted networks. Once you have settled into your sector, your key task is to build trusted networks. This is the true meaning of the phrase hearts and minds, which comprises two separate components. Hearts means persuading people their best interests are served by your success; minds means convincing them that you can protect them, and that resisting you is pointless. Note that neither concept has anything to do with whether people like you. Calculated self-interest, not emotion, is what counts. Over time, if you successfully build networks of trust, these will grow like roots into the population, displacing the enemy's networks, bringing him out into the open to fight you, and letting you seize the initiative. These networks include local allies, community leaders, local security forces, nongovernmental organizations (NGOs) and other friendly or neutral nonstate actors in your area, and the media. Conduct village and neighborhood surveys to identify needs in the community, and

then follow through to meet them. Build common interests and mobilize popular support. This is your true main effort; everything else is secondary. Actions that help build trusted networks serve your cause. Actions—even killing high-profile targets that undermine trust or disrupt your networks—help the enemy.

14. Start easy. If you were trained in maneuver warfare you know about surfaces and gaps. This applies to counterinsurgency as much as any other form of maneuver. Don't try to crack the hardest nut first—don't go straight for the main insurgent stronghold, try to provoke a decisive showdown, or focus efforts on villages that support the insurgents. Instead, start from secure areas and work gradually outwards. Do this by extending your influence through the locals' own networks. Go with, not against, the grain of local society. First win the confidence of a few villages and see who they trade, intermarry, or do business with. Now win these people over. Soon enough the showdown with the insurgents will come. But now you have local allies, a mobilized population, and a trusted network at your back. Do it the other way around and no one will mourn your failure.
15. Seek early victories. In this early phase, your aim is to stamp your dominance in your sector. Do this by seeking an early victory. This will probably not translate into a combat victory over the enemy. Looking for such a victory can be overly aggressive and create collateral damage—especially since you really do not yet understand your sector. Also, such a combat victory depends on the enemy being stupid enough to present you with a clear-cut target, which is a rare windfall in counterinsurgency. Instead, you may achieve a victory by resolving long-standing issues your predecessors have failed to address, or by co-opting a key local leader who has resisted cooperation with our forces. Like any other form of armed propaganda, achieving even a small victory early in the tour sets the tone for what comes later and helps seize the initiative, which you have probably lost due to the inevitable hiatus entailed by the handover-takeover with your predecessor.
16. Practice deterrent patrolling. Establish patrolling methods that deter the enemy from attacking you. Often our patrolling approach seems designed to provoke, then defeat, enemy attacks. This is counterproductive; it leads to a raiding, day-tripping mindset or, worse, a bunker mentality. Instead, practice deterrent patrolling. There are many methods for this, including multiple patrolling in which you flood an area with numerous small patrols working together. Each is too small to be a worthwhile target, and the insurgents never know where all the patrols are—making an attack on any one patrol extremely risky. Other methods include so-called blue-green patrolling, where you mount daylight, overt humanitarian patrols, which go covert at night and hunt specific targets. Again, the aim is to keep the enemy off balance, and the population reassured through constant and unpredictable activity, which, over time, deters attacks and creates a more permissive environment. A reasonable rule of thumb is that one—to two-thirds of your force should be on patrol at any time, day or night.
17. Be prepared for setbacks. Setbacks are normal in counterinsurgency, as in every other form of war. You will make mistakes, lose people, or occasionally kill or detain the wrong person. You may fail in building or expanding networks. If this happens, don't lose heart, simply drop back to the previous phase of your game plan and recover your balance. It is normal in company counterinsurgency operations for some

platoons to be doing well while others do badly. This is not necessarily evidence of failure. Give local commanders the freedom to adjust their posture to local conditions. This creates elasticity that helps you survive setbacks.

18. Remember the global audience. One of the biggest differences between the counterinsurgencies our fathers fought and those we face today is the omnipresence of globalized media. Most houses in Iraq have one or more satellite dishes. Web bloggers; print, radio, and television reporters; and others are monitoring and reporting your every move. When the insurgents ambush your patrols or set off a car bomb, they do so not to destroy one more track, but because they want graphic images of a burning vehicle and dead bodies for the evening news. Beware of the scripted enemy who plays to a global audience and seeks to defeat you in the court of global public opinion. You counter this by training people to always bear in mind the global audience, to assume that everything they say or do will be publicized, and to befriend the media. Get the press on-side—help them get their story, and trade information with them. Good relationships with nonembedded media, especially indigenous media, dramatically increase your situational awareness and help get your message across to the global and local audience.
19. Engage the women, beware of the children. Most insurgent fighters are men. But in traditional societies, women are hugely influential in forming the social networks that insurgents use for support. Co-opting neutral or friendly women, through targeted social and economic programs, builds networks of enlightened self-interest that eventually undermine the insurgents. You need your own female counterinsurgents, including interagency people, to do this effectively. Win the women, and you own the family unit. Own the family, and you take a big step forward in mobilizing the population. Conversely, though, stop your people from fraternizing with the local children. Your troops are homesick; they want to drop their guard with the kids, but children are sharp-eyed, lacking in empathy, and willing to commit atrocities their elders would shrink from. The insurgents are watching: They will notice a growing friendship between one of your people and a local child, and either harm the child as punishment, or use them against you. Similarly, stop people from throwing candies or presents to children. It attracts them to our vehicles, creates crowds the enemy can exploit, and leads to children being run over. Harden your heart and keep the children at arm's length.
20. Take stock regularly. You probably already know that a body count tells you little, because you usually can't know how many insurgents there were to start with, how many moved into the area, how many transferred from supporter to combatant status, or how many new fighters the conflict has created. But you still need to develop metrics early in the tour and refine them as the operation progresses. They should cover a range of social, informational, military, and economic issues. Use metrics intelligently to form an overall impression of progress—not in a mechanistic traffic-light fashion. Typical metrics include percentage of engagements initiated by our forces versus those initiated by insurgents; longevity of friendly local leaders in positions of authority; number and quality of tip-offs on insurgent activity that originate spontaneously from the population; and economic activity at markets and shops. These mean virtually nothing as a snapshot; it is trends over time that help you track progress in your sector.

## Groundhog Day

Now you are in "steady state." You are established in your sector, and people are settling into that "groundhog day" mentality that hits every unit at some stage during every tour. It will probably take you at least the first third of your tour to become effective in your new environment, if not longer. Then in the last period you will struggle against the short-timer mentality. So this middle part of the tour is the most productive—but keeping the flame alive, and bringing the local population along with you, takes immense leadership.

21. Exploit a "single narrative." Since counterinsurgency is a competition to mobilize popular support, it pays to know how people are mobilized. In most societies there are opinion makers—local leaders, pillars of the community, religious figures, media personalities, and others who set trends and influence public perceptions. This influence, including the pernicious influence of the insurgents, often takes the form of a "single narrative": a simple, unifying, easily expressed story or explanation that organizes people's experience and provides a framework for understanding events. Nationalist and ethnic historical myths, or sectarian creeds, provide such a narrative. The Iraqi insurgents have one, as do Al-Qaeda and the Taliban. To undercut their influence you must exploit an alternative narrative, or better yet, tap into an existing narrative that excludes the insurgents. This narrative is often worked out for you by higher headquarters—but only you have the detailed knowledge to tailor the narrative to local conditions and generate leverage from it. For example, you might use a nationalist narrative to marginalize foreign fighters in your area or a narrative of national redemption to undermine former regime elements that have been terrorizing the population. At the company level, you do this in baby steps by getting to know local opinion-makers, winning their trust, learning what motivates them, and building on this to find a single narrative that emphasizes the inevitability and rightness of your ultimate success. This is art, not science.
22. Local forces should mirror the enemy, not the Americans. By this stage, you will be working closely with local forces, training or supporting them and building indigenous capability. The natural tendency is to build forces in the US image, with the aim of eventually handing our role over to them. This is a mistake. Instead, local indigenous forces need to mirror the enemy's capabilities and seek to supplant the insurgent's role. This does not mean they should be irregular in the sense of being brutal or outside proper control. Rather, they should move, equip, organize like the insurgents, but have access to your support and be under the firm control of their parent societies. Combined with a mobilized population and trusted networks, this allows local forces to hard-wire the enemy out of the environment, under top-cover from you. At the company level, this means that raising, training, and employing local indigenous auxiliary forces (police and military) are valid tasks. This requires high-level clearance, of course, but if support is given, you should establish a company training cell. Platoons should aim to train one local squad, and then use that squad as a nucleus for a partner platoon. Company headquarters should train an indigenous leadership team. This mirrors the growth process of other trusted networks and tends to emerge naturally as you win local allies who want to take up arms in their own defense.

23. Practice armed civil affairs. Counterinsurgency is armed social work, an attempt to redress basic social and political problems while being shot at. This makes civil affairs a central counterinsurgency activity, not an afterthought. It is how you restructure the environment to displace the enemy from it. In your company sector, civil affairs must focus on meeting basic needs first, and then progress up Maslow's hierarchy as each successive need is met. You need intimate cooperation with interagency partners here—national, international, and local. You will not be able to control these partners—many NGOs, for example, do not want to be too closely associated with you because they need to preserve their perceived neutrality. Instead, you need to work on a shared diagnosis of the problem, building a consensus that helps you self-synchronize. Your role is to provide protection, identify needs, facilitate civil affairs, and use improvements in social conditions as leverage to build networks and mobilize the population. Thus, there is no such thing as impartial humanitarian assistance or civil affairs in counterinsurgency. Every time you help someone, you hurt someone else—not least the insurgents—so civil and humanitarian assistance personnel will be targeted. Protecting them is a matter not only of close-in defense, but also of creating a permissive operating environment by co-opting the beneficiaries of aid (local communities and leaders) to help you help them.
24. Small is beautiful. Another natural tendency is to go for large-scale, mass programs. In particular, we have a tendency to template ideas that succeed in one area and transplant them into another, and we tend to take small programs that work and try to replicate them on a larger scale. Again, this is usually a mistake: Often programs succeed because of specific local conditions of which we are unaware, or because their very smallness kept them below the enemy's radar and helped them flourish unmolested. At the company level, programs that succeed in one district often also succeed in another (because the overall company sector is small), but small-scale projects rarely proceed smoothly into large programs. Keep programs small; this makes them cheap, sustainable, low-key, and (importantly) recoverable if they fail. You can add new programs—also small, cheap and tailored to local conditions— as the situation allows.
25. Fight the enemy's strategy, not his forces. At this stage, if things are proceeding well, the insurgents will go over to the offensive. Yes, the offensive, because you have created a situation so dangerous to the insurgents (by threatening to displace them from the environment) that they have to attack you and the population to get back into the game. Thus it is normal, even in the most successful operations, to have spikes of offensive insurgent activity late in the operation. This does not necessarily mean you have done something wrong (though it may, it depends on whether you have successfully mobilized the population). At this point the tendency is to go for the jugular and seek to destroy the enemy's forces in open battle. This is rarely the best choice at company level, because provoking major combat usually plays into the enemy's hands by undermining the population's confidence. Instead, attack the enemy's strategy. If he is seeking to recapture the allegiance of a segment of the local population, then co-opt them against him. If he is trying to provoke a sectarian conflict, go over to peace-enforcement mode. The permutations are endless, but the principle is the same: Fight the enemy's strategy, not his forces.

26. Build your own solution—only attack the enemy when he gets in the way. Try not to be distracted or forced into a series of reactive moves by a desire to kill or capture the insurgents. Your aim should be to implement your own solution, the game plan you developed early in the operation and then refined through interaction with local partners. Your approach must be environment-centric (based on dominating the whole district and implementing a solution to its systemic problems) rather than enemy-centric. This means that particularly late in the operation you may need to learn to negotiate with the enemy. Members of the population that supports you also know the enemy's leaders. They may have grown up together in the small district that is now your company sector, and valid negotiating partners sometimes emerge as the operation progresses. Again, you need close interagency relationships to exploit opportunities to co-opt segments of the enemy. This helps you wind down the insurgency without alienating potential local allies who have relatives or friends in the insurgent movement. At this stage, a defection is better than a surrender, a surrender is better than a capture, and a capture is better than a kill.

## Getting Short

Time is short, and the tour is drawing to a close. The key problem now is keeping your people focused, maintaining the rage on all the multifarious programs, projects, and operations that you have started, and preventing your people from dropping their guard. In this final phase, the previous articles still stand, but there is an important new one.

27. Keep your extraction plan secret. The temptation to talk about home becomes almost unbearable toward the end of a tour. The locals know you are leaving, and probably have a better idea than you of the generic extraction plan. Remember, they have seen units come and go. But you must protect the specific details of the extraction plan, or the enemy will use this as an opportunity to score a high-profile hit, recapture the population's allegiance by scare tactics that convince them they will not be protected once you leave, or persuade them that your successor unit will be oppressive or incompetent. Keep the details secret within a tightly controlled compartment in your headquarters.

## Four "What Ifs"

The articles above describe what should happen, but we all know that things go wrong. Here are some what ifs to consider:

- What if you get moved to a different area? You prepared for ar-Ramadi and studied Dulaim tribal structures and Sunni beliefs. Now you are going to Najaf and will be surrounded by al-Hassani tribes and Shi'a communities. But that work was not wasted. In mastering your first area, you learned techniques you can apply: how to "case" an operational area and how to decide what matters in the local societal structure. Do the same again, and this time the process is easier and faster, since you have an existing mental structure and can focus on what is different. The same applies if you get moved frequently within a battalion or brigade area.

- What if higher headquarters doesn't "get" counterinsurgency? Higher headquarters is telling you the mission is to "kill terrorists," or pushing for high-speed armored patrols and a base-camp mentality. They just don't seem to understand counterinsurgency. This is not uncommon, since company-grade officers today often have more combat experience than senior officers. In this case, just do what you can. Try not to create expectations that higher headquarters will not let you meet. Apply the adage "first do no harm." Over time, you will find ways to do what you have to do. But never lie to higher headquarters about your locations or activities—they own the indirect fires.
- What if you have no resources? You have no linguists, the aid agencies have no money for projects in your area, and you have a low priority for civil affairs. You can still get things done, but you need to focus on self-reliance: Keep things small and sustainable and ruthlessly prioritize effort. The local population are your allies in this: They know what matters to them more than you do. Be honest with them; discuss possible projects and options with community leaders; get them to choose what their priorities are. Often they will find the translators, building supplies, or expertise that you need, and will only expect your support and protection in making their projects work. And the process of negotiation and consultation will help mobilize their support and strengthen their social cohesion. If you set your sights on what is achievable, the situation can still work.
- What if the theater situation shifts under your feet? It is your worst nightmare—Everything has gone well in your sector, but the whole theater situation has changed and invalidates your efforts. Think of the first battle of Fallujah, the Askariya shrine bombing, or the Sadr uprising. What do you do? Here is where having a flexible, adaptive game plan comes in. Just as the insurgents drop down to a lower posture when things go wrong, now is the time for you to drop back a stage, consolidate, regain your balance, and prepare to expand again when the situation allows. But see article 28: If you cede the initiative, you must regain it as soon as the situation allows, or you will eventually lose.

This, then, is the tribal wisdom, the folklore that those who went before you have learned. Like any folklore it needs interpretation and contains seemingly contradictory advice. Over time, as you apply unremitting intellectual effort to study your sector, you will learn to apply these ideas in your own way and will add to this store of wisdom from your own observations and experience. So only one article remains, and if you remember nothing else, remember this:

28. Whatever else you do, keep the initiative. In counterinsurgency, the initiative is everything. If the enemy is reacting to you, you control the environment. Provided you mobilize the population, you will win. If you are reacting to the enemy, even if you are killing or capturing him in large numbers, then he is controlling the environment and you will eventually lose. In counterinsurgency, the enemy initiates most attacks, targets you unexpectedly, and withdraws too fast for you to react. Do not be drawn into purely reactive operations: Focus on the population, build your

own solution, further your game plan, and fight the enemy only when he gets in the way. This gains and keeps the initiative.

## Appendix D. Twenty-Seven Articles of T. E. Lawrence

T. E. Lawrence wrote that he expressed these notes, published in *The Arab Bulletin*, 20 August 1917, "*in commandment form*" for clarity and brevity:

They are, however, only my personal conclusions, arrived at gradually while I worked in the Hejaz and now put on paper as stalking horses for beginners in the Arab armies. They are meant to apply only to Bedu; townspeople or Syrians require totally different treatment. They are of course not suitable to any other person's need, or applicable unchanged in any particular situation. Handling Hejaz Arabs is an art, not a science, with exceptions and no obvious rules. At the same time we have a great chance there; the Sharif trusts us, and has given us the position (towards his Government) which the Germans wanted to win in Turkey. If we are tactful, we can at once retain his goodwill and carry out our job, but to succeed we have got to put into it all the interest and skill we possess.

1. Go easy for the first few weeks. A bad start is difficult to atone for, and the Arabs form their judgments on externals that we ignore. When you have reached the inner circle in a tribe, you can do as you please with yourself and them.
2. Learn all you can about your Ashraf and Bedu. Get to know their families, clans and tribes, friends and enemies, wells, hills and roads. Do all this by listening and by indirect inquiry. Do not ask questions. Get to speak their dialect of Arabic, not yours. Until you can understand their allusions, avoid getting deep into conversation or you will drop bricks. Be a little stiff at first.
3. In matters of business deal only with the commander of the army, column, or party in which you serve. Never give orders to anyone at all, and reserve your directions or advice for the C.O., however great the temptation (for efficiency's sake) of dealing with his underlings. Your place is advisory, and your advice is due to the commander alone. Let him see that this is your conception of your duty, and that his is to be the sole executive of your joint plans.
4. Win and keep the confidence of your leader. Strengthen his prestige at your expense before others when you can. Never refuse or quash schemes he may put forward; but ensure that they are put forward in the first instance privately to you. Always approve them, and after praise modify them insensibly, causing the suggestions to come from him, until they are in accord with your own opinion. When you attain this point, hold him to it, keep a tight grip of his ideas, and push them forward as firmly as possibly, but secretly, so that no one but himself (and he not too clearly) is aware of your pressure.
5. Remain in touch with your leader as constantly and unobtrusively as you can. Live with him, that at meal times and at audiences you may be naturally with him in his tent. Formal visits to give advice are not so good as the constant dropping of ideas in casual talk. When stranger sheikhs come in for the first time to swear allegiance and

offer service, clear out of the tent. If their first impression is of foreigners in the confidence of the Sharif, it will do the Arab cause much harm.

6. Be shy of too close relations with the subordinates of the expedition. Continual intercourse with them will make it impossible for you to avoid going behind or beyond the instructions that the Arab C.O. has given them on your advice, and in so disclosing the weakness of his position you altogether destroy your own.
7. Treat the sub-chiefs of your force quite easily and lightly. In this way you hold yourself above their level. Treat the leader, if a Sharif, with respect. He will return your manner and you and he will then be alike, and above the rest. Precedence is a serious matter among the Arabs, and you must attain it.
8. Your ideal position is when you are present and not noticed. Do not be too intimate, too prominent, or too earnest. Avoid being identified too long or too often with any tribal sheikh, even if C.O. of the expedition. To do your work you must be above jealousies, and you lose prestige if you are associated with a tribe or clan, and its inevitable feuds. Sharifs are above all blood-feuds and local rivalries, and form the only principle of unity among the Arabs. Let your name therefore be coupled always with a Sharif's, and share his attitude towards the tribes. When the moment comes for action put yourself publicly under his orders. The Bedu will then follow suit.
9. Magnify and develop the growing conception of the Sharifs as the natural aristocracy of the Arabs. Intertribal jealousies make it impossible for any sheikh to attain a commanding position, and the only hope of union in nomad Arabs is that the Ashraf be universally acknowledged as the ruling class. Sharifs are half-townsmen, half-nomad, in manner and life, and have the instinct of command. Mere merit and money would be insufficient to obtain such recognition; but the Arab reverence for pedigree and the Prophet gives hope for the ultimate success of the Ashraf.
10. Call your Sharif 'Sidi' in public and in private. Call other people by their ordinary names, without title. In intimate conversation call a Sheikh 'Abu Annad', 'Akhu Alia' or some similar by-name.
11. The foreigner and Christian is not a popular person in Arabia. However friendly and informal the treatment of yourself may be, remember always that your foundations are very sandy ones. Wave a Sharif in front of you like a banner and hide your own mind and person. If you succeed, you will have hundreds of miles of country and thousands of men under your orders, and for this it is worth bartering the outward show.
12. Cling tight to your sense of humor. You will need it every day. A dry irony is the most useful type, and repartee of a personal and not too broad character will double your influence with the chiefs. Reproof, if wrapped up in some smiling form, will carry further and last longer than the most violent speech. The power of mimicry or parody is valuable, but use it sparingly, for wit is more dignified than humor. Do not cause a laugh at a Sharif except among Sharifs.
13. Never lay hands on an Arab; you degrade yourself. You may think the resultant obvious increase of outward respect a gain to you, but what you have really done is to build a wall between you and their inner selves. It is difficult to keep quiet when everything is being done wrong, but the less you lose your temper the greater your advantage. Also then you will not go mad yourself.

14. While very difficult to drive, the Bedu are easy to lead, if— have the patience to bear with them. The less apparent your interferences the more your influence. They are willing to follow your advice and do what you wish, but they do not mean you or anyone else to be aware of that. It is only after the end of all annoyances that you find at bottom their real fund of goodwill.
15. Do not try to do too much with your own hands. Better the Arabs do it tolerably than that you do it perfectly. It is their war, and you are to help them, not to win it for them. Actually, also, under the very odd conditions of Arabia, your practical work will not be as good as, perhaps, you think it is.
16. If you can, without being too lavish, forestall presents to yourself. A well-placed gift is often most effective in winning over a suspicious sheikh. Never receive a present without giving a liberal return, but you may delay this return (while letting its ultimate certainty be known) if you require a particular service from the giver. Do not let them ask you for things, since their greed will then make them look upon you only as a cow to milk.
17. Wear an Arab head cloth when with a tribe. Bedu have a malignant prejudice against the hat, and believe that our persistence in wearing it (due probably to British obstinacy of dictation) is founded on some immoral or irreligious principle. A thick head cloth forms a good protection against the sun, and if you wear a hat your best Arab friends will be ashamed of you in public.
18. Disguise is not advisable. Except in special areas, let it be clearly known that you are a British officer and a Christian. At the same time, if you can wear Arab kit when with the tribes, you will acquire their trust and intimacy to a degree impossible in uniform. It is, however, dangerous and difficult. They make no special allowances for you when you dress like them. Breaches of etiquette not charged against a foreigner are not condoned to you in Arab clothes. You will be like an actor in a foreign theatre, playing a part day and night for months, without rest, and for an anxious stake. Complete success, which is when the Arabs forget your strangeness and speak naturally before you, counting you as one of themselves, is perhaps only attainable in character: while half-success (all that most of us will strive for; the other costs too much) is easier to win in British things, and you yourself will last longer, physically and mentally, in the comfort that they mean. Also then the Turks will not hang you, when you are caught.
19. If you wear Arab things, wear the best. Clothes are significant among the tribes, and you must wear the appropriate, and appear at ease in them. Dress like a Sharif, if they agree to it.
20. If you wear Arab things at all, go the whole way. Leave your English friends and customs on the coast, and fall back on Arab habits entirely. It is possible, starting thus level with them, for the European to beat the Arabs at their own game, for we have stronger motives for our action, and put more heart into it than they. If you can surpass them, you have taken an immense stride toward complete success, but the strain of living and thinking in a foreign and half-understood language, the savage food, strange clothes, and stranger ways, with the complete loss of privacy and quiet, and the impossibility of ever relaxing your watchful imitation of the others for months on end, provide such an added stress to the ordinary difficulties of dealing

with the Bedu, the climate, and the Turks, that this road should not be chosen without serious thought.

21. Religious discussions will be frequent. Say what you like about your own side, and avoid criticism of theirs, unless you know that the point is external, when you may score heavily by proving it so. With the Bedu, Islam is so all-pervading an element that there is little religiosity, little fervor, and no regard for externals. Do not think from their conduct that they are careless. Their conviction of the truth of their faith, and its share in every act and thought and principle of their daily life is so intimate and intense as to be unconscious, unless roused by opposition. Their religion is as much a part of nature to them as is sleep or food.
22. Do not try to trade on what you know of fighting. The Hejaz confounds ordinary tactics. Learn the Bedu principles of war as thoroughly and as quickly as you can, for till you know them your advice will be no good to the Sharif. Unnumbered generations of tribal raids have taught them more about some parts of the business than we will ever know. In familiar conditions they fight well, but strange events cause panic. Keep your unit small. Their raiding parties are usually from one hundred to two hundred men, and if you take a crowd they only get confused. Also their sheikhs, while admirable company commanders, are too 'set' to learn to handle the equivalents of battalions or regiments. Don't attempt unusual things, unless they appeal to the sporting instinct Bedu have so strongly, unless success is obvious. If the objective is a good one (booty) they will attack like fiends, they are splendid scouts, their mobility gives you the advantage that will win this local war, they make proper use of their knowledge of the country (don't take tribesmen to places they do not know), and the gazelle-hunters, who form a proportion of the better men, are great shots at visible targets. A sheikh from one tribe cannot give orders to men from another; a Sharif is necessary to command a mixed tribal force. If there is plunder in prospect, and the odds are at all equal, you will win. Do not waste Bedu attacking trenches (they will not stand casualties) or in trying to defend a position, for they cannot sit still without slacking. The more unorthodox and Arab your proceedings, the more likely you are to have the Turks cold, for they lack initiative and expect you to. Don't play for safety.
23. The open reason that Bedu give you for action or inaction may be true, but always there will be better reasons left for you to divine. You must find these inner reasons (they will be denied, but are none the less in operation) before shaping your arguments for one course or other. Allusion is more effective than logical exposition: they dislike concise expression. Their minds work just as ours do, but on different premises. There is nothing unreasonable, incomprehensible, or inscrutable in the Arab. Experience of them and knowledge of their prejudices will enable you to foresee their attitude and possible course of action in nearly every case.
24. Do not mix Bedu and Syrians, or trained men and tribesmen. You will get work out of neither, for they hate each other. I have never seen a successful combined operation, but many failures. In particular, ex-officers of the Turkish army, however Arab in feelings and blood and language, are hopeless with Bedu. They are narrow minded in tactics, unable to adjust themselves to irregular warfare, clumsy in Arab etiquette, swollen-headed to the extent of being incapable of politeness to a tribesman for more than a few minutes, impatient, and, usually, helpless without their

troops on the road and in action. Your orders (if you were unwise enough to give any) would be more readily obeyed by Bedouins than those of any Mohammedan Syrian officer. Arab townsmen and Arab tribesmen regard each other mutually as poor relations, and poor relations are much more objectionable than poor strangers.

25. In spite of ordinary Arab example, avoid too free talk about women. It is as difficult a subject as religion, and their standards are so unlike our own that a remark, harmless in English, may appear as unrestrained to them, as some of their statements would look to us, if translated literally.
26. Be as careful of your servants as of yourself. If you want a sophisticated one you will probably have to take an Egyptian, or a Sudani, and unless you are very lucky he will undo on trek much of the good you so laboriously effect. Arabs will cook rice and make coffee for you, and leave you if required to do unmanly work like cleaning boots or washing. They are only really possible if you are in Arab kit. A slave brought up in the Hejaz is the best servant, but there are rules against British subjects owning them, so they have to be lent to you. In any case, take with you an Ageyli or two when you go up country. They are the most efficient couriers in Arabia, and understand camels.
27. The beginning and ending of the secret of handling Arabs is unremitting study of them. Keep always on your guard; never say an unnecessary thing: watch yourself and your companions all the time: hear all that passes, search out what is going on beneath the surface, read their characters, discover their tastes and their weaknesses and keep everything you find out to yourself. Bury yourself in Arab circles, have no interests and no ideas except the work in hand, so that your brain is saturated with one thing only, and you realize your part deeply enough to avoid the little slips that would counteract the painful work of weeks. Your success will be proportioned to the amount of mental effort you devote to it."

# GLOSSARY

## Section I—Acronyms and Abbreviations

| | |
|---|---|
| **3C** | cross-cultural competency |
| **AAR** | after-action review |
| **AI** | area of interest |
| **AO** | area of operations |
| **ASCOPE** | areas, structures, capabilities, organizations, people, and events |
| **BCT** | brigade combat team |
| **BDA** | battle damage assessment |
| **C2** | command and control |
| **CA** | civil affairs |
| **CALL** | Center for Army Lessons Learned |
| **CARE** | Cooperative for Assistance and Relief Everywhere |
| **CARVER-P** | Criticality<br>Accessibility |

| | |
|---|---|
| | Recoverability |
| | Vulnerability |
| | Effect |
| | Recognizability |
| | Psychological impact |
| **CAS** | close air support |
| **CASEVAC** | casualty evacuation |
| **CAT** | civil affairs team |
| **CBRN** | chemical, biological, radiological, or nuclear |
| **CCIR** | commander's critical information requirement |
| **CI** | civilian internee |
| **CID** | combat identification |
| **COA** | course of action |
| **COIN** | counterinsurgency |
| **COP** | common operational picture |
| **CORDS** | civil operations and revolutionary (rural) development support |
| **CWIED** | command-wired improvised explosive device |
| **D3A** | decide, detect, deliver, and assess |
| **DA** | Department of the Army |
| **DC** | dislocated civilian |
| **DCP** | detention control point |
| **DHA** | detainee holding area |
| **DNA** | deoxyribonucleic acid |
| **DOD** | Department of Defense |
| **DOMEX** | document and media exploitation |
| **DOTMLPF** | Doctrine |
| | Organization |
| | Training |
| | Materiel |
| | Leadership and Education |
| | Personnel |
| | Facilities |
| **DTG** | date-time group |
| **DVD** | digital videodisk |
| **EEFI** | essential elements of friendly information |
| **EOD** | explosive ordnance disposal |
| **EOF** | escalation of force |
| **EPW** | enemy prisoner of war |
| **FARC** | Revolutionary Armed Forces of Colombia (Fuerzas Armadas Revolucionarias de Colombia) |
| **FCM** | fire-control measure |
| **FFIR** | friendly force information requirement |
| **FID** | foreign internal defense |
| **FOB** | forward operating base |
| **FRAGO** | fragmentary order |
| **GIS** | geographic information system |

| | |
|---|---|
| **HCT** | HUMINT collection team |
| **HMMWV** | high-mobility, multipurpose wheeled vehicle |
| **HN** | host nation |
| **HOPE-P** | higher, operational, planning, enemy, and populace |
| **HPT** | high-payoff target |
| **HQ** | headquarters |
| **HUMINT** | human intelligence |
| **IAW** | in accordance with |
| **ID** | identification |
| **IDAD** | internal defense and development |
| **IDP** | internally displaced person |
| **IE** | information engagement |
| **IED** | improvised explosive device |
| **IGO** | intergovernmental organization |
| **IPB** | intelligence preparation of the battlefield |
| **IPI** | indigenous population and institutions |
| **IRA** | Irish Republican Army |
| **ISR** | intelligence, surveillance, reconnaissance |
| **JDAM** | joint direct attack munition |
| **JLENS** | Joint Land Attack Cruise Missile Defense Elevated Netted Sensor System |
| **JTACMS** | Joint Tactical Attack Cruise Missile System |
| **LEP** | law enforcement professionals |
| **LN** | local nationals |
| **LOA** | limit of advance |
| **LOC** | line of communications |
| **LOE** | line of effort |
| **M** | meter(s) |
| **MASINT** | measurement and signature intelligence |
| **MDMP** | military decision-making process |
| **METT-TC** | mission, enemy, terrain and weather, troops and support available, time available, and civilian considerations |
| **MILVAN** | military van (container) |
| **MOE** | measure of effectiveness |
| **MOP** | measure of performance |
| **MORTEAM** | Measure (assess)<br>Organize<br>Rebuild/build facilities<br>Train<br>Equip<br>Advise<br>Mentor |
| **MWD** | military working dog |
| **NATO** | North Atlantic Treaty Organization |
| **NCO** | noncommissioned officer |
| **NCOIC** | noncommissioned officer in charge |

| | |
|---|---|
| **NGA** | National Geospatial-Intelligence Agency |
| **NGO** | nongovernmental organization |
| **OAKOC** | observation and fields of fire, avenues of approach, key and decisive terrain, obstacles, cover and concealment. |
| **OBJ** | objective |
| **OE** | operational environment |
| **OP** | observation post |
| **OPORD** | operation order |
| **OPSEC** | operations security |
| **ORP** | objective rally point |
| **OXFAM** | Oxford Committee for Famine Relief |
| **PAO** | public affairs office |
| **PBIED** | person-borne improvised explosive device |
| **PCC** | precombat check |
| **PCI** | precombat inspection |
| **PID** | positively identify (a target) |
| **PIR** | priority intelligence requirement |
| **PKC** | Russian-made machine gun; also called *PK* or *PKS* |
| **PMESII-PT** | Political<br>Military<br>Economic<br>Social<br>Information<br>Infrastructure physical environment, time |
| **PRC** | populace and resource control |
| **PSYOP** | psychological operations |
| **QRF** | quick reaction force |
| **RAM** | random antiterrorism measure |
| **RCIED** | radio-controlled improvised explosive device |
| **RCT** | route clearance team |
| **RIP** | relief in place |
| **ROE** | rules of engagement |
| **ROI** | rules of interaction |
| **RPG** | rocket-propelled grenade |
| **RPK** | a light machine gun |
| **RSTA** | reconnaissance, surveillance, target acquisition |
| **S-1** | Personnel Staff Officer |
| **S-2** | Intelligence Staff Officer |
| **S-3** | Operations Staff Officer |
| **S-5** | Plans Staff Officer |
| **S-7** | Information Engagement Staff Officer |
| **S-9** | Civil Affairs Operations Staff Officer |
| **SCT** | small capture team |
| **SE** | site exploitation |
| **SEC** | second(s) |
| **SIGINT** | signals intelligence |

| | |
|---|---|
| **SITEMP** | situation template |
| **SKT** | small kill team |
| **SOF** | special operations forces |
| **SOP** | standing operating procedures |
| **SOT** | small observation team |
| **SVD** | Soviet semiautomatic sniper rifle, common throughout the former Eastern Bloc |
| **SWAT** | special weapons and tactics |
| **SWEAT-MSO** | sewers, water, electrical, academic, trash, medical facilities, safety, and other considerations |
| **TACSAT** | tactical satellite |
| **TM** | team |
| **TERP** | interpreter |
| **THT** | tactical human intelligence team |
| **TLP** | troop-leading procedures |
| **TM** | team |
| **TOA** | transfer of authority |
| **TPT** | tactical psychological operations team |
| **TRP** | target reference point |
| **TV** | television |
| **UAS** | Unmanned Aircraft System |
| **UN** | United Nations |
| **UNHCR** | United Nations High Commissioner for Refugees |
| **USAF** | United States Air Force |
| **USAID** | United States Agency for International Development |
| **USMC** | United States Marine Corps |
| **UXO** | unexploded ordnance |
| **VBIED** | vehicle-borne, improvised explosive device |
| **VIP** | very important person |
| **WARNORD** | warning order |
| **WFF** | warfighting function |
| **WIT** | weapons intelligence team |
| **XO** | executive officer |

## Section II—Terms

### Advisor

A military member who conducts operations that train Host Nation military individuals and units in tactical employment, sustainment, and integration of land, air, and maritime skills; provide advice and assisstance to military leaders, and provide training on tactics, techniques, and procedures required to protect the HN from subversion, lawlessness, and insurgency, and develop indigenous individual, leader, organizational skills.

**Ambush**

A form of attack by fire or other destructive means from concealed positions on a moving or temporarily halted enemy (FM 3-0).

**Area Defense**

A type of defensive operation that concentrates on denying enemy forces access to designated terrain for a specific time rather than destroying the enemy outright (FM 3-0).

**Area of Interest**

(Joint) That area of concern to the commander, including the area of influence, areas adjacent thereto, and extending into enemy territory to the objectives of current or planned operations. This area also includes areas occupied by enemy forces who could jeopardize the accomplishment of the mission (JP 1-02).

**Area of Operations**

An operational area defined by the joint force commander for land and naval forces. Areas of operations do not typically encompass the entire operational area of the joint force commander, but should be large enough for component commanders to accomplish their missions and protect their forces (JP 1-02).

**Area Security**

A form of security operations conducted to protect friendly forces, installation routes, and actions within a specific area (FM 3-90).

**Assessment**

(Army) The continuous monitoring and evaluation of the current situation and progress of an operation (FMI 5-0.1).

***Asymmetric Warfare**

Conflict in which a weaker opponent uses unorthodox or surprise tactics to attack weak points of a stronger opponent, especially if the tactics include terrorism, guerrilla warfare, criminal activity, subversion, or propaganda.

**Auxiliary**

In unconventional warfare, that element of the resistance force established to provide the organized civilian support of the resistance movement. (AR 310-25).

**Avenue of Approach**

An air or ground route of an attacking force of a given size leading to its objective or to key terrain in its path (JP 1-02).

**Capacity Building**

The process of creating an environment that fosters Host Nation institutional development, community participation, human resources development, and strengthening managerial systems (FM 3-07).

**Center of Gravity**

(Joint) The source of power that provides moral or physical strength, freedom of action, or will to act (JP 3-0).

**Civil Considerations**

How the manmade infrastructure, civilian institutions, and attitudes and activities of the civilian leaders, populations, organizations within an area of operations influence the conduct of military operations (FM 6-0). See also METT-TC.

**Civil-Military Operations**

(FM 3-07) The activities of a commander that establish, maintain, influence, or exploit relations between military forces, governmental and nongovernmental civilian organizations and authorities, and the civilian populace in a friendly, neutral, or hostile operational area in order to facilitate military operations, to consolidate and achieve operational US objectives. Civil-military operations may include performance by military forces of activities and functions normally the responsibility of the local, regional, or national government. These activities may occur prior to, during, or subsequent to other military actions. They may also occur, if directed, in the absence of other military operations. Civil-military operations may be performed by designated civil affairs, by other military forces, or by a combination of civil affairs and other forces (JP 3-57).

**Civil War**

A war between opposing groups of citizens of the same country.

**Clear**

(Army) A tactical mission task that requires the commander to remove all enemy forces and eliminate organized resistance in an assigned area (FM 3-90).

**Close Air Support**

(CAS) Air action by fixed—and rotary-wing aircraft against hostile targets that are in close proximity to friendly forces and which require detailed integration of each air mission with the fire and movement of those forces (DOD).

**Close Combat**

Combat carried out with direct-fire weapons, supported by indirect fires, air-delivered fires, and nonlethal engagement means. Close combat defeats or destroys enemy forces or seizes and retains ground (FM 3-0).

**Coalition**

(Joint) An ad hoc arrangement between two or more nations for common action. (JP 5-0)

**Collateral Damage**

Unintended and undesirable civilian personnel injuries or material damage adjacent to a target produced by the effects of demolition weapons.

**Combat Patrol**

(NATO): For ground forces, a tactical unit sent out from the main body to engage in independent fighting; detachment assigned to protect the front, flank, or rear of the main body by fighting if necessary.

**Combat Power**

(Army) The total means of destructive, constructive, and information capabilities that a military unit/formation can apply at a given time. Army forces generate combat power by converting potential into effective action (FM 3-0).

**Combined Arms**

The synchronized and simultaneous application of the elements of combat power—to achieve an effect greater than if each element of combat power was used separately or sequentially (FM 3-0).

**Command**

(Joint) 1. The authority that a commander in the armed forces lawfully exercises over subordinates by virtue of rank or assignment. Command includes the authority and responsibility for effectively using available resources and for planning the employment of, organizing, directing, coordinating, and controlling military forces for the accomplishment of assigned missions. It also includes responsibility for health, welfare, morale, and discipline of assigned personnel.

2. An order given by a commander; that is, the will of the commander expressed for the purpose of bringing about a particular action.

3. A unit or units, an organization, or an area under the command of one individual (JP 1).

**Command and Control**

(Joint) The exercise of authority and direction by a properly designated commander over assigned and attached forces in the accomplishment of a mission. Command and control functions are performed through an arrangement of personnel, equipment, communications, facilities, and procedures employed by a commander in planning, directing, coordinating, and controlling forces and operations in the accomplishment of the mission (JP 1). (Army) The exercise of authority and direction by a properly designated commander over assigned and attached forces in the accomplishment of a mission. Commanders perform command and control functions through a command and control system (FM 6-0).

**Commander's Critical Information Requirement**

(Joint) An information requirement identified by the commander as being critical to facilitating timely decisionmaking. The two key elements are friendly force information requirements and priority intelligence requirements (JP 3-0).

**Commander's Intent**

(Joint) A concise expression of the purpose of the operation and the desired end state. It may also include the commander's assessment of the adversary commander's intent and an assessment of where and how much risk is acceptable during the operation (JP 3-0). (Army)

A clear, concise statement of what the force must do and the conditions the force must establish with respect to the enemy, terrain, and civil considerations that represent the desired end state (FM 3-0).

**Commander's Visualization**

The mental process of developing situational understanding, determining a desired end state, and envisioning the broad sequence of events by which the force will achieve that end state (FM 3-0).

**Comprehensive Approach**

An approach that integrates the cooperative efforts of the departments and agencies of the United States Government, intergovernmental and nongovernmental organizations, multinational partners, and private sector entities to achieve unity of effort toward a shared goal (FM 3-07)

**Concept of Operations**

(Joint) A verbal or graphic statement that clearly and concisely expresses what the joint force commander intends to accomplish and how it will be done using available resources. The concept is designed to give an overall picture of the operation. (JP 5-0, Army) A statement that directs the manner in which subordinate units cooperate to accomplish the mission and establishes the sequence of actions the force will use to achieve the end state. It is normally expressed in terms of decisive, shaping, and sustaining operations (FM 3-0).

**Condition**

(DOD) Those variables of an operational environment or situation in which a unit, system, or individual is expected to operate and may affect performance.

**Control**

(Joint) 1. Authority that may be less than full command exercised by a commander over part of the activities of subordinate or other organizations.

2. In mapping, charting, and photogrammetry, a collective term for a system of marks or objects on the Earth or on a map or a photograph, whose positions or elevations (or both) have been or will be determined.

3. Physical or psychological pressures exerted with the intent to assure that an agent or group will respond as directed. 4. An indicator governing the distribution and use of documents, information, or material. Such indicators are the subject of intelligence community agreement and are specifically defined in appropriate regulations (JP 1-02).

(Army) 1. In the context of command and control, the regulation of forces and warfighting functions to accomplish the mission in accordance with the commander's intent (FM 3-0).

2. A tactical mission task that requires the commander to maintain physical influence over a specified area to prevent its use by an enemy (FM 3-90).

3. An action taken to eliminate a hazard or reduce its risk (FM 5-19).

4. In the context of stability mechanisms, to impose civil order (FM 3-0).

**Conventional Forces**

Those forces capable of conducting operations using nonnuclear weapons (JP 1-02).

**Counterinsurgency**

(Joint) Those military, paramilitary, political, economic, psychological, and civic actions taken by a government to defeat insurgency (JP 1-02).

**Counterterrorism**

(Joint) operations that include the offensive measures taken to prevent, deter, preempt, and respond to terrorism (JP 1-02).

**Course of Action**

1. Any sequence of activities that an individual or a unit may follow.
2. A possible plan open to an individual or a commander that would accomplish or is related to the accomplishment of a mission.
3. The scheme adopted to accomplish a job or mission.
4. A line of conduct in an engagement.
5. A product of the Joint Operation Planning and Execution System concept development phase (JP 1-02).

**Deception**

Those measures designed to mislead the enemy by manipulation, distortion, or falsification of evidence to induce him to react in a manner prejudicial to his interests (JP 1-02).

**Decision Point**

(Army) An event, area, or point in the battle where and when the friendly commander will make a critical decision (FM 3-0).

**Decisive Operation**

The operation that directly accomplishes the mission. It determines the outcome of a major operation, battle, or engagement. The decisive operation is the focal point around which commanders design the entire operation (FM 3-0).

**Decisive Point**

(Joint) A geographic place, specific key event, critical factor, or function that, when acted upon, allows commanders to gain a marked advantage over an adversary or contribute materially to achieving success. (JP 3-0) [Note: In this context, adversary also refers to enemies.]

**Defeat**

A tactical mission task that occurs when an enemy force has temporarily or permanently lost the physical means or the will to fight. The defeated force's commander is unwilling or unable to pursue his adopted course of action, thereby yielding to the friendly commander's will, and can no longer interfere to a significant degree with the actions of friendly forces. Defeat can result from the use of force or the threat of its use (FM 3-90).

**Displaced Person**

(Army) A civilian who is involuntarily outside the national boundaries of his or her country or as an internally displaced person is a civilian involuntarily outside his area or region within his country (FM 2-0).

**Disrupt**

A tactical mission task in which a commander integrates direct and indirect fires, terrain, and obstacles to upset an enemy's formation or tempo, interrupt his timetable, or cause his forces to commit prematurely or attack in piecemeal fashion (FM 3-90).

**Doctrine**

Fundamental principles by which the military forces or elements thereof guide their actions in support of national objectives. It is authoritative but requires judgment in application (JP 1-02).

**Defensive Operations**

Combat operations conducted to defeat an enemy attack, gain time, economize forces, and develop conditions favorable for offensive or stability operations (FM 3-0).

**Demobilization**

(Joint) The process of transitioning a conflict or wartime military establishment and defense-based civilian economy to a peacetime configuration while maintaining national security and economic vitality (JP 4-05).

**Destroy**

In the context of defeat mechaisms, to apply lethal combat power on an enemy capability so that it can no longer perform any function and cannot be restored to a usable condition without being entirely rebuilt (FM 3-0).

2. A tactical mission task that physically renders an enemy force combat-ineffective until it is reconstituted (FM 3-90).

**Effect**

(Army) A result, outcome, or consequence of an action (FMI 5-0.1).

**Engagement**

A tactical conflict, usually between opposing lower echelon maneuver forces (JP 1-02).

**End State**

(Joint) The set of required conditions that defines achievement of the commander's objectives (JP 3-0).

**Enemy**

A party identified as hostile against which the use of force is authorized (FM 3-0).

**Enemy Combatant**

A person engaged in hostilities against the United States or its coalition partners during an armed conflict. This term includes both "enemy combatants" and "unlawful enemy combatants" (DODD 2310.01E. September 5, 2006).

**Execution**

Putting a plan into action by applying combat power to accomplish the mission and using situational understanding to assess progress and make execution and adjustment decisions (FM 3-0).

**Exfiltration**

The removal of personnel or units from areas under enemy control by stealth, deception, surprise, or clandestine means (JP 1-02).

**Exploitation**

1. Taking full advantage of success in military operations, following up initial gains, and making permanent the temporary effects already achieved.

2. An offensive operation that usually follows a successful attack and is designed to disorganize the enemy in depth (JP 1-02).

**Explosive Ordnance Disposal**

The detection, identification, on-site evaluation, rendering safe, recovery, and final disposal of unexploded ordnance. It may also include explosive ordnance which has become hazardous by damage or deterioration (JP 1-02).

**Fires**

The effects of lethal or nonlethal weapons (JP 1-02).

**Fix**

(Army) 1. A tactical mission task where a commander prevents the enemy from moving any part of its force from a specific location for a specific period of time. 2. An engineer obstacle effect that focuses fire planning and obstacle effort to slow an attacker's movement within a specified area, normally an engagement area (FM 3-90).

**Foreign Internal Defense**

(Joint) Participation by civilian and military agencies of a government in any of the action programs taken by another government or other designated organization to free and protect its society from subversion, lawlessness, and insurgency (JP 1-02).

**Forward Operating Base**

An area used to support tactical operations without establishing full support facilities (FM 0).

**Full-Spectrum Operations**

The Army's operational concept: Army forces combine offensive, defensive, and stability or civil support operations simultaneously as part of an interdependent joint force to seize, retain, and exploit the initiative, accepting prudent risk to create opportunities to achieve decisive results. They employ synchronized action—lethal and nonlethal—proportional to the mission and informed by a thorough understanding of all variables of the operational environment. Mission command that conveys intent and an appreciation of all aspects of the situation guides the adaptive use of Army forces (FM 3-0).

**Goals (Governmental Planning)**

Goals are more specific statements than objectives, they represent the actions or things to be accomplished in order to achieve the established objectives. A goal is an observable and measurable end result.

**Guerrilla**

(DOD definition): A combat participant in guerrilla warfare. Dictionary definition: A member of an irregular, usually indigenous military or paramilitary unit that operates in small units and uses guerrilla warfare. Source: The Spanish diminutive form of *guerra* (war) that means "small" or "little war." The word developed in reference to the tactics that the Spanish resistance used against Napoleon's forces in Spain.

**Guerrilla Force**

(DOD) A group of irregular, predominantly indigenous personnel organized along military lines to conduct military and paramilitary operations in enemy-held, hostile, or denied territory.

**Guerrilla Warfare**

(GW, DOD, NATO) Military and paramilitary operations conducted in enemy held or hostile territory by irregular, predominantly indigenous forces.

**Host Nation**

(Joint) A nation which permits, either by written agreement or official invitation, government representatives and/or agencies of another nation to operate, under specified conditions, within its borders. (JP 1-02).

**Host Nation Support**

Civil or military assistance rendered by a nation to foreign forces within its territory during peacetime, crises or emergencies, or war based on agreements mutually concluded between nations (JP 1-02).

**Human Intelligence**

(Army) The collection of information by a trained human intelligence collector from people and their associated documents and media sources to identify elements, intentions, composition, strength, dispositions, tactics, equipment, personnel, and capabilities (FM 2-22.3). [*Note*: trained HUMINT collectors are Soldiers holding military occupational specialties 97E, 351Y {formerly 351C}, 351M {formerly 351E}, 35E, and 35F, and Marines holding the specialty 0251.]

**Imagery Intelligence**

Intelligence derived from the exploitation of collection by visual photography, infrared sensors, lasers, electro-optics, and radar sensors such as synthetic aperture radar wherein images of objects are reproduced optically or electronically on film,electronic display devices, or other media (JP 1-02).

**Indigenous**

Existing, born, or produced in a lor region. Synonym: Native.

**Infiltration**

(Army) A form of maneuver in which an attacking force conducts undetected movement through or into an area occupied by enemy forces to occupy a position of advantage in the enemy rear while exposing only small elements to enemy defensive fires (FM 3-90).

**Information Engagement**

The integrated employment of public affairs to inform US and friendly audiences; psychological operations, combat camera, US Government strategic communication and defense support to public diplomacy, and other means necessary to influence foreign audiences; and, leader and Soldier engagements to support both efforts (FM 3-0).

**Information Warfare**

Information Operations conducted during time of crisis or conflict to achieve or promote specific objectives over a specific adversary or adversaries (JP 1-02).

**Insurgency**

(Joint) An organized movement aimed at the overthrow of a constituted government through the use of subversion and armed conflict (JP 1-02).

**Insurgent**

(DOD) Member of a political party who rebels against established leadership.

**Intelligence**

1. The product resulting from the collection, processing, integration, analysis, evaluation, and interpretation of available information concerning foreign countries or areas.

2. Information and knowledge about an adversary obtained through observation, investigation, analysis, or understanding (JP 1-02).

**Intelligence Preparation of the Battlefield**

The systematic, continuous process of analyzing the threat and environment in a specific geographic area. Intelligence preparation of the battlefield (IPB) is designed to support the running estimate and military decision-making processes. Most intelligence requirements are generated as a result of the IPB process and its interrelation with the decision-making process (FM 34-130).

**Intelligence, Surveillance, and Reconnaissance**

(Army) An activity that synchronizes and integrates the planning and operation of sensors, assets, and processing, exploitation, and dissemination systems in direct support of current and future operations. This is an integrated intelligence and operations function. For Army forces, this activity is a combined arms operation that focuses on priority intelligence requirements while answering the commander's critical information requirements (FM 3-0).

**Interagency**

(Joint) United States Government agencies and departments, including the Department of Defense (JP 3-08).

**Interagency Coordination**

(Joint) Within the context of Department of Defense involvement, the coordination that occurs between elements of Department of Defense and engaged US Government agencies for the purpose of achieving an objective (JP 1-02).

**Interdict**

A tactical mission task where the commander prevents, disrupts, or delays the enemy's use of an area or route (FM 3-90).

**Intergovernmental Organization**

(Joint) An organization created by a formal agreement, such as a treaty, between two or more governments. It may be established on a global, regional, or functional basis for wide-ranging or narrowly defined purposes. Formed to protect and promote national interests shared by member states. Examples include the United Nations, North Atlantic Treaty Organization, and the African Union (JP 3-08).

**Internal Defense and Development**

(IDAD, DOD) The full range of measures taken by a nation to promote its growth and to protect itself from subversion, lawlessness, and insurgency. It focuses on building viable institutions (political, economic, social, and military) that respond to the needs of society.

**Irregular Forces**

(DOD) Armed individuals or groups who are not members of the regular armed forces, police, or other internal security forces (FM 3-07).

**Irregular Warfare**

A broad form of conflicts in which insurgency, counterinsurgency, and unconventional warfare are the principle activities (FM 3-0).

**Isolate**

In the context of defeat mechanisms, to deny an enemy or adversary access to capabilities that enable the exercise of coercion, influence, potential advantage, and freedom of action (FM 3-0).

**Joint**

Connotes activities, operations, organizations, and so on, in which elements of two or more Military Departments participate (JP 1-02).

**Joint Force**

A general term applied to a force composed of significant elements, assigned or attached, of two or more Military Departments, operating under a single joint force commander (JP 1-02).

**Joint Operations**

A general term to describe military actions conducted by joint forces, or by Service forces in relationships, such as support or coordinating authority which, of themselves,do not create joint forces (JP 1-02).

**Key Terrain**

Any locality, or area, the seizure or retention of which affords a marked advantage to either combatant (JP 1-02).

**Leadership**

The process of influencing people by providing purpose, direction, and motivation, while operating to accomplish the mission and improving the organization (FM 6-22).

**Line of Effort**

A line that links multiple tasks and missions using the logic of purpose—cause and effect—to focus efforts toward establishing operational and strategic conditions (FM 3-0).

**Line of Communications**

(Joint) A route, either land, water, or air, that connects an operating military force with a base of operations and along which supplies and military forces move (JP 1-02).

**Major Operation**

A series of tactical actions (battles, engagements, strikes) conducted by various combat forces of a single or several Services, coordinated in time and place, to accomplish operational and, sometimes, strategic objectives in an operational area. These actions are conducted simultaneously or sequentially in accordance with a common plan and are controlled by a single commander (JP 1-02).

**Maneuver**

(Joint) Employment of forces in the operational area through movement in combination with fires to achieve a position of advantage in respect to the enemy in order to accomplish the mission (JP 3-0).

**Measure of Effectiveness**

(Joint) A criterion used to assess changes in system behavior, capability, or operational environment that is tied to measuring the attainment of an end state, achievement of an objective, or creation of an effect (JP 3-0).

**Measure of Performance**

(Joint) A criterion used to assess friendly actions that is tied to measuring task accomplishment (JP 3-0).

**METT-TC**

A memory aid used in two contexts:

1. Information Management— the major subject categories into which relevant information is grouped for military operations: mission, enemy, terrain and weather, troops and support available, time available, civil considerations (FM 6-0). 2. In the context of tactics, major variables considered during mission analysis (mission variables, FM 3-90).

**Milestone**

A significant event in a project.

**Military Decision-Making Process**

A process that integrates the activities of the commander, staff and subordinate commanders in developing and operation plan or order. It establishes procedures for analyzing a mission; developing, analyzing, and comparing courses of action; selecting the best course of action; and producing an operation plan or order (FMI 5-0.1).

**Mission**

(Joint) 1. The task, together with the purpose, that clearly indicates the action to be taken and the reason therefore. 2. In common usage, especially when applied to lower military units, a duty assigned to an individual or unit; a task. 3. The dispatching of one or more aircraft to accomplish one particular task (JP 1-02).

**Mission Command**

The conduct of military operations through decentralized execution based upon mission orders for effective mission accomplishment. Successful mission command results from subordinate leaders at all echelons exercising disciplined initiative within the commander's intent to accomplish missions. It requires an environment of trust and mutual understanding (FM 6-0).

**Movement to Contact**

A form of the offensive designed to develop the situation and to establish or regain contact (JP 1-02).

**Multinational Operations**

(Joint) A collective term to describe military actions conducted by forces of two or more nations, usually undertaken within the structure of a coalition or alliance (JP 3-16).

**Neutral**

(Army) A party identified as neither supporting nor opposing friendly or enemy forces (FM 3-0).

**Nongovernmental Organization**

(Joint) A private, self-governing, not-for-profit organization dedicated to alleviating human suffering; or promoting education, health care, economic development, environmental protection, human rights, and conflict resolution; or encouraging the establishment of democratic institutions and civil society (JP 3-08).

**Nonlethal Fires**

Any fires that do not directly seek the physical destruction of the intended target and are designed to impair, disrupt, or delay the performance of enemy operational forces, functions, and facilities. Psychological operations, electronic warfare (jamming), and other command and control countermeasures are all nonlethal fire options (FM 6-20).

**Objective**

(Army) A location on the ground used to orient operations, phase operations, facilitate changes of direction, and provide for unity of effort (FM 3-90).

**Objective Area**

(DOD, NATO) A defined geographical area within which is located an objective to be captured or reached by the military forces. This area is defined by competent authority for purposes of command and control.

**Offensive Operations**

Combat operations conducted to defeat and destroy enemy forces and seize terrain, resources, and population centers. They impose the commander's will on the enemy (FM 3-0).

**Operation**

1. A military action or the carrying out of a strategic, operational, tactical, service, training, or administrative military mission. 2. The process of carrying on combat, including movement, supply, attack, defense, and maneuvers needed to gain the objectives of any battle or campaign (JP 1-02).

**Operation Order**

(OPORD, DOD) A directive issued by a commander to subordinate commanders for the purpose of effecting the coordinated execution of an operation. Also called the five paragraph field order.

**Operation Plan**

(DOD) Any plan for the conduct of military operations. Operation plans are prepared in either a complete format (OPLAN) or as a concept plan (CONPLAN).

**Operating Tempo**

The annual operating miles or hours for the major equipment system in a battalion-level or equivalent organization. Commanders use operating tempo to forecast and allocate funds for fuel and repair parts for training events and programs (FM 7-0).

**Operations Process**

The major command and control activities performed during operations: planning, preparing, executing, and continuously assessing the operation. The commander drives the operations process (FM 3-0).

**Operations Security**

A process of identifying critical information and subsequently analyzing friendly actions attendant to military operations and other activities to— a. identify those actions that can be observed by adversary intelligence systems; b. determine indicators hostile intelligence systems might obtain that could be interpreted or pieced together to derive critical information in time to be useful to adversaries; and c. select and execute measures that eliminate or reduce to an acceptable level the vulnerabilities of friendly actions to adversary exploitation (JP 1-02).

**Order**

A communication that is written, oral, or by signal, which conveys instructions from a superior to a subordinate. In a broad sense, the terms "order" and "command" are synonymous. However, an order implies discretion as to the details of execution, whereas a command does not (JP 1-02). Combat orders pertain to operations and their service support.

**Paramilitary Forces**

Forces or groups distinct from the regular armed forces of any country, but resembling them in organization, equipment, training, or mission (JP 1-02).

**Partner Unit**

A unit that shares all or a portion of an area of operation with an HN security force unit. US forces operating as partner units to HN security forces need to be prepared to make some organizational changes.

**Patrol**

A detachment of ground, sea, or air forces sent out for the purpose of gathering information or carrying out a destructive, harassing, mopping-up, or security mission (JP 1-02).

**Phase**

A planning and execution tool used to divide an operation in duration or activity. A change in phase usually involves a change of mission, task organization, or rules of engagement. Phasing helps in planning and controlling and may be indicated by time, distance, terrain, or an event (FM 3-0).

**Plan**

A design for a future or anticipated operation (FM 5-0).

**Planning**

The process by which commanders (and the staff, if available) translate the commander's visualization into a specific course of action for preparation and execution, focusing on the expected results (FM 3-0).

**PMESII-PT**

A memory aid for the varibles used to describe the operational environment (operational variables, FM 3-0):

**P**olitical
**M**ilitary
**E**conomic
**S**ocial
**I**nformation
**I**nfrastructure
**P**hysical (environment)
**T**ime

**Populace Controls**

Controls that provide security for the populace, mobilize human resources, deny personnel to the guerrilla, and detect and reduce the effectiveness of guerrilla agents.

**Preparation**

Activities performed by units to improve their ability to execute an operation. Preparation includes, but is not limited to, plan refinement; rehearsals; intelligence, surveillance, and reconnaissance; coordination; inspections; and movement (FM 3-0).

**Procedures**

Standard and detailed courses of action that describe how to perform a task.

**Program (Governmental Planning)**

A program supports the plan and is a more detailed determination of specific objectives established in the plan.

**Projects (Governmental Planning)**

Projects comprise the components of programs and are the specific actions and tasks to be accomplished.

**Propaganda**

Any form of communication in support of national objectives designed to influence the opinions, emotions, attitudes, or behavior of any group in order to benefit the sponsor, either directly or indirectly (JP 1-02).

**Psychological Operations**

(PSYOP) Planned operations to convey selected information and indicators to foreign audiences to influence their emotions, motives, objective reasoning, and ultimately the behavior of foreign government, organizations, groups, and individuals. The purpose of psychological operations is to induce or reinforce foreign attitudes and behavior favorable to the originator's objectives (JP 1-02).

**Public Affairs**

(PA) Those public information, command information, and community relations activities directed toward both the external and internal publics with interest in the Department of Defense (JP 1-02).

**Pursuit**

An offensive operation designed to catch or cut off a hostile force attempting to escape, with the aim of destroying it (JP 1-02).

**Quick Reaction Force**

(QRF) A designated organization for any immediate response requirement that occurs in a designated area of operation (FM 3-90.6).

**Raid**

(Joint) An operation to temporarily seize an area in order to secure information, confuse an adversary, capture personnel or equipment, or to destroy a capability. It ends with a planned withdrawal upon completion of the assigned mission (JP 3-0) [Note: In this context, *adversary* also refers to *enemies*].

**Rebellion**

Open, armed, and usually unsuccessful defiance of or resistance to an established government.

**Regular Forces**

Members of a nation's armed forces, police, or other internal security forces.

**Reintegration**

The process through which former combatants, belligerents, and dislocated civilians receive amnesty, reenter civil society, gain sustainable employment, and become contributing members of the local populace (FM 3-07).

**Relief in Place**

(Army) An tactical enabling operation in which, by direction of higher authority, all or part of a unit is replaced in an area by the incoming unit (FM 3-90).

**Reserve**

Portion of a body of troops which is kept to the rear or withheld from action at the beginning of an engagement, in order to be available for a decisive movement (JP 1-02).

**Resistance Movement**

An organized effort by some portion of the civil population of a country to resist the legally established government or an occupying power and to disrupt civil order and stability (JP 1-02).

**Resource Controls**

Controls that regulate the movement or consumption of materiel resources, mobilize materiel resources, and deny materiel to the guerrilla.

**Revolution**

The overthrow or renunciation of one government or ruler and the substitution of another by the governed.

**Risk**

(DOD) Probability and severity of loss linked to hazards (JP 1-02).

**Risk Management**

The process of identifying, assessing, and controlling risk arising from operational factors, and making decisions that balance risk cost with mission benefits (JP 1-02).

**Rule of Law**

A principle under which all persons, institutions, and entities, public and private, including the state itself, are accountable to laws that are publicly promulgated, equally enforced, and independently adjudicated, and that are consistent with international human rights principles (FM 3-07).

**Rules of Engagement**

(Joint) Directives issued by competent military authority that delineate the circumstances and limitations under which United States forces will initiate or continue combat engagement with other forces encountered (JP 1-02).

**Running Estimate**

A staff section's continuous assessment of current and future operations to determine if the current operation is proceeding according to the commander's intent and if future operations are supportable (FMI 5-0.1).

**Search and Attack**

A technique of conducting a movement to contact that shares many of the characteristics of an area security mission (FM 3-0).

**Sabotage**

(DOD) An act or acts with intent to injure, interfere with, or obstruct the national defense of a country by willfully injuring or destroying, or attempting to injure or destroy, any national defense or war materiel, premises, or utilities, to include human and natural resources.

**Security**

(Joint) 1. Measures taken by a military unit, an activity or installation to protect itself against all acts designed to, or which may, impair its effectiveness. 2. A condition that results from the establishment and maintenance of protective measures that ensure a state of inviolability from hostile acts or influences (JP 1-02).

**Security Force Assistance**

The unified action to generate, employ, and sustain local, Host Nation, or regional security forces in support of a legitimate authority (FM 3-07).

**Security Operations**

Those operations undertaken by a commander to provide early and accurate warning of enemy operations, to provide the force being protected with time and maneuver space within which to react to the enemy, and to develop the situation to allow the commander to effectively use the protected force.

**Security Sector Reform**

The set of policies, plans, programs, and activities that a government undertakes to improve the way it provides safety, security, and justice (FM 3-07).

**Shaping Operation**

An operation at any echelon that creates and preserves conditions for the success of the decisive operation (FM 3-0).

**Situational Awareness**

Knowledge of the immediate present environment, including knowledge of the factors of METT-TC (FMI 5-0.1).

**Situational Understanding**

(Army) The product of applying analysis and judgment to the common operational picture to determine the relationship among the factors of METT-TC (FM 3-0).

**Special Operations**

Operations conducted in hostile, denied, or politically sensitive environments to achieve military, political, economic, or informational objectives employing military capabilities for which there is no broad conventional force requirement. These operations often require covert, clandestine, or low visibility capabilities. Special operations are applicable across the rnage of military operations. They can be conducted independently or in conjunction with operations of conventional forces or other government agencies and may include operations through, with, or by indigenous or surrogate forces. Special operations differ from conventional operations in degree of physical and political risk, operational techniques, mode of employment, independence from friendly support, and dependence on detailed operational intelligence and indigenous assets (JP 1-02).

**Stability Operations**

(Joint) An overarching term encompassing various military missions, tasks, and activities conducted outside the United States in coordination with other instruments of national power to maintain or reestablish a safe and secure environment, provide essential governmental services, emergency infrastructure reconstruction, and humanitarian relief (JP 3-0).

**Strike**

(Joint) An attack to damage or destroy an objective or capability (JP 1-02).

**Support**

(Joint) The action of a force that aids, protects, complements, or sustains another force in accordance with a directive requiring such action.

2. A unit that helps another unit in battle. 3. An element of a command that assists, protects, or supplies other forces in combat (JP 1, Army) In the context of stability mechanisms, to establish, reinforce, or set the conditions necessary for the other instruments of national power to function effectively (FM 3-0).

**Supporting Range**

The distance one unit may be geographically separated from a second unit yet remain within the maximum range of the second unit's weapons systems (FM 3-0).

**Subversion**

(DOD) Action designed to undermine the military, economic, psychological, political strength, or morale of a regime.

**Synchronization**

1. The arrangement of military actions in time, space, and purpose to produce maximum relative combat power at a decisive place and time (FM 3-0).

2. In the intelligence context, application of intelligence sources and methods in concert with the operation plan (JP 1-02).

**Tactical Questioning**

(DOD) Direct questioning by any DOD personnel of a captured or detained person to obtain time-sensitive tactical intelligence, at or near the point of capture or detention and consistent with applicable law.

**Tactical Mission Task**

The specific activity a unit performs while conducting a form of tactical operation or form of maneuver. It may be expressed in terms of either actions by a friendly force or effects on an enemy force (FM 7-15).

**Tactics**

(Joint) The employment ordered arrangement of forces in relation to each other. (CJCSI 5120.02A)

**Target**

(DOD) 1. An entity or object considered for possible engagement or other action.

2. In intelligence usage, a country, area, installation, agency, or person against which intelligence operations are directed. 3. An area designated and numbered for future firing.

**Task-Organizing**

(Army) The act of designing an operating force, support staff, or logistic package of specific size and composition to meet a unique task or mission. Characteristics to examine when task-organizing the force include, but are not limited to— training, experience, equipage, sustainability, operating environment, enemy threat, and mobility. For Army forces, it includes allocating available assets to subordinate commanders and establishing their command and support relationships (FM 3-0).

**Techniques**

(Army/Marine Corps) The general and detailed methods used by troops or commanders to perform assigned missions and functions, specifically, the methods of using equipment and personnel (FM 3-90).

**Tempo**

(Army) The rate of military action (FM 3-0).

(Marine Corps) The relative speed and rhythm of military operations over time with respect to the enemy (MCRP 5-12A).

**Terrorism**

(Joint) The calculated use of unlawful violence or threat of unlawful violence to inculcate fear; intended to coerce or to intimidate governments or societies in the pursuit of goals that are generally political, religious, or ideological (JP 3-07.2).

**Terrorist**

(DOD) An individual who uses violence, terror, and intimidation to achieve a result.

**Troop-Leading Procedures**

A dynamic process used by small unit leaders to analyze amission, develop a plan, and prepare for an operation (FM 5-0).

**Unconventional Warfare**

A broad spectrum of military and paramilitary operations, normally of long duration, predominantly conducted through, with, or by indigenous or surrogate forces who are organized, trained, equipped, supported, and directed in varying degrees by an external source. It includes, but is not limited to, guerrilla warfare, subversion, sabotage, intelligence activities, and unconventional assisted recovery (JP 1-02).

**Underground**

A covert unconventional warfare organization established to operate in areas denied to the guerrilla forces or conduct operations not suitable for guerrilla forces FM 1-02.

**Unexploded Ordnance**

Explosive ordnance which has been primed, fuzed, armed, or otherwise prepared for action, and which has then been fired, dropped, launched, projected, or placed in such a manner as to constitute a hazard to operations, installations, personnel, or material, and remains unexploded either by malfunction or design, or for any other cause (JP 1-02).

**Unified Action**

(Joint) The synchronization, coordination, or integration of the activities of governmental and nongovernmental entities with military operations to achieve unity of effort (JP 1).

**Unity of Effort**

(Joint) The coordination and cooperation toward common objectives, even if the participants are not necessarily part of the same common organization—the product of successful unified action (JP 1).

**Vulnerable State**

A nation either unable or unwilling to provide adequate security and essential services to significant portions of the population (FM 3-07)

**Warfighting Function**

A group of tasks and systems (people, organizations, information, and processes) united by a common purpose that commanders use to accomplish missions and training objectives (FM 3-0).

**Weapons of Mass Destruction**

Weapons that are capable of a high order of destruction or of being used in such a manner as to destroy large numbers of people. Weapons of mass destruction can be high explosives or nuclear, biological, chemical, or radiological weapons, but exclude the means of transporting or propelling the weapon where such means is a separable and divisible part of the weapon (JP 1-02).

## REFERENCES

Army Regulation

AR 350-30, *Code of Conduct/Survival, Evasion, Resistance and Escape (SERE) Training*. 10 December 1985.

Field Manuals

FM 2-0, *Intelligence*. 17 May 2004.

### Sources Used

These are the sources quoted or paraphrased in this publication.

Call Handbooks

For Center for Army Lessions Learned (CALL) Handbooks go to http://usacac.army.mil/cac2/call/archives.asp

07-06, *Southern Afghanistan Counterinsurgency Operations*. November 2006.

07-08, *CF/SOF Integration and Interoperability*. January 2007.

Field Manuals

FM 1-02, *Operational Terms and Graphics*. 21 September 2004.

FM 3-0, *Operations*. 27 February 2008.

FM 3-05.40, *Civil Affairs Operations*. 29 September 2006.

FM 3-05.120 (S/NF), *Army Special Operations Forces Intelligence*. 15 July 2007 (U).
FM 3-05.202, *Special Forces Foreign Internal Defense Operations*. 2 February 2007.
FM 3-05.301, *Psychological Operations Process Tactics, Techniques, and Procedures*. 30 August 2007.
FM 3-05.302, *Tactical Psychological Operations Tactics, Techniques, and Procedures*. 28 October 2005.
FM 3-06.11, *Combined Arms Operations in Urban Terrain*. 28 February 2002.
FM 3-06.20, *Multi-Service Tactics, Techniques, and Procedures for Cordon and Search Operations*. 25 April 2006.
FM 3-07, *Stability Operations*. 6 October 2008.
FM 3-21.10, *The Infantry Rifle Company*. 27 July 2006.
FM 3-21.11, *The SBCT Infantry Rifle Company*. 23 January 2003.
FM 3-21.20, *The Infantry Battalion*. 13 December 2006.
FM 3-21.21, *The Stryker Brigade Combat Team Infantry Battalion*. 8 April 2003.
FM 3-21.75, *The Warrior Ethos and Soldier Combat Skills*. 28 January 2008.
FM 3-24, *Counterinsurgency*. 15 December 2006.
FM 3-90, *Tactics*. 4 July 2001.
FM 3-90.5, *The Combined Arms Battalion*. 7 April 2008.
FM 3-90.6, *The Brigade Combat Team*. 4 August 2006.
FM 3-90.15, *Sensitive Site Operations*. 25 April 2007.
FM 3-90.119, *Combined Arms Improvised Explosive Device Defeat Operations*. 21 September 2007.
FM 5-0, *Army Planning and Orders Production*. 20 January 2005.
FM 5-19, *Composite Risk Management*. 21 August 2006.
FM 5-103, *Survivability*. 10 June 1985.
FM 6-0, *Mission Command: Command and Control of Army Forces*. 11 August 2003.
FM 7-98, *Operations in a Low-Intensity Conflict*. 19 October 1992.
FM 90-8, *Counterguerrilla Operations*. 29 August 1986.
FM 3-34.210, *Explosive Hazards Operations*. 27 March 2007.
FM 34-130, *Intelligence Preparation of the Battlefield*. 8 July 1994.

Joint Publications

[44] JP 1, *Doctrine for the Armed Forces of the United States*. 14 May 2007.
[45] JP 3-07.1, *Joint Tactics, Techniques, and Procedures for Foreign Internal Defense (FID)*. 30 April 2004.
[46] JP 3-07.2, *Antiterrorism*. 14 April 2006.
[47] JP 3-07.3, *Peace Operations*. 17 October 2007.
[48] JP 3-13, *Information Operations*. 13 February 2006.
[49] JP 3-53, *Joint Doctrine for Psychological Operations*. 5 September 2003.

Training Circulars
TC 7-98-1, *Stability and Support Operations Training Support Package*. 5 June 1997.

## Other Publications

[1] NAVMC 2890 (FMFRP 12-15), *Small Wars Manual (Reprint of1940 USMC)*. 1 April 1987. http://www.au.af/mil/au/awc/awcgate/swm/full.pdf.

[2] *Counterinsurgency Operations Southern & Western Afghanistan*. Task Force 31, January 2007.

[3] *Counterinsurgency Handbook. Multi-National Force—Iraq*. 1st ed. Camp Taji, Iraq: Counterinsurgency Center for Excellence, May, 2006.

[4] *Commander's Handbook for Security Force Assistance*. Joint Center for International Security Force Assistance, Fort Leavenworth, Kansas. 14 July 2008.

[5] *Developing Iraq's Security Sector: The Coalition Provisional Authority's Experience*. RAND National Defense Institute. 2005.

## Internet Web Sites

Some of the documents listed in these References may be downloaded from Army websites:

Air Force Doctrine https://www.doctrine.af.mil/.
Air Force Pubs http://afpubs.hq.af.mil/.
Army Formshttp://www.apd.army.mil/usapa _PUB _formrange _f.asp.
Army Knowledge Online https://akocomm.us.army.mil/usapa/doctrine/index.html.
NATO ISAshttp://www.nato.int/docu/standard.htm.
Reimer Digital Library http://www.train.army.mil.

## Civilian Publications

These books can be located in libraries or purchased at bookstores or online. The Combined Arms Research Library at___http://www-cgsc.army.mil/carl/resources biblio/mildep.asp has some of these publications.

[1] Barber, Noel. (1971). *The War of the Running Dogs: Malaya, 1948-1960*. New York: Weybright and Tulley.

[2] Boot, Max. (2002). *The Savage Wars of Peace: Small Wars and the Rise of American Power*. New York: Basic Books.

[3] Callwell, Charles E. *Small Wars: Their Principles and Practice*. Lincoln, NE: University of Nebraska.

[4] Ellis, John. (1995). *From the Barrel of a Gun: A History of Guerrilla, Revolutionary, and Counterinsurgency Warfare from the Romans to the Present*. London: Greenhill.

[5] Galula, David. (1964). *Counterinsurgency Warfare—Theory and Practice*. London: Praeger.

[6] Hammes, T. X. (1977). *The Sling and the Stone: On War in the 21st Century*. Osceola, WI: Zenith Press, 2004. Horne, Alistair. *A Savage War of Peace*. New York: Viking.

[7] Krepinevich, Andrew Jr. (1962). *The Army and Vietnam*. Baltimore: Johns Hopkins University Press, 1986. Larteguy, Jean. *The Centurions*. New York: Dutton.

[8] Linn, Brian McAllister. (2002). *The Philippine War, 1899--1902*. Lawrence, KS: University Press of Kansas.
[9] Mao Zedong. (1965). *On Guerrilla Warfare*. London: Cassell.
[10] McCuen, John J. (2005). *The Art of Counter-Revolutionary War*. St. Petersburg, FL: Hailer Publishing.
[11] Nagl, John A. (2005). *Learning to Eat Soup with a Knife: Counterinsurgency Lessons from Malaya and Vietnam*. Chicago: University of Chicago Press.
[12] O'Neill & Bard E. (1996). *Insurgency and Terrorism: From Revolution to Apocalypse*. Dulles, VA: Potomac Press
[13] Taber, Robert. (2002). *War of the Flea: The Classic Study of Guerrilla Warfare*. Dulles, VA: Potomac Books.
[14] Trinquier, Roger. (1964). *Modern Warfare—A French View of Counterinsurgency*. New York: Praeger.
[15] United States Marine Corps. (1972). *Small Wars Manual*. Washington, D.C.: Government Printing Office, 1987. West, Bing. *The Village*. New York: Pocket Books.

## Military Publications

[1] Military Review articles are located at http://usacac.army.mil/CAC2/MilitaryReview /repository/MREditions2006-English.xml
[2] Baker, Ralph O. (May-Jun 2006). *The Decisive Weapon: A Brigade Combat Team Commander's Perspective on the Conduct of Information Operations. Military Review 86, 3,* 1 3—32.
[3] Chiarelli, Peter W. & Patrick R. Michaelis. (Jul-Aug 2005). *Winning the Peace: The Requirement for Full-Spectrum Operations*, Military Review, *85*, 4, 4—17.
[4] Kilcullen, David. (May-Jun 2006). *'Twenty-Eight Articles': Fundamentals of Company-level Counterinsurgency. Military Review, 86*, 3, 103—108.
[5] Petraeus, David. (Jan-Feb 2006). *Learning Counterinsurgency: Observations from Soldiering in Iraq*. Military Review, *86*, 1, 2—12.
[6] Lawrence, T. E. (20 Aug 1917). *The 27 Articles of T. E. Lawrence*. The Arab Bulletin. http://www.gwpda.org/1917/27arts.html
[7] Sepp, Kalev I. (May-Jun 2005). *Best Practices in Counterinsurgency. Military Review 85*, 3, 8—12.

# CHAPTER SOURCES

The Following chapters have been previously published:

Chapter 1 – This is an edited, reformatted and augmented version of a United States Government, Interagency Counterinsurgency Initiative Report, dated January 13, 2009.

Chapter 2 – This is an edited, reformatted and augmented version of a Headquarters, Department of the Army Report, dated April 21, 2009.

# INDEX

**B**



**C**

## F

## G



## H

## I

## J



## K



## L

## M

## Q



## R

## S

## T

## U



## V

## W



## Y